Practical Statistical Consulting

This book presents a practical introduction to statistical consulting through 36 carefully designed hypothetical case studies drawn from realistic problems that resemble those commonly encountered in statistical consulting and applied research using data. Every case study is self-contained and focuses on the essential ideas needed to understand and apply the relevant methods. Discussions are intentionally concise and centered on key concepts, modelling choices, and interpretation of results. A standardized workflow is used across different cases so that readers can quickly orient themselves when approaching a new problem. The case studies collectively cover a broad range of statistical methods commonly used in practice.

The book can serve both as a teaching resource and as a practical reference. Students can use it as an exercise-oriented companion for developing applied statistical skills through many case studies, while researchers and practitioners from diverse fields may consult it to find a case study resembling their own analytical situation and adapt the accompanying codes and workflow to their own work.

Key Features:

- Hypothetical yet realistic case studies allow comprehensive coverage of topics and practical issues.
- Familiar everyday cases remove unnecessary mental load while focusing on essential ideas and concepts.
- The example-to-method approach leads to growth of abilities while deepening understanding of methods.
- A standardized workflow makes complex problems approachable and abstract problem-solving skills learnable.
- Complete R codes make learned methods immediately applicable and adaptable to readers' own work.

Lei Hua is a professor in the School of Mathematical and Statistical Sciences at the Northern Illinois University. His research focuses on multivariate non-Gaussian, dependence modelling, and quantitative risk management. He earned a Ph.D. in statistics from the University of British Columbia.

Practical Statistical Consulting

Learning from Everyday Cases Using R

Lei Hua

CRC Press
Taylor & Francis Group
Boca Raton London New York

CRC Press is an imprint of the
Taylor & Francis Group, an **informa** business
A CHAPMAN & HALL BOOK

Designed cover image: Lei Hua

First edition published
2027 by CRC Press
2385 NW Executive Center Drive, Suite 320, Boca Raton FL 33431

and by CRC Press
4 Park Square, Milton Park, Abingdon, Oxon, OX14 4RN

CRC Press is an imprint of Taylor & Francis Group, LLC

ISBN: 978-1-041-15789-2 (hbk)
ISBN: 978-1-041-16305-3 (pbk)
ISBN: 978-1-003-68382-7 (ebk)

DOI: 10.1201/9781003683827

Typeset in Nimbus font
by KnowledgeWorks Global Ltd.

Publisher's note: This book has been prepared from camera-ready copy provided by the author.

To Michelle, Claire, and Daniel

Contents

List of Notes

Preface

The main motivation for writing this book is to have a book that I wish I had four years ago when I started to supervise graduate students in statistical consulting as the director of the Statistical Consulting Services at Northern Illinois University. I needed a book to train students with basic statistical knowledge to be able to handle various statistical consulting requests from clients with different backgrounds. There are many excellent books on teaching statistics from methods to examples, but there are very few books on teaching statistics from examples to methods, which is exactly the situation that most researchers and practitioners face when they need to use statistics for their specific research questions. I found that most of the existing books on statistical consulting are either too theoretical or too advanced for students with basic statistical knowledge, and the topics covered in those books are very limited and not close to our daily consulting practice. I wanted a book that has many simple case studies that are relevant to everyday consulting scenarios, and that can teach students how to select appropriate methods for different cases, and how to communicate results effectively to clients with different levels of statistical knowledge. I also wanted a book that can be used as a primary textbook for courses on statistical consulting, as well as a reference book for researchers and practitioners from other fields who are potentially our statistical consulting clients. I hope this book can be a useful exercise book for students to learn statistics through case studies, and a practical guide for researchers and practitioners to use statistics effectively in their own projects.

During the multiple-year preparation of this book, I have witnessed rapid evolution in the field of artificial intelligence (AI). I acknowledge that earlier free-access versions of generative AI tools, including Claude, ChatGPT, and Grok, and Writefull from Overleaf were attempted as assistive tools. These tools were used to help develop desired hypothetical case studies, assist with the preparation of R code examples and notes, and improve language and readability. The exact model versions used at the time were not recorded, and therefore cannot be reliably reported. I have thoroughly reviewed, verified, edited, and substantially revised all AI-assisted content, in many cases through multiple rounds of revision. Although reasonable efforts were made to ensure the accuracy and quality of the content, errors or inaccuracies may still exist, and your feedback is welcome and appreciated.

Lei Hua
DeKalb, Illinois, USA

1

Introduction

This chapter provides an overview of the book, including intended audience, why to use hypothetical case studies, how to use the book, commonly used statistical jargons, and conventions and styles used throughout the book.

1.1 Intended Audience

The book is designed for those who seek practical guidance on statistical consulting or applied statistics using case studies. The intended audience includes instructors and students in need of a textbook for a one-semester statistical consulting course, as well as practitioners or researchers who have research questions and data but unsure about how to analyze them and draw conclusions. The book is also suitable for those who have taken some beginner-level statistics courses and want to widen their knowledge of applied statistics through case studies.

It is assumed that the readers have a basic understanding of statistics, such as descriptive statistics, hypothesis testing, confidence intervals, and regression analysis. The book is not intended to be a comprehensive textbook on statistical theory or methods, but rather a practical guide that emphasizes the use of case studies to develop problem-solving skills in using statistics.

1.2 Why Hypothetical Case Study

To prepare a comprehensive set of case studies that cover a wide range of topics, scenarios, and statistical methods, hypothetical cases are carefully created for this book. All the case studies discussed in this book are hypothetical, as accessible real-world data can rarely cover so many different aspects that the book aims to cover. Therefore, any conclusions drawn from the case studies are not meant to be generalizable to real-world situations, but rather serve as examples for educational purposes. Using hypothetical cases is to ensure that some specific data or model features are included in the cases so that readers have an opportunity to learn how to identify and handle those situations. For example, some cases are designed to have violations of certain assumptions, and we need to learn how to identify them and how to find alternative methods to solve the problems.

The choice of the cases is based on everyday scenarios that are commonly encountered in statistical consulting and in various fields, such as healthcare, social sciences, business, and education. On the one hand, using everyday cases can reduce the mental barrier for readers to spend much effort in understanding the background of the cases, and instead focus more on the statistical aspects. On the other hand, these cases are designed to be relatable and relevant to a wide range of readers, making it easier for them to see how the statistical methods can be applied in their own contexts. Although the cases are hypothetical, they are based on real-world scenarios and data structures, and

DOI: 10.1201/9781003683827-1

therefore, the case studies presented in the book can be used as templates for readers to analyze their own data with similar characteristics.

1.3 How to Use the Book: From Case Studies to Statistical Methods

There are mainly two different ways of using this book. One is to read it sequentially, following the chapters in order. This approach is suitable for those who want a comprehensive understanding of statistical consulting and applied statistics through case studies. The other way is to use it as a reference guide, focusing on specific case studies or topics of interest. In particular, the summary table in Section 5.1 provides an overview of all the case studies included in the book, along with their key features, so that readers can quickly identify which case studies are relevant to their interests or needs.

The book is particularly structured to serve as a textbook for a one-semester statistical consulting course for graduate students or advanced undergraduate students in statistics or related fields. The case studies are organized in a way that allows instructors to select specific cases that align with the learning objectives of their course, while also providing a comprehensive set of cases so that students can systematically learn how to apply statistical methods to real-world problems.

Through my experience teaching statistical consulting courses and mentoring students in real-world consulting projects, I observed a common challenge faced by many students and practitioners: the difficulty in identifying appropriate statistical methods to address real-world problems, even after they have learned various statistical models in their coursework. Many existing statistical textbooks and courses focus heavily on statistical theory, and the typical approach of teaching or learning statistical methods is from theory to examples. However, in statistical practices, we are facing the opposite situation: we often start from research questions and data, and then identify suitable statistical methods to address them. This gap between theory and practice can be frustrating for students and practitioners, as there is often a lack of easy-to-follow steps to help them navigate through the process of analyzing real-world data and drawing conclusions. To address this issue, a standard set of initial questions is designed to guide readers step by step in understanding the case studies and identifying appropriate methods; see Section 2.2 for a general discussion on these initial questions and Chapter 5 for the application of these questions for all the case studies. Through a standardized process of answering these initial questions, students and practitioners can feel less overwhelmed and more confident in identifying appropriate statistical methods to analyze real-world data and draw conclusions. For each case study, the relevant statistical methods are introduced in the form of notes. As this book covers a wide range of topics, the notes are kept to the point and focuses on the key aspects of the methods. For readers who want to learn more about the methods, references are provided for further reading.

The key to using this book effectively, especially for those who have already learned various statistical methods but need improving their ability to apply them in practice, is to start with the case study descriptions without looking at the analyses first, while trying to answer the initial questions on their own; a worksheet with the initial questions is provided in Section 8.2. To facilitate this process, all cases are first presented in Chapter 5 without the statistical analyses initially, mimicking real-world scenarios where statisticians must assess situations without predefined solutions. Formal analyses of these case studies are then organized in Chapters 6 and 7 based on data types and statistical methods. Each case study includes detailed explanations of the statistical methods used, along with R codes to facilitate hands-on learning. Moreover, a companion website for this book is available at `https://larryleihua.github.io/PSCBook/`, where datasets and printable worksheets including case descriptions and initial questions and thoughts for all the case studies are available.

TABLE 1.1: Statistical Jargons: Definitions and Other Names.

Term	Definition	Other Names
Subject	A participant in a study, usually a person (but can also be an animal, object, or experimental unit) from whom data are collected.	Participant, Experimental unit, Sample unit.
Response Variable	The outcome of interest that you are trying to explain, predict, or model in a study. It is the variable that may depend on other variables.	Dependent variable, Outcome variable, Target variable, Labels.
Explanatory Variable	A variable that is used to explain or predict the response variable. It is the one that could explain changes in the response variable.	Independent variable, Predictors, Exposure variable, Risk factor, Input variable, Features.
Covariates	A variable that is not the main focus of the study but is included in the analysis because it may influence the response variable. Adjusting for covariates helps isolate the effect of the main explanatory variable. In practice, covariates and explanatory variables are often used interchangeably, unless a clear distinction is made.	Control variable, Background variable, Nuisance variable, Moderator.
Interaction Term	A product (or combination) of two or more explanatory variables and covariates included in a statistical model to explicitly capture their interaction effect on the response variable.	Product term.
Confounding Variable	A variable that is related to both the explanatory variable and the response variable, creating a false or biased association between the response variable and the explanatory variable if not properly controlled. Confounders can make it seem like there is a relationship when there isn't, or hide a true relationship.	Confounder.
Power of a Study/Test	The chance of detecting a real effect if one exists.	Statistical power, Sensitivity.

1.4 Statistical Jargons

Statistical jargons are those words and phrases that are frequently used without explanations, either through face to face communications or in written materials, such as textbooks, research articles, and reports. These jargons are often used by statisticians and other researchers to communicate complex statistical concepts and methods efficiently. However, they can be confusing and intimidating for non-statisticians who may not be familiar with the terminology. To help readers overcome this challenge, Table 1.1 provides definitions and alternative names for commonly used statistical terms throughout the book. Some other more specific terms will be defined in specific chapters when they are first introduced.

1.5 Conventions and Styles

The following conventions and styles are used throughout the book:

- p-values are rounded values reported in the text, and their exact values are provided in the R code chunks, and equal signs ("=") or words ("is", "are") are used for rounded values, even if they are not exactly equal, for simplicity and readability.
- In notes that include subject-level formulas, such as in Note 7.26, when there is no confusion, $Y|\mathbf{x}$ instead of Y_i or Y_j is used to keep the notation easier to read. Unless specified otherwise, all subjects are assumed to be independent of each other, and therefore, the subscript for subjects is omitted for simplicity.
- R code chunks are presented in a gray background for easy identification.
- In R code chunks, comments are preceded by the hash symbol (#) and R outputs are preceded by the colon (:) which is not the default setting of R but modified for better readability.
- R plots are attempted to be generated by `ggplot2` package for better aesthetics, unless specified otherwise.
- R plots are attempted to be generated in gray color scheme for better visibility when printed in black and white. If a plot is created by `ggplot2`, a colorful version can be obtained by replacing those color controlling functions, such as `scale_fill_grey()`, with their colorful counterparts.
- The R packages used in the code examples are assumed to be installed. If not, please install them using the `install.packages("packageName")` command before running the code.

2

The Consultation Process

This chapter outlines the typical workflow of statistical consulting, from the initial client request to project completion. It also provides a set of initial questions to ask during the initial meeting with clients, along with practical considerations for effective consulting.

2.1 Overview of the Processes

1. **Client's request:** Upon receiving a client request form, reach out to schedule an initial meeting promptly. To enhance the effectiveness of the initial meeting, request the client to prepare certain key information in advance: the research objective, statistical questions, questionnaires used, a sample dataset, etc. Additionally, prior to sharing any data, remind the client to eliminate any identifying information, such as names or medical ID numbers, from the data.
2. **Initial meeting:** The initial meeting serves as a critical opportunity to collect essential information about the case, including research objectives, specific statistical questions, and details about the data or data collection plans, etc. At this moment, clients may or may not have collected the data, but they might have already formulated the study and data collection plan. If data have not been collected, clients might inquire about sample size calculation and power analysis. If a preliminary sample has been obtained, it is time to carefully review it with the client and to clarify the scope of inference and potential limitations of the analysis; a set of standard initial questions provided in Section 2.2 can be used to guide the discussion. In addition to these critical questions, the following practical aspects should also be addressed: the project timeline, budgeting, the software to be used, the statistical consultant's duties, and expectations for post-meeting steps.
3. **Analysis plan:** Following the initial meeting, an analysis plan including estimated hours and costs should be sent to the client prior to investing considerable time in the actual analysis. Once clients have reviewed and approved the plan along with the projected costs, proceed with the principal work.
4. **Exploratory data analysis:** Before conducting a formal analysis, an exploratory data analysis should be performed to understand the data, including summarizing key statistics, identifying any missing values, detecting outliers, exploring the distributions and correlations among variables, and considering candidate statistical methods that may be appropriate for answering the research questions.
5. **Formal analysis:** After the exploratory data analysis, start exploring the candidate methods and choose the most suitable ones for the analysis. The formal analysis includes, but is not limited to, hypothesis testing, fitting regression models, model diagnosing, and interpreting the outcomes. Model assumptions should be checked, and if they are violated, alternative methods should be considered. Sometimes, it may be necessary to try multi-

DOI: 10.1201/9781003683827-2

ple methods and compare the results to ensure robustness. Interpretation of the results is also a critical step, and it is important to avoid overinterpretation or misinterpretation.

6. **Reporting and documentation:** Prepare a well-organized report that clearly details the objectives, methodologies, analytical findings, underlying assumptions, limitations, suggestions, etc. The report should be written in a manner that fits the client's background in statistics, and avoid unnecessary technical jargon.
7. **Follow-up and refinement:** Upon presenting the findings, allocate time for client feedback and inquiries. Clients might seek further explanation or suggest other analyses based on the initial results. Based on this input, reassess the analysis, adjust the models, or conduct more analyses as needed. But it is important to manage client expectations regarding the scope of work and the time required for additional analyses, especially when they involve significant changes to the original plan. Moreover, it is important to avoid biasing the analysis by trying to fit the results to the client's expectations, especially when the client has a strong expectation for a particular outcome. Instead, the analysis should be guided by the data and the research questions, and any discrepancies between the results and the client's expectations should be discussed openly and objectively.
8. **Ongoing support:** Clients may need ongoing support as new data or additional questions arise. It often benefits both parties to establish a long-term consulting relationship, which can reduce the time and effort required for future consultations, as the consultant will already be familiar with the client's research context and data.
9. **Invoicing/case closure:** Once the client is certain that no additional tasks are required, the case may be closed and invoices can be sent. Prior to formally dispatching the invoice, it is advisable to provide the client with an activity sheet detailing the hours worked and a comprehensive cost breakdown to avoid any confusion or disputes regarding the billing.

In Figure 2.1, we present a visual representation of the workflow of statistical consulting, which can be used as a reference for consultants to navigate through the various stages of the consulting process effectively.

2.2 One-Pagers of Initial Questions

It may be difficult for clients to articulate their research objectives and statistical questions clearly, especially when they are not familiar with statistical concepts. To facilitate the initial meeting and ensure that critical information is collected effectively, a set of standard questions can be used as a guide. These questions cover essential aspects of the research design and data structures, which are crucial for selecting appropriate statistical methods and for interpreting results accurately. From the consultant's perspective, using a standard set of questions can help to simplify the initial meeting, ensure that important information is effectively collected from various real-world cases that are often non-standard and even appear to be messy.

The following is a list of standard questions that can be used during the initial meeting with clients. A blank worksheet with these questions is also provided in Section 8.2 for convenience. Although these questions are not exhaustive, they cover most of the critical aspects that are needed to understand the research context and to guide the selection of appropriate statistical methods. These questions are meant to be a starting point for discussion, and additional questions may arise based on specific cases.

These questions are initially designed for the initial meeting in a consulting setting, but they are perfectly applicable to those who want to analyze their own data. In fact, it is recommended for

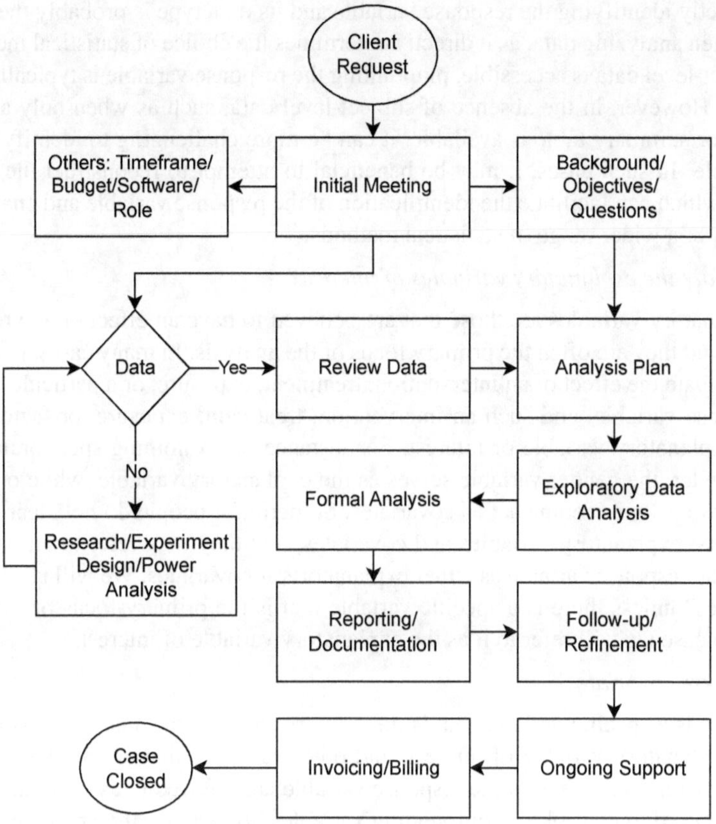

FIGURE 2.1: Workflow of Statistical Consulting.

anyone who wants to analyze data to ask themselves these questions before starting the analysis. For each case study in this book, we will provide answers to these questions, as they are very helpful to guide us in choosing the right methods and to interpret the results correctly.

1. *What are the research objectives?*

   Research objectives are qualitative, non-technical statements that outline the goals of the study. They are often more broad and less specific than statistical questions, and they may not even be directly answerable through data analysis. However, they are important, as any statistical questions asked and any statistical methods used should serve the purpose of addressing the research objectives.

2. *What are the statistical questions?*

   Statistical questions are specific and quantitative questions that can be answered through data analysis. They are often more technical than research objectives and involve specific hypotheses and parameters of interest. Statistical questions directly guide the choice of statistical methods and serve the purpose of addressing the research objectives. Therefore, it is crucial to formulate these questions thoughtfully to align with the research objectives.

3. *What is the response variable, and what is the data type of the response variable?*

   Correctly identifying the response variable and its data type is probably the first thing to do when analyzing data, as it directly determines the choice of statistical methods. When subject-level data is accessible, pinpointing the response variable is typically straightforward. However, in the absence of subject-level data, such as when only a contingency table or summary table is available, it can be more challenging to identify the response variable. In such cases, it may be beneficial to attempt to reconstruct the subject-level data, which can facilitate the identification of the response variable and enable the application of a wider range of statistical methods.

4. *What are the explanatory variables of interest?*

   Explanatory variables are those that are believed to have an effect on the response variable, and they are often the primary focus of the analysis. In many cases, researchers aim to evaluate the effect of an intervention, treatment, exposure, or a particular factor on the response variable, and such an intervention, treatment, exposure, or factor is typically the explanatory variable of interest. For instance, in examining sport brand preference by gender, the gender variable serves as the explanatory variable, while other variables such as age and income act as covariates. Sometimes, people do not clearly distinguish between explanatory variables and covariates, and they may refer to all variables other than the response variable as either explanatory or covariates. We will use the word "covariates" unless there is a specific variable that is the primary focus of the analysis, in which case we will refer to it as the explanatory variable of interest.

5. *Are there covariates?*

   If there is a specific variable that is the primary focus of the analysis, it is referred to as the explanatory variable of interest, and other variables that are not the primary focus but may have an effect on the response variable are referred to as covariates. Covariates can be used to control for confounding variables, to increase the precision of estimates, and to increase the power of statistical tests. Therefore, it is important to include any covariates that are available in the data and are relevant to our analysis.

6. *What is the population of interest?*

   Population of interest is the group of individual or entities that the sample is drawn from and to which we want to generalize any findings from the sample. It is important to clearly define the population of the study, as it directly impacts the choice of statistical methods and the scope of inference. After identifying the population of interest, a key aspect is to determine whether the sample being analyzed is able to represent the population of interest.

7. *What is the subject, and what is the number of distinct subjects?*

   In statistical analysis, the subject is the unit that interventions or treatments are applied to, and the response variable is measured on. The subject can be an individual, a group, etc., depending on the context of the study. Accurate identification of the subject is crucial as it determines the structure of data and impacts our selection of suitable statistical techniques. The number of distinct subjects affects the choice of methods, representativeness of the sample, and the power of the analysis. In certain instances, it is possible for the number of distinct subjects to be just one, even when there are many data rows available.

8. *Are there subject-level data?*

   The granularity of data is important, with the optimal scenario being the availability of comprehensive information for each subject. If direct subject-level data is unavailable and only a contingency table or summary table is present, we can strive to reconstruct the subject-level data, potentially enabling the application of a broader range of statistical methods.

9. *Are response variables dependent (repeated measures/clustered subjects)?*

   Once the response variable and the subject are correctly identified, we need to identify whether the response variables are dependent, and if so, we need to use statistical methods to account for the dependence, such as using mixed-effects models. There are various scenarios that can lead to dependence among response variables, such as when the response variable is measured multiple times for each subject or when subjects are clustered within groups. For the former, we refer to it as repeated measures, and note that, repeated measures should be only counted for the response variable and not for the other variables. For example, in a case study for innovative teaching on test scores (Section 5.6), the posttest score or the score gain is the response variable. Although each student has two scores (pretest and posttest), they are not repeated measures, as the pretest score is not the response variable and it is actually a covariate.

10. *Are the subjects selected randomly?*

    This question examines whether the sample being analyzed is able to represent the population of interest. When the sample is not randomly chosen, such as via volunteer participation or convenience, any findings from the analysis may not be generalized to the population of interest, as there may be selection bias in the sample. Such limitations should be clearly communicated to clients early, and remedies such as redesigning the study to use random sampling can be suggested for future studies.

11. *Are the subjects randomly assigned to different groups?*

    Random assignment of subjects can be very helpful for reducing confounding variables and thus establishing causality. However, many everyday cases are observational studies where subjects are not randomly assigned to groups, and often, control groups are absent. In the absence of random assignment to treatment and control groups, it becomes challenging to eliminate confounding variables, and thus, causality cannot be established. However, it is still meaningful to conduct observational studies, as they may suggest potential associations between variables, and suggest potential hypotheses for future studies.

2.3 Initial Thoughts

After gathering the necessary information from the client, it is beneficial to communicate with clients about initial thoughts regarding critical aspects of the analysis, such as which statistical methods could be relevant, the scope of inference, and potential limitations of the study. In particular, many everyday cases are observational studies, and it is important to communicate early and openly with clients that causality cannot be established from observational studies, as well as the potential confounding variables that may exist in the study. This can help to manage client expectations and to avoid overinterpretation of the results. Other considerations, such as data quality, missing data, and sample size, can also be discussed at this stage, as they may have a significant impact on the choice of methods and reliability of the results.

2.4 Practical Considerations

In addition to the above critical aspects that may directly impact the choice of methods and the interpretation of results, there are some practical considerations that are important for effective statistical consulting. Here we discuss some of these considerations, which are not exhaustive but cover many common issues that may arise in statistical consulting.

2.4.1 Data Privacy and Confidentiality

Preserving privacy and maintaining confidentiality are of utmost importance in statistical consulting. Clients should be informed to conceal all identifying details, such as names and medical IDs, before sharing any data. At this moment, it is also important to remind clients to use the same pseudo ID for the same subject and to keep a record of the mapping between real and pseudo IDs for their own reference, as we need to keep the original data structure while removing identifying information. It is crucial to inform clients of such protocol early to prevent unintentional submission of identifiable data. For example, this can be highlighted in the client service request form and/or during the scheduling of the initial consultation meeting. Even without explicit identification details, clients' data may still contain a wealth of sensitive and confidential information. Consultants must apply appropriate data security measures, such as only using institutionally approved secure platforms for data storage and analysis, to ensure the security and confidentiality of all client data throughout the consulting process.

2.4.2 Ethical Considerations in Statistical Consulting

Consultants must ensure honesty, transparency, and objectivity in all phases of analysis, including data collection, processing, modeling, interpretation, and reporting, as the role of the consultant often directly influences research integrity, policy decisions, and stakeholders' trust. A key ethical responsibility is to avoid manipulating data or results to fit client expectations or personal biases. Consultants should be transparent about the limitations of the data and methods used as well as any uncertainties and potential biases in the analysis.

2.4.3 Artificial Intelligence in Statistical Consulting

Artificial intelligence (AI) has evolved rapidly and has the potential to significantly enhance the efficiency of statistical consulting. However, consultants must maintain the privacy and confidentiality of clients' data, and therefore, uploading any client data to AI platforms should be prohibited. AI may provide useful resources for consultants to identify relevant statistical methods and to generate code snippets, but consultants should critically evaluate the suggestions provided by AI and use it mainly as an informational resource rather than a decision-making tool, as AI may not always provide accurate or appropriate suggestions for specific cases.

2.4.4 Working with Diverse Clients

Statistical consultants often collaborate with clients from different backgrounds, including health, business, social sciences, public policy, etc. Try to communicate effectively using language that is accessible and meaningful to non-statisticians. In addition, clients may have a variety of cultural or political backgrounds. It is important for statistical consultants to respect these differences, maintain professionalism, and concentrate on the technical aspects of the research rather than the potential opinions derived from it.

2.4.5 Statistical Tests and Statistical Procedures

A statistical test is used for testing a specific hypothesis of interest, and it typically produces a test statistic and a p-value to guide decisions about rejecting or accepting the null hypothesis. Although useful to some extent, a statistical test should not be relied upon as the sole basis for decision-making. Instead, it should be considered as one component of a comprehensive statistical procedure, which includes not only hypothesis testing but also estimation, construction of confidence intervals, calculation of effect sizes, interpretations, etc.

2.4.6 Avoiding Misuse and Misinterpretation

A critical responsibility of statistical consultants is to prevent the misuse of statistical methods and the misinterpretation of results. This includes ensuring that the models used are appropriate for the data and research questions, that the model assumptions are met, and that the results are interpreted correctly.

2.4.7 Dealing with Missing Data

Missing data are very common in applied research and, if not handled appropriately, can lead to biased estimates, reduced statistical power, and misleading conclusions. Therefore, it is important to carefully examine the pattern and mechanism of missing data, and to use appropriate methods for either imputing missing values or choosing statistical methods that can handle missing data directly. For instance, in Section 6.1.11.3, we will discuss how to use a linear mixed-effects model for a longitudinal study that has missing data.

2.4.8 Handling Small Sample Sizes

Small sample sizes present unique challenges in statistical analysis, including increased variability, reduced statistical power, and limited generalizability of results. It is important to choose methods that are appropriate for small sample sizes. For example, techniques such as Student t procedures, exact tests, and bootstrapping are usually more reliable for small sample sizes than methods that rely on large sample approximations. Consultants should clearly communicate with clients the limitations of using a small sample for analysis and avoid overinterpretation of the results.

2.4.9 Working Under Time Constraints

Time constraints can be a challenge in statistical consulting. During the initial meeting, consultants should clarify project timelines and manage client expectations regarding the time required for different stages of the consulting work. During the analysis phase, consultants should prioritize tasks that are most critical to clients' primary research questions while maintaining transparency about time constraints. Although expediency is important, consultants must resist the pressure to compromise analytical integrity, ensuring that results remain reliable, reproducible, and interpretable even when working under tight deadlines.

3

Understanding the Basics

In this chapter, we introduce some fundamental concepts and tools that are essential for statistical consulting. We begin with an introduction to the R software, which is widely used for data analysis and visualization, followed by a brief overview of a typical workflow in R. We then discuss the fundamentals of data, including common data types, formats, and manipulations. Next, we cover the basics of data visualization, highlighting commonly used plots for exploratory data analysis. Finally, we review some fundamental statistical concepts, including common parametric distributions, descriptive statistics, and how to choose appropriate statistical methods based on data types and research questions in a general sense.

3.1 Introduction to the R Software

R [107] is a powerful open-source programming language widely used for statistical computing, data analysis, and graphical representation. One of the greatest strengths of R lies in its comprehensive collection of packages that cover many aspects of statistical analyses, including more current and advanced methods. Due to its accessibility, adaptability, and emphasis on reproducibility, R has become a cornerstone of modern statistical practice and a valuable skill for researchers, analysts, and consultants.

A typical workflow in R consists of several key steps that guide the user from raw data to interpretable results. The process often includes data import, data manipulation, statistical analysis, visualization, and reporting. The following is an example of a typical workflow of using R.

3.2 A Typical Workflow in R

To illustrate a typical workflow in R, consider a simple dataset that contains information on the age, sex, and income of individuals. The following steps demonstrate how to import, manipulate, analyze, visualize data, and write report in R.

Step 1: Data Import

Data can be imported from `.csv` files, Excel files, databases, and other formats using various R functions and packages. For example:

```
data <- read.csv("data.csv")
```

For demonstration, we create a small dataset directly:

DOI: 10.1201/9781003683827-3

```
data <- data.frame(age = c(25, 32, 47, 51, 62),
                   sex = c("F", "M", "F", "M", "F"),
                   income = c(40000, 52000, 61000, 58000, 72000))
```

Step 2: Data Manipulation

The dplyr package provides convenient functions for handling data. The following is just a brief illustration, and the dplyr package will be frequently employed in the majority of the case studies discussed in the book.

```
library(dplyr)

# Calculate average income by sex
summary_data <- data %>% group_by(sex) %>%
                         summarise(mean_income = mean(income))
print(summary_data)
: # A tibble: 2 x 2
:   sex   mean_income
:   <chr>       <dbl>
: 1 F          57667.
: 2 M          55000
```

Step 3: Statistical Analysis

As an example, a simple linear regression can be used to examine whether income increases with age.

```
model <- lm(income ~ age, data = data)
summary(model)
:
: Call:
: lm(formula = income ~ age, data = data)
:
: Residuals:
:     1     2     3     4     5
: -2657  4039  1672 -4359  1305
:
: Coefficients:
:             Estimate Std. Error t value Pr(>|t|)
: (Intercept)  23711.7     6030.7   3.932   0.0293 *
: age            757.8      132.9   5.704   0.0107 *
: ---
: Signif. codes:  0 '***' 0.001 '**' 0.01 '*' 0.05 '.' 0.1 ' ' 1
:
: Residual standard error: 3953 on 3 degrees of freedom
: Multiple R-squared:  0.9156, Adjusted R-squared:  0.8874
: F-statistic: 32.54 on 1 and 3 DF,  p-value: 0.01069
```

Step 4: Data Visualization

The `ggplot2` package is widely used for creating quality graphics, and we try to use `ggplot()` as much as possible throughout the book to give a coherent plot style. In `ggplot2` package, `scale_color_grey()` and `scale_fill_grey()` are functions that control the color and fill aesthetics using grayscale palettes. We use these functions purposefully to display only monochrome plots to make all the plots readable on certain display media such as black-and-white printouts. When needed, we can simply remove these functions to get colored plots.

```
library(ggplot2)
ggplot(data, aes(x = age, y = income, color = sex)) +
  geom_point(size = 2) + theme_bw() +
  geom_smooth(method = "lm", se = FALSE, formula = 'y ~ x',
              linewidth = 0.5) + scale_color_grey(start = 0.3, end = 0.8) +
  labs(title = "Income vs. Age by Sex", x = "Age", y = "Income")
```

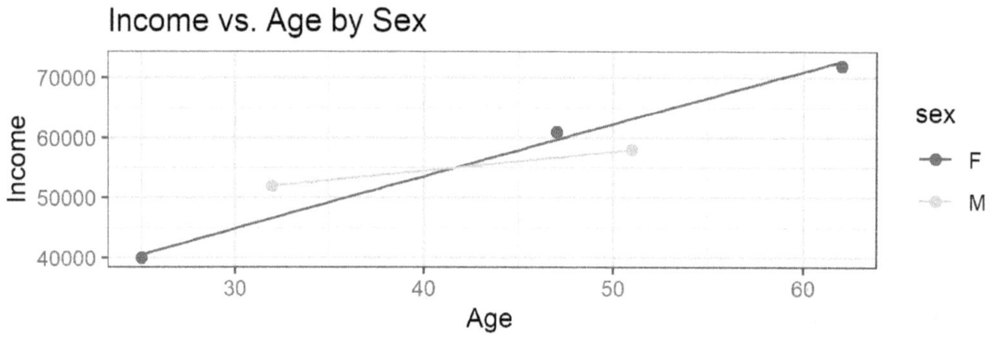

Step 5: Reproducible Reporting

The entire workflow, including narrative text, code, and results, can be combined into a single document for transparent and reproducible reporting using R `Markdown` or `Sweave`. A more flexible and powerful alternative is LaTeX, a typesetting system for producing high-quality documents. When combined with the `knitr` package in R, it creates `.Rtex` files that allow users to incorporate R codes directly into LaTeX documents for creating customized and professionally formatted reports. The majority of the book is actually written in such `.Rtex` files.

3.3 Fundamentals of Data

In this section, we introduce some fundamental concepts related to data, including common data types, formats, and manipulations.

3.3.1 Data Types

We begin with an introduction to different types of data, which is often the determining factor in selecting appropriate statistical methods for analysis. Figure 3.1 illustrates the common data types used in statistical analysis, which can be broadly categorized into continuous and discrete data.

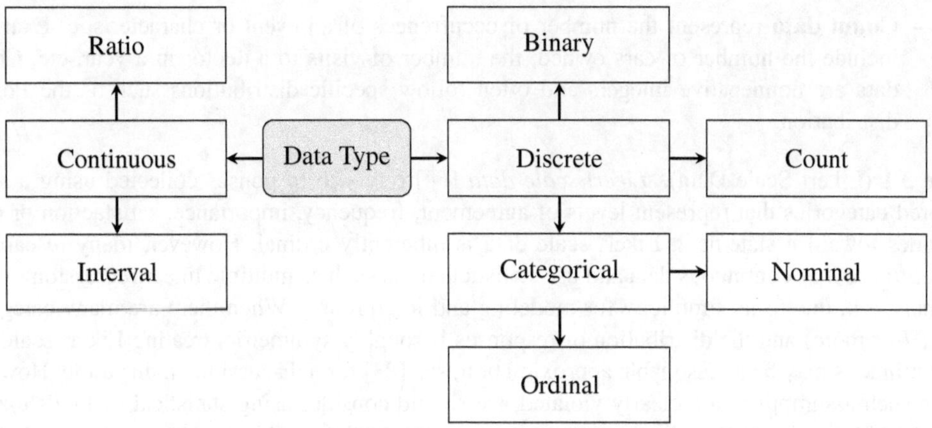

FIGURE 3.1: Common Data Types in Statistical Analysis.

- **Continuous data** can take any value within a given range and can be measured with infinite precision. Examples include height, weight, and temperature. Based on whether there exists a true "zero", continuous data can be divided into the following two categories.
  - **Interval data** are quantitative measurements in which the differences between values are meaningful and consistent, but there is no true zero point; that is, the zero is arbitrary and does not represent the complete absence of the measured attribute. For example, temperature is interval data, and we cannot say that a zero temperature means "no temperature". Interval data can be positive, negative, or zero.
  - **Ratio data** are quantitative measurements in which both the differences and the ratios between values are meaningful, and there is a true zero that represents the absence of the measured attribute. For example, income is ratio data as we can say that person A earns twice as much as person B, and time to event is ratio data. Ratio data are usually nonnegative.

  Most standard statistical methods for continuous data, such as linear regression models, rely on the equal spacing property of continuous data, not necessarily the presence of a true zero, and thus can be applied for both interval and ratio data. However, in order to select models and interpret them correctly, some attention should be paid. For example, a logarithm transformation cannot be applied for temperatures that can be negative, and a regression model may wrongly predict a negative income and a logarithm transformation on income may be needed to avoid this.

- **Discrete data** can only take certain values within a finite range and often represent counts or categories. Discrete data can be further classified into the following categories.
  - **Binary data** consist of only two possible values, typically represented as 0 and 1. Examples include yes or no responses and presence or absence of a characteristic.
  - **Categorical data** represent characteristics or attributes and can take on values in a finite set of categories. They include both nominal and ordinal data, as described below.
    * **Nominal data** represent categories or labels without any inherent order. Examples include sex, ethnicity, or brands of cars.
    * **Ordinal data** also represent categories, but these categories have a meaningful order and the differences between the categories are unknown and not necessarily equal. Examples include education levels (e.g., high school, college, graduate), Likert scale responses (e.g., strongly agree, agree, neutral, disagree, strongly disagree), etc.

- **Count data** represent the number of occurrences of an event or characteristic. Examples include the number of cars owned, the number of visits to a doctor in a year, etc. Counts data are nonnegative integers and often follow specific distributions such as the Poisson distribution.

Note 3.1 (Likert Scale Data). *Likert scale data* [68] refers to responses collected using a set of ordered categories that represent levels of agreement, frequency, importance, satisfaction or other attitudes toward a statement. Likert scale data is inherently ordinal. However, many researchers prefer to use it as continuous data and use statistical tools such as multiple linear regression models to analyze it, due to its simplicity for modeling and interpreting. When there are many categories (e.g., 7 or more) and the distribution of responses is roughly symmetric, treating Likert scale data as continuous may be a reasonable approximation; see [83] for a discussion on this topic. However, when such assumptions are clearly violated, we should consider using statistical methods that are designed for dealing with ordinal responses, such as proportional odds models (see Note 7.16). For detailed discussions on statistical methods for analyzing Likert scale data, please refer to [3] and [60].

Choosing a relatively simple approach like a linear regression model offers numerous advantages, and it is recommended to directly use numerical responses instead of Likert scales when designing the survey. For example, in the survey questionnaire on satisfaction in service, we can ask the following "On a scale of 1 to 10, where 1 = Not at all satisfied and 10 = Extremely satisfied, how satisfied are you with our service?" Then we can more comfortably treat the response as continuous data and use statistical methods designed for continuous data accordingly.

3.3.2 Common Data Formats in Statistical Analysis

Data in statistical analysis can often be represented in multiple formats according to how the data is structured and organized. Four of the most common formats include wide format, long format, contingency tables, and summary statistics tables. Each provides a different balance of detail, structure, and usability.

- **Wide-format data** is commonly used in experimental or longitudinal studies where each subject has multiple measurements at different conditions or time points. In this structure, each subject occupies a single row and has repeated measures across multiple columns of the row. This format is convenient for human inspection, but is often less flexible for statistical modeling in R.

- **Long-format data** restructures wide data so that each measurement is a row, and columns specify the subject ID, the condition (or time point), and the observed value. This format is preferred by most modeling functions in R.

- **Contingency tables** summarize categorical data by cross-classifying observations into counts across two (or more) categorical variables. These tables are central to classical categorical data analysis methods such as the Chi-square test, Fisher's exact test, and McNemar's test.

- **Summary statistics tables**, in contrast, present aggregate statistics such as counts and proportions, but without the joint structure of contingency tables. Although useful for descriptive summaries, they may discard critical subject-level information that is often needed for statistical analyses.

The following are examples of these data formats. We use a toy example to illustrate the different data formats, and the dataset contains ratings for three users evaluating two products: A and B.

```
# wide format
dat_wide <- data.frame(user = 1:3, A = c(4, 5, 3), B = c(2, 5, 4))
dat_wide
:   user A B
: 1    1 4 2
: 2    2 5 5
: 3    3 3 4
```

```
# Convert to long format
library(tidyr)

dat_long <- pivot_longer(dat_wide, cols = c(A, B),
                         names_to = "product", values_to = "rating")
dat_long
: # A tibble: 6 x 3
:    user product rating
:   <int> <chr>    <dbl>
: 1     1 A            4
: 2     1 B            2
: 3     2 A            5
: 4     2 B            5
: 5     3 A            3
: 6     3 B            4
```

```
# Contingency table: distribution of ratings between products
table(dat_long$product, dat_long$rating)
:
:      2 3 4 5
:   A 0 1 1 1
:   B 1 0 1 1
```

```
# Summary statistics table: sum of ratings per product
aggregate(rating ~ product, data = dat_long, FUN = sum)
:   product rating
: 1       A     12
: 2       B     11
```

3.3.3 Common Data Manipulations

R has some base functions for data manipulation, and alternatively, we can use R package `dplyr` for more convenient data manipulation. Below are some commonly used data manipulation tasks with examples using both base R functions and functions in `dplyr`.

```
library(dplyr)

# Create a sample data frame
df <- data.frame(id = 1:5, group = c("A", "A", "B", "B", "C"),
```

```
  score = c(10, 15, 8, 20, 18))
df
:   id group score
: 1  1     A    10
: 2  2     A    15
: 3  3     B     8
: 4  4     B    20
: 5  5     C    18
```

```
# Select rows based on conditions
df[df$score > 10, ]
:   id group score
: 2  2     A    15
: 4  4     B    20
: 5  5     C    18
df %>% filter(score > 10)
:   id group score
: 1  2     A    15
: 2  4     B    20
: 3  5     C    18
```

```
# Select columns
df[c("id", "score")]
:   id score
: 1  1    10
: 2  2    15
: 3  3     8
: 4  4    20
: 5  5    18
df %>% select(id, score)
:   id score
: 1  1    10
: 2  2    15
: 3  3     8
: 4  4    20
: 5  5    18
```

```
# Add a new column
df$score2 <- df$score * 2
df <- df %>% mutate(score2 = score * 2)
df
:   id group score score2
: 1  1     A    10     20
: 2  2     A    15     30
: 3  3     B     8     16
: 4  4     B    20     40
: 5  5     C    18     36
```

```
# Aggregate by group
aggregate(score ~ group, data = df, FUN = mean)
:   group score
: 1     A  12.5
: 2     B  14.0
: 3     C  18.0
df %>% group_by(group) %>% summarise(mean_score = mean(score))
: # A tibble: 3 x 2
:   group mean_score
:   <chr>      <dbl>
: 1 A           12.5
: 2 B           14
: 3 C           18
```

```
# Count observations by group
table(df$group)
:
: A B C
: 2 2 1
df %>% count(group)
:   group n
: 1     A 2
: 2     B 2
: 3     C 1
```

```
# Arrange (sort)
df[order(-df$score), ]
:   id group score score2
: 4  4     B    20     40
: 5  5     C    18     36
: 2  2     A    15     30
: 1  1     A    10     20
: 3  3     B     8     16
df %>% arrange(desc(score))
:   id group score score2
: 1  4     B    20     40
: 2  5     C    18     36
: 3  2     A    15     30
: 4  1     A    10     20
: 5  3     B     8     16
```

```
# Rename columns
df0 <- df
names(df0)[names(df) == "id"] <- "identifier"
df0
:   identifier group score score2
: 1          1     A    10     20
: 2          2     A    15     30
```

```
: 3            3    B     8     16
: 4            4    B    20     40
: 5            5    C    18     36
df %>% rename(identifier = id)
:   identifier group score score2
: 1            1    A    10     20
: 2            2    A    15     30
: 3            3    B     8     16
: 4            4    B    20     40
: 5            5    C    18     36
```

The R package `tidyr` provides some convenient functions for reshaping data between wide and long formats.

```
library(tidyr)

# Wide to long
df_long <- df %>% pivot_longer(cols = score:score2, names_to = "metric",
                values_to = "value")
df
:   id group score score2
: 1  1     A    10     20
: 2  2     A    15     30
: 3  3     B     8     16
: 4  4     B    20     40
: 5  5     C    18     36
df_long
: # A tibble: 10 x 4
:        id group metric value
:     <int> <chr> <chr>  <dbl>
:  1      1 A     score     10
:  2      1 A     score2    20
:  3      2 A     score     15
:  4      2 A     score2    30
:  5      3 B     score      8
:  6      3 B     score2    16
:  7      4 B     score     20
:  8      4 B     score2    40
:  9      5 C     score     18
: 10      5 C     score2    36
```

```
# Long to wide
df_wide <- df_long %>% pivot_wider(names_from = metric, values_from = value)
df_wide
: # A tibble: 5 x 4
:      id group score score2
:   <int> <chr> <dbl>  <dbl>
: 1     1 A        10     20
: 2     2 A        15     30
: 3     3 B         8     16
```

```
: 4       4 B           20      40
: 5       5 C           18      36
```

Datasets can be efficiently merged with different join functions offered by the R package `dplyr`.

The following `left_join` only keeps rows that have matching keys (`id`) in the left dataset (`df1`).

```
library(dplyr)
df1 <- data.frame(id = 1:3, score = c(10, 20, 30))
df2 <- data.frame(id = 2:4, grade = c("B", "A", "C"))
df1
:   id score
: 1  1    10
: 2  2    20
: 3  3    30
df2
:   id grade
: 1  2     B
: 2  3     A
: 3  4     C
# Left join
left_join(df1, df2, by = "id")
:   id score grade
: 1  1    10  <NA>
: 2  2    20     B
: 3  3    30     A
```

The following `right_join` only keeps rows that have matching keys (`id`) in the right dataset (`df2`).

```
# Right join
right_join(df1, df2, by = "id")
:   id score grade
: 1  2    20     B
: 2  3    30     A
: 3  4    NA     C
```

The following `inner_join` only keeps rows that have matching keys (`id`) in both datasets.

```
# Inner join
inner_join(df1, df2, by = "id")
:   id score grade
: 1  2    20     B
: 2  3    30     A
```

The following `full_join` keeps rows that have matching keys (`id`) in either dataset.

```
# Full join
full_join(df1, df2, by = "id")
:   id score grade
: 1  1    10  <NA>
```

```
: 2  2    20     B
: 3  3    30     A
: 4  4    NA     C
```

The above examples illustrate some of the most commonly used data manipulation tasks in R, and these will be frequently employed throughout the case studies in this book.

3.4 Basics of Data Visualization

In this section, we introduce some commonly used plots for exploratory data analysis, and we will frequently employ these across all case studies in this book. The following examples use the built-in `mtcars` dataset in R, which contains information on various car models, including miles per gallon (mpg), number of cylinders (cyl), horsepower (hp), weight (wt), etc.

- **Histogram:** Examine distribution, skewness, and modality.

```
library(ggplot2)
ggplot(mtcars, aes(x = mpg)) +
  geom_histogram(binwidth = 2) + theme_bw() +
  labs(title = "Histogram of MPG", x = "Miles per Gallon", y = "Count")
```

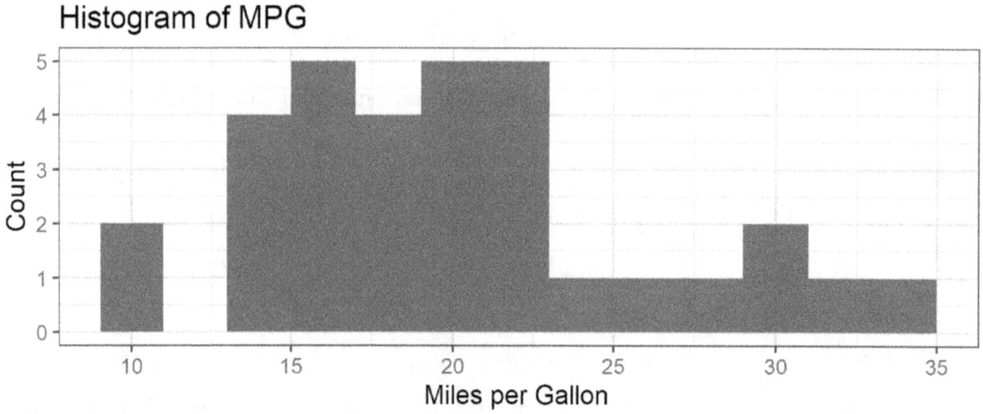

- **Box plot:** Compare distributions across groups and detect outliers. The box plot displays the median, interquartile range (IQR), and potential outliers for each group. The mean, indicated as a diamond shape, is also added as a point for reference. For example, for 4 cylinders, the mean is larger than the median, indicating a right-skewed distribution.

```
ggplot(mtcars, aes(y = factor(cyl), x = mpg)) +
  geom_boxplot() + theme_bw() +
  stat_summary(fun = "mean", geom = "point", shape = 23, size = 2) +
  labs(title = "Box plot of MPG by Cylinders", y = "Cylinders",
       x = "Miles per Gallon")
```

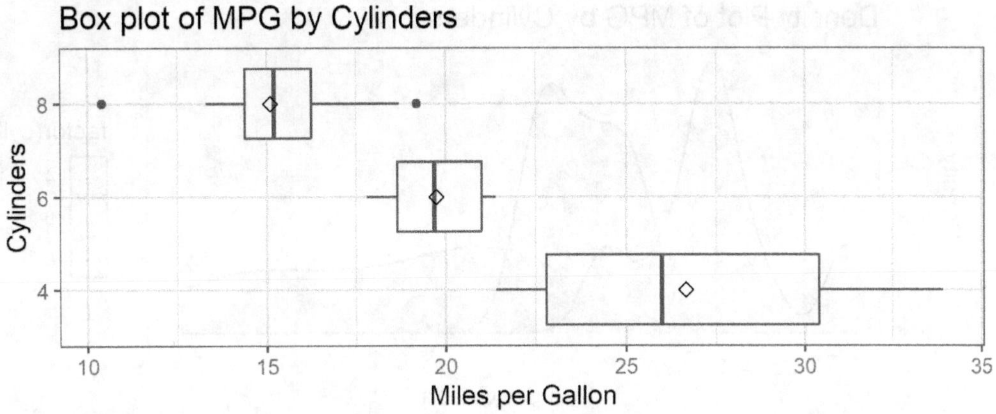

- **Scatter plot:** Explore relationships between two variables, and an optional regression line can be added to assess linear trends.

```
ggplot(mtcars, aes(x = wt, y = mpg)) +
  geom_point(color="grey", size = 2) + theme_bw() +
  labs(title = "Scatter Plot of MPG vs. Weight",
       x = "Weight", y = "Miles per Gallon") +
  geom_smooth(method = "lm", se = FALSE, color = "black", formula = y ~ x)
```

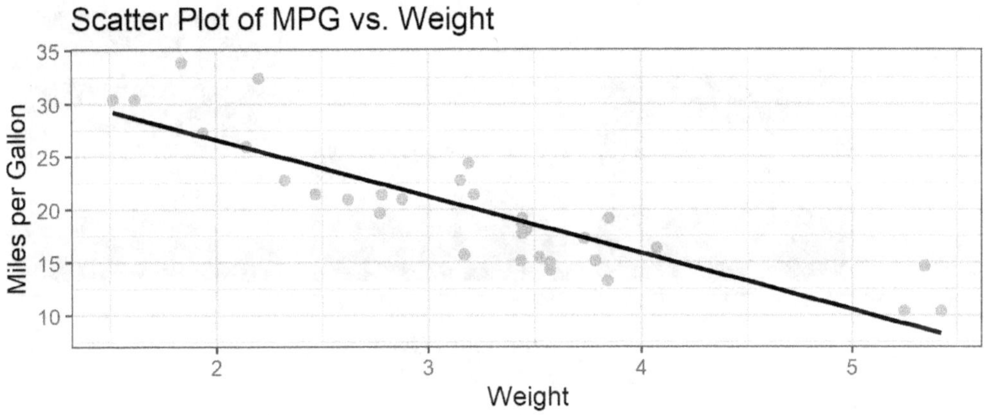

- **Density plot:** Smoothed distribution, good for obtaining the basic shape of a distribution.

```
ggplot(mtcars, aes(x = mpg, fill = factor(cyl))) +
  geom_density(alpha = 0.5) + theme_bw() +
  scale_fill_grey(start = 0.3, end = 0.8) +
  labs(title = "Density Plot of MPG by Cylinders",
       x = "Miles per Gallon", y = "Density")
```

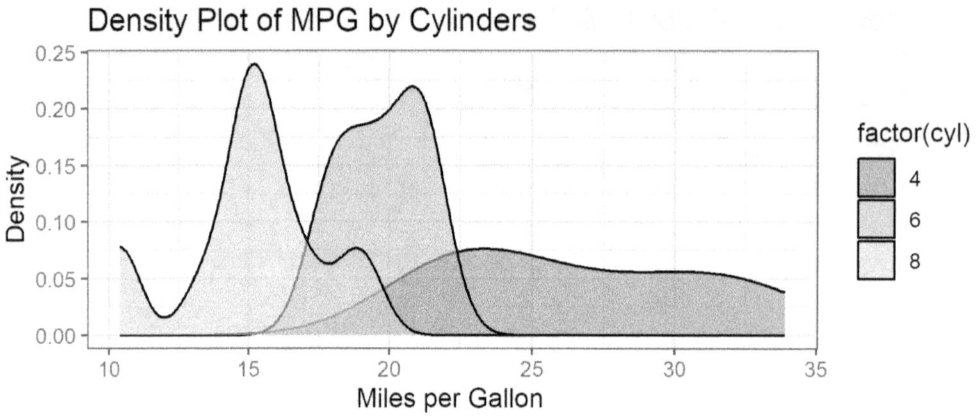

- **Bar plot:** Frequency of categorical variables.

```
ggplot(mtcars, aes(x = factor(cyl))) + geom_bar() + theme_bw() +
  labs(title = "Bar Plot of Cylinders", x = "Number of Cylinders",
       y = "Count")
```

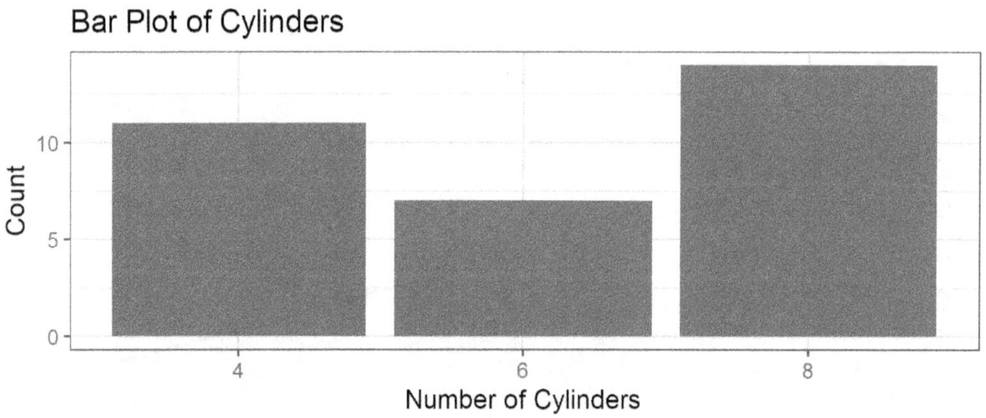

- **Pair plot (scatterplot matrix):** Examine relationships across multiple variables.

```
library(GGally)
ggpairs(mtcars[, c("mpg", "wt", "hp")], lower=list(continuous="smooth"),
        progress = F) + theme_bw()
```

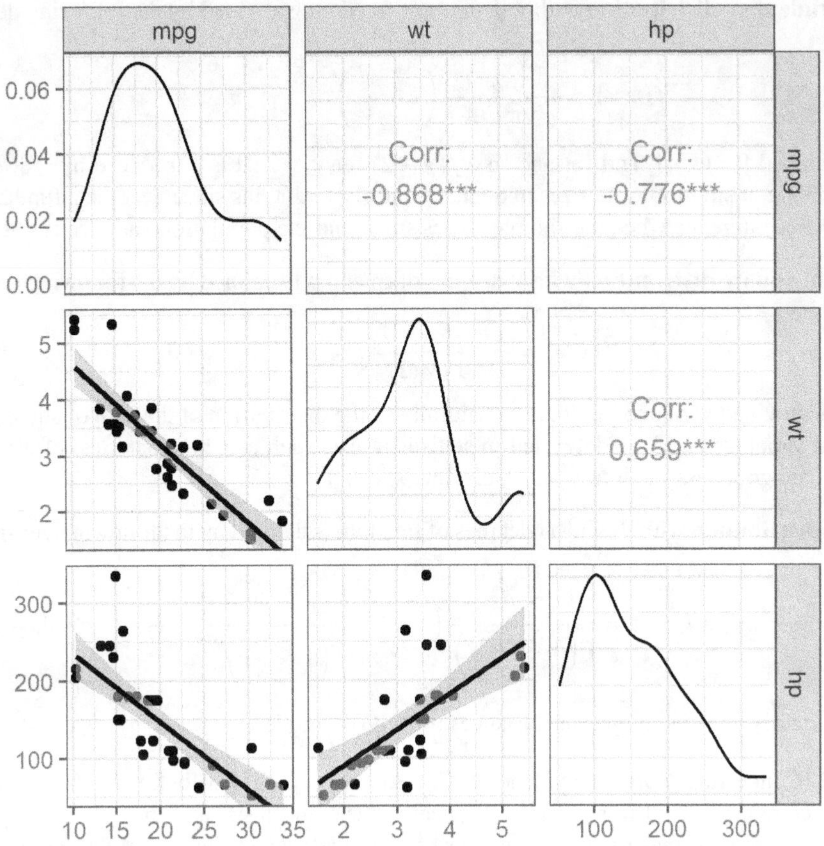

3.5 Fundamentals of Statistics

In this section, we review some fundamental statistical concepts, including common parametric distributions, descriptive statistics, and general guidelines on choosing appropriate statistical methods.

3.5.1 Parametric Distributions in Statistical Analysis

In statistical analysis, various parametric probability distributions are often employed to represent data based on specific assumptions. The following are the most frequently used for standard models discussed in this book. We first list the definitions of these distributions in terms of their probability density functions, and then we discuss the connections between them in Note 3.2.

- **The normal (Gaussian) distribution** is defined by the following density function

$$f(x;\mu,\sigma^2) = \frac{1}{\sqrt{2\pi\sigma^2}} \exp\left(-\frac{(x-\mu)^2}{2\sigma^2}\right), \quad x \in (-\infty,\infty),$$

  with mean $\mathrm{E}(X) = \mu$ and variance $\mathrm{Var}(X) = \sigma^2$. It is the most fundamental distribution in statistics, widely used in regression modeling and hypothesis testing, such as ANOVA and linear regression models.

- **The Student's *t* distribution** with ν degrees of freedom is defined by the following density function

$$f(x;\nu) = \frac{\Gamma(\frac{\nu+1}{2})}{\sqrt{\nu\pi}\Gamma(\frac{\nu}{2})}\left(1+\frac{x^2}{\nu}\right)^{-\frac{\nu+1}{2}}, \quad x \in (-\infty, \infty).$$

Its mean is 0 for $\nu > 1$, and variance is $\nu/(\nu-2)$ for $\nu > 2$. It is introduced by [106] to address the problem of small sample sizes in quality controls, and it has since become a fundamental tool in statistics inference when the sample size is small and the population variance is unknown.

- **The Chi-square distribution** with ν degrees of freedom is defined by the following density function

$$f(x;\nu) = \frac{1}{2^{\nu/2}\Gamma(\nu/2)} x^{\nu/2-1} e^{-x/2}, \quad x > 0,$$

with mean ν and variance 2ν. It arises naturally as the distribution of the sum of squared standard Normal variables and is widely used in statistical tests such as Chi-square test of independence, Chi-square goodness-of-fit test, and McNemar's test.

- **The *F* distribution** with d_1 and d_2 degrees of freedom is defined by the following density function

$$f(x;d_1,d_2) = \frac{\Gamma\left(\frac{d_1+d_2}{2}\right)}{\Gamma\left(\frac{d_1}{2}\right)\Gamma\left(\frac{d_2}{2}\right)}\left(\frac{d_1}{d_2}\right)^{d_1/2}\frac{x^{d_1/2-1}}{\left(1+\frac{d_1}{d_2}x\right)^{(d_1+d_2)/2}}, \quad x > 0.$$

It is primarily used in comparing variances in ANOVA models.

Note 3.2 (Connecting Normal, Student's *t*, Chi-Square, and *F* Distributions)**.** These distributions are most commonly used in standard statistical models and tests, and they are closely related to each other. Knowing these connections can help us understand the underlying assumptions and properties of various statistical methods. For example, the Student's *t* distribution can be derived from the Normal distribution when the variance is estimated from the data, and the *F* distribution can be derived from the ratio of two independent Chi-square distributions.

More specifically, if $Z_i \overset{iid}{\sim} \mathscr{N}(0,1), i = 1, \ldots, n$, the standard Normal distribution, then

$$\sum_{i=1}^{n} Z_i^2 \sim \chi^2(n),$$

the Chi-square distribution with n degrees of freedom. If $Z \sim \mathscr{N}(0,1)$ and $U \sim \chi^2(\nu)$ are independent, then

$$T = \frac{Z}{\sqrt{U/\nu}} \sim t(\nu),$$

a Student's *t* distribution with ν degrees of freedom. As $\nu \to \infty$, Student's *t* distribution and the Normal distribution converge, and this is why in practice, when the sample size is large, the Normal distribution can be used but when the sample size is small, the Student's *t* distribution should be used to account for the additional uncertainty in estimating the variance. If $U_1 \sim \chi^2(d_1)$ and $U_2 \sim \chi^2(d_2)$ are independent, then

$$F = \frac{U_1/d_1}{U_2/d_2} \sim F(d_1,d_2),$$

a *F* distribution with d_1 and d_2 degrees of freedom. These connections are fundamental to understanding the properties of various statistical tests, such as the two-sample *t* test, ANOVA, and Chi-square tests, which are based on these distributions. We refer readers to [52] and [21] for details on more probability distributions and their applications in statistical models.

3.5.2 Commonly Used Descriptive Statistics

In statistical analysis, descriptive statistics provide numerical summaries of data, capturing essential features such as central tendency, dispersion, and distributional shape. Below are the most commonly used descriptive statistics with their definitions and formulas. For these quantities, we focus on sample versions, which are computed from observed data, and there are corresponding population versions that describe the entire population.

- **Sample mean**, or arithmetic average of a sample, is a measure of central tendency defined as $\bar{x} = (\sum_{i=1}^{n} x_i)/n$, where $x_1, x_2, \ldots, x_n$ are observations and n is the sample size.
- **Sample variance** measures the dispersion of observations around the sample mean: $s^2 = (\sum_{i=1}^{n}(x_i - \bar{x})^2)/(n-1)$. Using $n-1$ in the denominator provides an unbiased estimator of the population variance when the data are independent and identically distributed.
- **Sample standard deviation** is defined as the square root of the sample variance.
- **Median** is the middle value when observations are ordered, and it is a robust measure of central tendency and is less sensitive to outliers than the mean.
- **Mode** is the value that occurs most frequently in the dataset.
- **The *p*th percentile** is the value below which $p\%$ of the data fall. Special cases include the three quartiles (25%, 50%, 75%).
- **Quartiles** divide the ordered data into four equal parts: Q_1 (25th percentile), Q_2 (median, 50th percentile), and Q_3 (75th percentile).
- **Interquartile range (IQR)** is the difference between the third and first quartiles: $\mathrm{IQR} = Q_3 - Q_1$, and measures the spread of the middle 50% of the data.
- **Sample skewness** measures the asymmetry of a sample: $\text{Sample skewness} = (\sum_{i=1}^{n}(x_i - \bar{x})^3)/(ns^3)$, where s is the sample standard deviation. A positive skewness indicates a right-skewed distribution, while a negative skewness indicates a left-skewed distribution.
- **Sample kurtosis** measures the tail heaviness of a sample: $\text{Sample kurtosis} = (\sum_{i=1}^{n}(x_i - \bar{x})^4)/(ns^4)$, and a higher kurtosis indicates heavier tails compared to a normal distribution. The kurtosis of a normal distribution is 3, and excess kurtosis is defined as the difference between the sample kurtosis and 3.
- **Sample covariance** measures the covariance of X and Y in a sample of paired observations (x_i, y_i)'s: $\mathrm{Cov}(X,Y) = (\sum_{i=1}^{n}(x_i - \bar{x})(y_i - \bar{y}))/(n-1)$, where $\bar{x}$ and $\bar{y}$ are the sample means of X and Y, respectively. The denominator $n-1$ provides an unbiased estimator of the population covariance when the data are independent and identically distributed.
- **Sample correlation** is a standardized measure of association between X and Y in a sample of paired data (x_i, y_i)'s: $r = \mathrm{Cov}(X,Y)/(s_X s_Y)$, where s_X and s_Y are the sample standard deviations of X and Y, respectively.
- **The *z*-score** standardizes an observation relative to the sample mean and sample standard deviation: $z_i = (x_i - \bar{x})/s$, and it indicates how many standard deviations an observation is from the sample mean.

3.5.3 Choosing the Right Methods

There are some general guidelines for choosing appropriate statistical methods for data analysis, and we summarize some of the most important ones below. These guidelines are not exhaustive, and there are many other important considerations in practice, but they provide a useful starting point.

- **Descriptive vs. Inferential Statistics: Descriptive statistics** refers to methods for summarizing and describing the main features of a dataset. This includes numerical measures such as the mean, median, variance, and standard deviation, as well as graphical representations such as histograms, box plots, or scatter plots. By contrast, **inferential statistics** focuses on drawing conclusions about a population based on a sample of data from that population. This involves estimating population parameters, testing hypotheses, and making predictions. For example, while descriptive statistics might report the average test score in a given sample of students, inferential statistics allows researchers to determine whether the average score differs significantly between two schools. In short, descriptive statistics provide a snapshot of the data at hand, whereas inferential statistics enable broader generalizations and conclusions beyond the observed data.

- **Hypothesis Testing vs. Regression Analysis: Hypothesis testing** is primarily concerned with evaluating specific hypotheses about population parameters, such as testing whether two proportions differ significantly. Its goal is to assess evidence against a null hypothesis using test statistics and p-values. In contrast, **regression analysis** focuses on modeling the relationship between the response variable Y and explanatory variables. However, many classical hypothesis tests, such as t tests and ANOVA, can be formulated as special cases of regression models. If covariates are available in the data, it is often more effective to use regression models to account for potential confounding variables, even when the primary interest is in testing group differences.

- **Parametric vs. Nonparametric Methods: Parametric methods** rely on specific assumptions about the underlying population distribution from which the data are drawn. These methods like t tests, ANOVA, and linear regression are generally more powerful than nonparametric approaches if the parametric assumptions are met, as they utilize the complete distributional details that interpolate and extrapolate beyond the sample data, unlike nonparametric methods that rely on distribution-free quantities such as ranks that summarize only partial information about the data. However, when the parametric assumptions are violated, parametric methods can yield misleading results. In contrast, **nonparametric methods** do not assume a particular distributional form and are therefore more flexible and robust to violations of parametric assumptions. Examples include the Mann-Whitney U test and the Kruskal-Wallis test. While nonparametric methods are generally less powerful under ideal parametric conditions, they provide valid inference in a wider range of scenarios, especially with small sample sizes or skewed distributions. Thus, the choice between parametric and nonparametric methods involves a trade-off between efficiency and robustness. When the assumptions for parametric methods clearly hold, it is generally better to use parametric methods. However, when there is uncertainty about the validity of these assumptions, using both parametric and nonparametric methods is advisable: when different methods yield consistent results, it boosts confidence in the findings, while inconsistent results can prompt further investigation, offering valuable insights into the data characteristics.

- **Observational Design vs. Experimental Design:** In an **observational design**, the investigator does not manipulate the exposure or treatment, and only observes the outcomes as they naturally occur. Common examples include cross-sectional studies, cohort studies, and case-control studies, where associations can be identified, but causal inference is often limited due to potential confounders. In contrast, an **experimental design** involves deliberate manipulation of one or more explanatory variables (e.g., treatment assignment) while controlling for other factors, often through randomization. Randomized controlled trials (RCTs) represent the gold standard of this approach, allowing stronger causal inference by minimizing bias and balancing observed and unobserved covariates across groups. In everyday scenarios, most studies are observational, and we should be cautious about drawing causal conclusions from them. When feasible, experimental designs are preferred for establishing causality.

4

From Data to Conclusions

The process of drawing conclusions from data is a fundamental aspect for statistical consulting or any data analysis project. With the rise of big data and advanced analytical techniques, the concept of "data-driven" decision-making has become increasingly popular in recent years, emphasizing the importance of using data and statistical analysis to inform conclusions and guide actions. However, making conclusions from data is often more complex than it appears. Statistical analysis can provide valuable insights, but it is crucial to understand the limitations and proper interpretations of the results. In this chapter, we discuss several important concepts and considerations when making conclusions from data, including discussing the difference between association versus causation, common study designs, confidence of evidence, limitations of p-values, effect size, and challenges in making conclusions.

4.1 Association and Causation

In statistical analysis, it is important to distinguish between association and causation. An **association** exists when two variables show a systematic relationship such that changes in one variable tend to coincide with changes in another. Mathematically, we can say that two variables X and Y are associated if the conditional distribution of Y given X differs from the marginal distribution of Y. However, such an association does not necessarily imply that X causes changes in Y, or vice versa. For example, if we observe that people who exercise more tend to have lower body weight, we can say that there is an association between exercise and body weight. However, this does not mean that exercise directly causes lower body weight, as there could be other confounding factors involved, such as diet or genetics. **Causation**, on the other hand, refers to a directional effect where variation in one variable causes changes in another. All causal relationships imply some form of association, but not all associations can be interpreted as causal relationships. Establishing causality typically requires more rigorous study designs, such as randomized controlled trials, to control for confounding variables. Most everyday cases in statistical consulting are observational studies, in which confounders remain uncontrolled, and therefore, it is crucial to avoid making causal claims based on associations observed in such studies.

4.2 Common Study Designs in Statistical Analysis

In statistical analysis, a study design is the overall plan or structured framework that specifies how data are collected, organized, and analyzed to answer a research question. Different study designs have different strengths and limitations, and they are suitable for addressing different types of research questions. The study design determines the random mechanism of subject selection and

DOI: 10.1201/9781003683827-4

group assignment, which in turn affects the scope of statistical inferences that can be made from the data. For more details on study designs, we refer to [19] and [93].

4.2.1 A List of Common Study Designs

In what follows, we outline several commonly used study designs.

- **Cross-sectional study**: subjects are measured for exposures and outcomes at the same time, allowing one to estimate prevalence and explore associations between exposures and outcomes. For example, a survey is conducted to assess the prevalence of smoking and its association with respiratory health in a population at a specific point in time.
- **Prospective cohort study**: a defined group is followed over time to examine associations between exposures and future outcomes. For instance, a group of individuals exposed to a certain risk factor is followed to see if they develop a specific disease compared to those not exposed.
- **Retrospective cohort study**: existing records are used to classify individuals by past exposures and then evaluate subsequent outcomes. For example, medical records are reviewed to identify patients who were exposed to a particular medication and their health outcomes are compared to those who were not exposed.
- **Case-control study**: subjects with an outcome are compared to those without to identify prior exposures. A typical example is comparing patients with a rare disease to matched controls without the disease to identify potential risk factors.
- **Randomized controlled trial (RCT)**: subjects are randomly assigned to treatment and control groups, and this study is considered the gold standard for assessing causal relationships.

4.2.2 Randomness Mechanism in Study Designs

Study designs vary in their random mechanisms, potentially affecting what can be achieved by statistical analysis. Study designs can be classified according to the random mechanisms in the following two aspects: (1) how subjects are selected from the population, and (2) how subjects are assigned to different groups (e.g., treatment vs. control). The random mechanisms in these two aspects determine the scope of statistical inference that can be made from the data.

In Table 4.1, we summarize the statistical inference allowed by different random mechanisms of subject selection and group assignment. When addressing a research question with a specific data set, it is crucial to identify the random mechanism to decide the scope of statistical inference.

TABLE 4.1: Statistical Inference Allowed by Random Mechanism in Study Designs.

How to Assign Subjects to Groups / How to Select Subjects	Randomly (Randomized Experiment)	Not Randomly (Observational Study)
Randomly	Causal inference Inference to population	Inference to population
Not randomly	Causal inference	None

Due to typically higher costs of randomized experiments, many researchers opt for observational studies. Although such studies cannot be used to establish causality, they can contribute to locating potential causal links and suggesting areas for future research. Given that many everyday cases are observational studies, it is important to resist the urge to improperly infer causal relationships based on associations observed in such studies.

4.2.3 Cross-Sectional, Prospective, Retrospective Study Designs

Depending on the timing of the subjects being selected and their exposures and outcomes being measured, study designs can be broadly classified into the following three categories: cross-sectional, prospective and retrospective study designs.

- In a **cross-sectional study**, both exposures and outcomes are assessed at the same time. This design is mainly used to assess the prevalence of outcomes for different exposure groups at a single time point.

- In a **prospective study**, subjects are identified and followed over time from exposure to outcome. For example, both the randomized controlled trial (RCT) and the prospective cohort study discussed above are prospective studies. This design allows better control of confounding variables and can establish temporal relationships between exposures and outcomes, making it more suitable for causal inference. However, prospective studies can be time-consuming and expensive to conduct.

- In a **retrospective study**, subjects are identified based on outcome and looked back to evaluate previous exposures. The case-control study and the retrospective cohort study discussed above are retrospective studies. This design is more cost-effective, especially for rare outcomes, but are more susceptible to biases, such as recall bias that may arise from inaccurate recollection of past exposures, and selection bias due to non-random selection of subjects.

4.2.4 Study Designs and Identifiable Measures of Association

Different study designs allow for different measures of association between exposures and outcomes. To simplify the discussion, we consider binary exposures (exposed vs. not exposed) and binary outcomes (outcome present vs. outcome absent), and the following is a contingency table for the exposures and outcomes.

	Outcome Present	Outcome Absent	Total
Exposed	a	b	$a+b$
Not Exposed	c	d	$c+d$
Total	$a+c$	$b+d$	n

- **Cross-sectional study**: *Prevalence* of outcome among exposed $(a/(a+b))$ and among unexposed $(c/(c+d))$ can be directly computed. Commonly used measures of association are the *prevalence ratio* $((a/(a+b))/(c/(c+d)))$ and the *prevalence odds ratio* $((a/b)/(c/d))$.

- **Cohort studies (prospective or retrospective) and RCT**: *Risk* of outcome among exposed $(a/(a+b))$ and among unexposed $(c/(c+d))$ can be directly computed. Both the *relative risk* $((a/(a+b))/(c/(c+d)))$ and the *odds ratio* $((a/b)/(c/d))$ can be used to assess the association between exposures and outcomes.

- **Case-control study**: Since subjects are selected based on outcome status, the marginal totals $(a+c)$ and $(b+d)$ are determined by design and do not represent the actual prevalence of outcomes in the population. Therefore, risk and thus relative risk cannot be directly estimated from a case-control study. In fact, after the total number of cases $(a+c)$ is fixed by design, the values of $(a/(a+b))$ and $(c/(c+d))$ depend on how many controls are selected for the study; the larger the number of controls $(b+d)$, the smaller the values of $(a/(a+b))$ and $(c/(c+d))$. However, the odds ratio of exposures $(a/c)/(b/d)$ is identifiable as changing the number of controls $(b+d)$ will change both b and d proportionally. Furthermore, the odds ratio of exposures is always the same as the odds ratio of outcomes $(a/b)/(c/d)$. Therefore, for a case-control study, the odds ratio

(ad/bc) can be used as the measure of association between exposures and outcomes. Actually, the odds ratio is the only measure of association that can be identified from a case-control study.

In practice, (prevalence) odds ratios are commonly used as long as one of the above five study designs is applied. Also, the odds ratios and the relative risk are approximately equal when the outcome is rare; that is, if ac is very small, then relative risk $= ((a/(a+b))/(c/(c+d))) \approx (a/c)/(b/d) =$ odds ratio.

4.3 Confidence of Evidence

Confidence of evidence refers to the degree of certainty or reliability associated with conclusions drawn from statistical analyses. It reflects how much trust we can place in the results based on the quality of the study design, data collection methods, sample size, and statistical techniques used. High confidence of evidence indicates that the findings are robust and likely to hold true in future studies, while low confidence suggests that the results may be less reliable and should be interpreted with caution. The following provides a general hierarchy of evidence from high to low confidence levels, but the specific context of each study should always be considered.

- Systematic Reviews and Meta-Analyses of Randomized Controlled Trials: These are at the top because of their comprehensive nature, synthesizing evidence from multiple high-quality studies, and rigorous methodology.
- Randomized Controlled Trials: Provide high-quality evidence for causal relationships.
- Cohort Studies: Provide moderate evidence, useful for studying the relationship between risk factors and outcomes.
- Case-Control Studies: Provide evidence but are more susceptible to biases, like recall bias, since the control group (those without the outcome) might be less motivated to remember exposures.
- Cross-Sectional Studies: Limited in providing evidence of causality, best for understanding prevalence.
- Case Reports: Provide lower levels of evidence, generally descriptive.
- Expert Opinion: Although useful, this is the lowest form of evidence, based on individual expertise rather than systematic research.

4.4 Limitations of p-Values and Hypothesis Testing

In 2016, the American Statistical Association published a formal statement on the use of hypothesis testing and p-values ([113]). The statement highlights that p-values are often misunderstood and misused, and it provides six key principles: (1) p-values indicate the degree of incompatibility between the observed data and a specified model; (2) they do not measure the probability that the hypothesis is true or that the data were produced by chance alone; (3) scientific conclusions should not be based solely on whether a threshold such as $p < 0.05$ is crossed; (4) sound inference demands transparency and full reporting; (5) p-values do not measure effect size or importance; and (6) by themselves, p-values are an inadequate measure of evidence. These principles emphasize that p-values can play a role in statistical inference but should be considered alongside effect sizes, interval estimates, study design, and substantive meaning.

Recent discussions in the statistical community emphasize that binary significance testing overlooks the complexity of real-world inference and promotes publication bias toward "significant" findings. The special issue of *The American Statistician* dedicated to "Statistical Inference in the 21st Century: A World Beyond $p < 0.05$" strongly advocates abandoning the use of statistical significance as a default decision criterion ([1]). Instead, researchers are urged to focus on effect sizes, confidence intervals, and transparency in reporting, while integrating statistical findings with substantive knowledge and considerations of study design. The shift away from rigid reliance on p-values reflects a broader recognition that scientific evidence cannot be reduced to a single numerical threshold, but requires nuanced interpretation.

In Table 4.2, we provide examples of interpreting p-values according to the ASA statement, including their literal meanings, common misinterpretations, and practical guidance for reporting and interpreting results responsibly.

TABLE 4.2: Examples of Interpreting p-values According to the ASA Statement.

p-value	Meaning	Misinterpretation	Practical Guidance
0.00423%	If the null hypothesis were true, data this extreme or more would occur about 0.00423% of the time.	Not the probability that the null is true; not a proof that the effect is large or important.	Report exact p, effect size, and measures of uncertainty; emphasize context and practical significance, not just small p.
4.6%	If the null were true, data this extreme or more would occur about 4.6% of the time.	Not a 4.6% chance the null is true; not evidence of a "real" effect simply because $p < 0.05$.	Avoid binary decision; report exact p, effect size, and measures of uncertainty; place result in scientific context.
5.8%	If the null were true, data this extreme or more would occur about 5.8% of the time.	Not a proof there is "no effect"; this p-value is not fundamentally different from 4.6%.	Report $p = 5.8\%$ without claiming it as "not significant"; report effect size, and measures of uncertainty; place result in scientific context.
43.7%	If the null were true, data this extreme or more would occur about 43.7% of the time.	Not a 43.7% chance the null is true; not a proof of no effect.	Report large p-value as weak evidence against the null; report exact p, effect size, measures of uncertainty, and study power.

4.5 Effect Size

Effect size quantifies the strength of a phenomenon, revealing practical significance beyond p-values, which do not indicate the magnitude of the effect. Effect size helps researchers comprehend the relevance of the results, particularly when large sample sizes can easily produce statistical significance while the effect size is actually very minimal. Effect size enables comparisons of effect magnitudes across various studies and is therefore commonly used in meta-analysis for combining findings from different studies. For power analysis and sample size calculation for a planned statistical study, the effect size is always involved.

In the following, we discuss some commonly used effect size measures in linear models. For other types of models, such as generalized linear models for binary or count data, different effect size measures may be more appropriate. We refer to [27], [39], and [66] for more comprehensive discussions on effect size measures in various contexts. In particular, page 41 of [39] provides a summary table for various effect sizes and their interpretations.

Many standard statistical methods can be treated as a special case of linear models, which has the following form:

$$\mathbf{Y} = \mathbf{X}\beta + \varepsilon, \quad \varepsilon \sim \mathcal{N}(0, \sigma^2\mathbf{I}),$$

where $\mathbf{Y}$ is the response variable, $\mathbf{X}$ is the predictors, β is the unknown regression coefficients, and ε are independent error terms.

- **Coefficient of determination (R^2).** It is used to quantify the proportion of variance explained by the predictors.

  $$R^2 = \frac{SS_{\text{model}}}{SS_{\text{total}}} = 1 - \frac{SS_{\text{error}}}{SS_{\text{total}}}, \quad 0 \le R^2 \le 1.$$

  An **adjusted** R^2 is often used to avoid R^2 being inflated simply by increasing the number of predictors:

  $$R^2_{\text{adj}} = 1 - \frac{(1-R^2)(n-1)}{n-p-1}, \quad -\infty < R^2_{\text{adj}} \le 1,$$

  where n is the sample size and p is the number of predictors. The adjusted R^2 may be negative in poor fit.

- **Eta-squared (η^2) and partial eta-squared (η_p^2).** For ANOVA / ANCOVA models, these two measures are commonly used to quantify the effect size of certain factors.

  $$\eta^2 = \frac{SS_{\text{effect}}}{SS_{\text{total}}}, \quad 0 \le \eta^2 \le 1; \qquad \eta_p^2 = \frac{SS_{\text{effect}}}{SS_{\text{effect}} + SS_{\text{error}}}, \quad 0 \le \eta_p^2 \le 1,$$

  where SS_{effect} is the sum of squares for the predictor and SS_{error} is the residual sum of squares. η^2 is more suitable for one-way ANOVA, and it shows the proportion of variance explained by the effect in relation to total variance explained. η_p^2 is often used in multi-way ANOVA or ANCOVA, to assess the unique contribution of a specific factor after accounting for other factors in the model.

- **Cohen's f^2.** It is used to quantify the local effect size of certain predictors, and it compares variance explained by the predictors of interest to the source of errors directly. It can be written in terms of either R^2

  $$f^2 = \frac{R^2_{\text{full}} - R^2_{\text{reduced}}}{1 - R^2_{\text{full}}}, \quad f^2 \ge 0,$$

  where R^2_{full} includes the predictors of interest, and R^2_{reduced} excludes them, or in terms of η_p^2 as follows:

  $$f^2 = \frac{\eta_p^2}{1-\eta_p^2} = \frac{SS_{\text{effect}}}{SS_{\text{error}}}, \quad f^2 \ge 0.$$

- **Cohen's d.** For two groups with sample means $\bar{y}_1, \bar{y}_2$ and common standard deviation s_p:

  $$d = \frac{\bar{y}_1 - \bar{y}_2}{s_p}, \quad s_p = \sqrt{\frac{(n_1-1)s_1^2 + (n_2-1)s_2^2}{n_1+n_2-2}}, \quad -\infty < d < \infty.$$

For small samples, a less biased **Hedges' g** ([50]) is used, and it applies a bias correction factor J to Cohen's d:

$$g = J \cdot d, \quad J = 1 - \frac{3}{4(n_1 + n_2 - 2) - 1}, \quad -\infty < g < \infty.$$

These measures provide different perspectives on the strength of effects in linear models. R^2 and η^2 quantify global model fit, η_p^2 and f^2 describe variance explained by factors, while Cohen's d and Hedges' g quantify pairwise effects. Together, they form the core toolbox of effect size measures commonly used in the context of linear models.

4.6 Challenges in Making Conclusions

Drawing valid conclusions is often one of the most difficult aspects of statistical consulting due to the complexity of real-world data and the limitations inherent in study designs. We must consider issues such as study designs, potential confounding variables, small sample sizes, and missing values, all of which can affect the robustness of statistical inference. In addition, clients may have specific expectations or goals that increase the risk of bias in interpreting results. We must carefully interpret the results, clearly communicate the uncertainty, and provide guidance on the strength and scope of the conclusions.

4.6.1 Interpreting Results

Interpreting results accurately can be more challenging than performing the analysis itself. The following are some common mistakes that should be avoided when interpreting statistical results:

- **Generalization to population:** A statistical hypothesis test is designed to make inference about the parameters of a population. It is necessary to explicitly reveal any conditions that might restrict applying the study to the population. For example, the sample studied may not be representative of the population due to selection bias or non-random sampling.
- **Causation:** Many everyday cases are observational studies in which confounders remain uncontrolled. In these situations, establishing causality is impossible, and we must be cautious not to confuse association with causation.
- **Alternative hypothesis:** Rejecting the null hypothesis does not mean that we accept the alternative, and it only suggests that the data are inconsistent with the null hypothesis. We should avoid over-interpreting the alternative hypothesis without sufficient evidence, as other explanations may exist that lead to the small p-value, such as model misspecification, unaccounted confounding variables, or alternative hypotheses that are not being tested.
- **Practical significance:** Statistical significance does not necessarily mean practical significance. For example, keeping increasing the sample size could help detect a subtle difference in population means among two groups, as the p-value can be made arbitrarily small by increasing the sample size. However, such a subtle difference may not be meaningful practically. Refer to Section 4.4 and the references therein for understanding the limitation of p-values and how to avoid making binary decisions about statistical significance.
- **Confirmation bias:** Driven by the desire of getting "significant" results or supporting one's preexisting beliefs, one may conduct many tests without adjusting for multiple comparisons, engaging in "p-hacking", or emphasizing results that align with prior beliefs while downplaying or ignoring

results that do not. This should be avoided by pre-specifying the analysis plan, adjusting for multiple comparisons when necessary, and transparently reporting all analyses conducted, regardless of the results.

4.6.2 Addressing Bias

Bias can arise at multiple stages of research, including study design, data collection, model selection, interpretation of results, etc. We should identify, assess, and mitigate potential sources of bias to ensure valid and reliable conclusions. This may involve advising clients on appropriate sampling strategies, addressing confounding variables by collecting relevant covariates, using robust statistical methods, and performing sensitivity analyses. Communicate transparently about the presence and implications of bias and interpret the findings with appropriate caution.

4.6.3 Addressing Confounding Factors

Confounding factors are critical concerns and common issues in statistical analysis as they can distort estimates and lead to invalid conclusions. To address this issue, randomly assigning subjects to groups and including more covariates in the model or stratifying analyses by confounders are commonly used to minimize bias. In addition, sensitivity analyses by systematically varying assumptions about unmeasured confounders can be used to assess the robustness of results to potential unmeasured confounders. Ultimately, transparent reporting of possible sources of bias and confounders is an essential part of a responsible statistical report.

4.6.4 Including Written Disclaimers

Including written disclaimers is a recommended practice in statistical consulting to clarify the scope, limitation, and intended use of the analyses. Disclaimers help manage client expectations, reduce the risk of misinterpretation, and reinforce the consultant's ethical responsibility to communicate uncertainty and methodological constraints. They may address aspects such as the assumptions underlying statistical models, the potential impact of missing or biased data, the limited generalizability of results, the lack of guarantee of specific outcomes, and the scope of the use of the deliverables.

5

Case Descriptions and Initial Steps

This chapter presents descriptions of the case studies, followed by a standard set of initial questions to provide a consistent framework for understanding the cases, while mimicking the initial steps of a statistical consulting project or analyzing a dataset on your own. Some initial thoughts are provided to help you get started. The detailed analyses for these case studies are intentionally deferred to later chapters to allow you to attempt the analyses on your own first. You can refer to the later chapters for detailed analyses of these case studies.

5.1 Overview of All Cases

To cover a wide range of data types, data structures, assumptions, and commonly used statistical methods, the hypothetical cases are carefully designed, drawing inspiration from real everyday cases for statistical consulting practices. Table 5.1 provides a summary of all case studies included in the book, along with their key characteristics, such as data types (continuous, binary, categorical, ordinal, contingency table, Likert scale), presence of covariates, repeated measures, and the corresponding sections where detailed analyses can be found.

Table 5.1 serves as a roadmap for readers to quickly identify cases that align with their interests or specific statistical methods they wish to learn. All case descriptions are followed by answers to the standard set of initial questions as well as some initial thoughts. A blank worksheet with the same set of initial questions is provided in Section 8.2, and you can use it to work through the cases on your own before referring to the answers and thoughts that are provided for each case.

DOI: 10.1201/9781003683827-5

TABLE 5.1: Overview of All Case Studies: Con, continuous response; Bin, binary response; Cat, categorical response; Ord, ordinal response or covariates; Tab, contingency table; Lik, Likert scale; Cnt, count response; Dep, response variables are dependent due to repeated measures or clustered subjects; Section, section where the case is analyzed.

#	Brief Description	Con	Bin	Cat	Ord	Tab	Lik	Cnt	Dep	Section
1	Effectiveness of learning formats	✓								6.1.1
2	Exercises on blood pressure	✓							✓	6.1.2
3	New blood pressure medication (sample size calculation)	✓								6.1.3
4	Dietary supplements on marathon performance	✓								6.1.4
5	Innovative teaching on posttest	✓								6.1.5
6	Innovative teaching on posttest (power analysis)	✓								6.1.6
7	Factors on housing prices	✓								6.1.7
8	Factors on job satisfaction	✓			✓					6.1.8
9	Study techniques on quiz performance	✓							✓	6.1.9
10	Chip thickness by machines	✓							✓	6.1.10
11	Student math performance	✓							✓	6.1.11.1
12	Student math performance with school effects	✓							✓	6.1.11.2
13	Stress reduction program	✓							✓	6.1.11.3
14	Environmental factors on antler length	✓							✓	6.1.12
15	Salary comparisons between sectors	✓								6.2.1
16	Sleep hygiene practices on sleep quality	✓							✓	6.2.2
17	Training methods on employee performance	✓								6.2.3
18	Teaching methods on students engagement	✓								6.2.4
19	Student psychological well-being survey	✓			✓		✓		✓	6.2.5
20	User satisfaction on software interfaces				✓		✓		✓	6.2.6
21	Cognitive training on test scores	✓								6.3
22	Sport brand preference by gender		✓			✓				7.1.1
23	Treatment protocol and adverse reactions		✓			✓				7.1.2
24	Educational intervention on smoking cessation		✓			✓			✓	7.1.3
25	Advertisements on purchase intention		✓			✓			✓	7.1.4
26	Drug effects among hospitals		✓			✓				7.1.5
27	Job satisfactory survey		✓							7.1.6
28	Customer satisfaction on product features		✓			✓			✓	7.1.7
29	Customer satisfaction on product features (subject-level)		✓			✓			✓	7.1.8
30	Career choice among college students			✓						7.2.1
31	Customer satisfaction survey				✓		✓			7.2.2
32	Patient satisfaction survey				✓		✓			7.2.3
33	Dietary intervention on cholesterol levels				✓	✓			✓	7.2.4
34	Environment noise level and exam confidence				✓		✓			7.2.5
35	Modeling property crimes with multiple predictors							✓		7.3.1
36	Assessing factors affecting customer complaints							✓		7.3.2
#	Brief Description	Con	Bin	Cat	Ord	Tab	Lik	Cnt	Dep	Section

5.2 Case Study: Effectiveness of Learning Formats

Case 1: Effectiveness of Learning Formats

This study aims to compare the effectiveness of two distinct learning formats: in-person and online. A random sample of 42 students is drawn from a large class consisting of two groups based on their choice of learning format. Both groups undergo the same final exam after one semester. The scores are listed in Table 5.2. Evaluate whether the final exam scores differ significantly between students in the two learning formats and whether the effectiveness of the two learning formats is significantly different.

TABLE 5.2: Learning Format and Final Score Comparison.

Format	Final Score	Format	Final Score
Online	70	In-Person	77.5
Online	74	In-Person	72
Online	68.5	In-Person	72.5
Online	69.5	In-Person	68.5
Online	74.5	In-Person	76
Online	68	In-Person	73.5
Online	72.5	In-Person	76.5
Online	69	In-Person	71
Online	73.5	In-Person	75
Online	71.5	In-Person	73
Online	71	In-Person	78
Online	72.5	In-Person	73.5
Online	68.5	In-Person	75.5
Online	71	In-Person	74.5
Online	70	In-Person	76.5
Online	73	In-Person	74
Online	71.5	In-Person	72
Online	70	In-Person	71.5
Online	69.5	In-Person	74.5
Online	68	In-Person	72.5
Online	70.5		
Online	69.5		

Initial Questions

1. What are the research objectives?

   Compare the effectiveness of two learning formats: in-person and online.

2. What are the statistical questions?

   Assess whether the final exam scores differ significantly between students in the two learning formats.

3. What is the response variable, and what is the data type of the response variable?

   The final exam score is the response variable and can be treated as continuous data.

4. What are the explanatory variables of interest?

The learning format is the only explanatory variable, and it is a categorical variable and has two levels: online and in person.

5. Are there covariates?

   No.

6. What is the population of interest?

   Based on the limited information from the description of the case study, the large class is the population of interest. It would be better to clarify whether a larger population is considered, aligned with the research objective.

7. What is the subject, and what is the number of distinct subjects?

   Each individual student is a subject: 22 subjects for the online learning format and 20 subjects for the in-person learning format.

8. Are there subject-level data?

   Yes.

9. Are response variables dependent (repeated measures/clustered subjects)?

   No.

10. Are the subjects selected randomly?

    Yes.

11. Are the subjects randomly assigned to different groups?

    No, the subjects are grouped based on their choice of learning format.

Initial Thoughts

Based on Question 11, this is an observational study, so causality cannot be established. But Question 10 indicates that we can make statistical inference to the population of interest. The response variable is continuous, the sample size for each group is small, and there are no repeated measures, so a two-sample t test would be appropriate for this case, assuming the normality assumption is satisfied.

In Section 6.1.1, a two-sample t test will be conducted for this case study.

5.3 Case Study: Exercises on Blood Pressures

Case 2: Exercises on Blood Pressures

This case study examines the effect of a 12-week exercise program on blood pressure. A group of 25 volunteers participate in the study. Table 5.3 shows the blood pressures measured before and after the program. Does the 12-week exercise program significantly reduce blood pressure?

TABLE 5.3: Blood Pressure Before and After the Exercise Program.

Patient ID	BP Before	BP After
1	139	120
2	143	143
3	161	136
4	146	125
5	146	146
6	162	139
7	150	132
8	132	143
9	138	143
10	141	142
11	157	141
12	149	140
13	149	134
14	146	132
15	139	132
16	163	129
17	150	133
18	125	124
19	152	155
20	140	146
21	134	125
22	143	131
23	135	131
24	138	142
25	139	134

Initial Questions

1. What are the research objectives?

   Evaluate the effect of a 12-week exercise program on blood pressure.

2. What are the statistical questions?

   Whether the blood pressures measured after the program are significantly lower than those measured before the program.

3. What is the response variable, and what is the data type of the response variable?

   The response variable is blood pressure and can be treated as continuous data.

4. What are the explanatory variables of interest?

   The time of measurement (before vs. after the program) is the only explanatory variable, and it is a categorical variable with two levels: before and after.

5. Are there covariates?

   No.

6. What is the population of interest?

   It is unknown, and it needs to be further clarified.

7. What is the subject, and what is the number of distinct subjects?

   The subject is each individual volunteer and there are 25 distinct subjects.

8. Are there subject-level data?

   Yes.

9. Are response variables dependent (repeated measures/clustered subjects)?

   Yes, each subject has been measured twice.

10. Are the subjects selected randomly?

    No, the subjects are volunteers and are not randomly selected from a larger population.

11. Are the subjects randomly assigned to different groups?

    No, there is only one group and no control group in this case study.

Initial Thoughts

From Items 10 and 11, it is an observational study and no causality can be established, and furthermore, subjects are not randomly selected and any statistical inference based on the data for a more general population than the sample is not applicable. Each subject has been measured twice, the response is continuous, and the sample size is fairly small. So, a paired t test would be appropriate for comparing the means of the measures before and after the program.

In Section 6.1.2, a paired t test will be conducted for this case study.

5.4 Case Study: Evaluating a New Blood Pressure Medication

Case 3: Evaluating a New Blood Pressure Medication (Sample Size Calculation)

A pharmaceutical company is developing a new blood pressure (BP) drug. They want to design a clinical trial to compare the effectiveness of their new drug with current standard treatment. The researchers hypothesize that the new drug will result in a greater reduction in systolic BP compared to standard treatment. The expected mean reduction in systolic BP with the new treatment is 12 mmHg, and the expected mean reduction in systolic BP with the standard treatment is 8 mmHg. The expected standard deviation of the reduction in systolic BP is 10 mmHg, and the minimum clinically significant difference is 4 mmHg. How many volunteers must be recruited to achieve at least 90% power of the study at a significance level of 2.5%?

Initial Questions

1. What are the research objectives?

   Compare the effectiveness of the new drug with the current standard treatment, with the hypothesis that the new drug will result in a greater reduction in systolic BP compared to standard treatment.

2. What are the statistical questions?

   The statistical question is to decide the minimum sample size to achieve at least 90% power to detect a 4 mmHg greater reduction in systolic BP at a significance level of 2.5%, for a test to assess whether the new drug results in a higher reduction in systolic BP than the standard treatment.

3. What is the response variable, and what is the data type of the response variable?

   The response variable is the reduction in systolic BP and can be treated as continuous data.

4. What are the explanatory variables of interest?

   The treatment group is the only explanatory variable, and it is a categorical variable with two levels: new drug and standard treatment.

5. Are there covariates?

   No.

6. What is the population of interest?

   Unknown.

7. What is the subject, and what is the number of distinct subjects?

   Each individual volunteer is a subject. The number of distinct subjects is to be determined.

8. Are there subject-level data?

   Subject-level data is expected, but not needed for sample size calculation.

9. Are response variables dependent (repeated measures/clustered subjects)?

   Unknown. It is possible that the response variable is measured only once for each subject after the treatment. This needs to be clarified with clients, as it will affect the choice of statistical method for the sample size calculation.

10. Are the subjects selected randomly?

Unknown.

11. Are the subjects randomly assigned to different groups?

Unknown.

Initial Thoughts

The answers to questions 9, 6, 10, and 11 are still unknown at this stage and need to be clarified further. However, the answers to Questions 6, 10, and 11 mainly influence how to interpret the result and do not directly affect the sample size calculation. To decide the minimum sample size required to achieve a certain level of power, we do need to know the answer of Item 9, as different statistical methods will lead to different sample sizes needed. In this case, we assume that the response variable "Reduction in systolic BP" is measured only once for each subject after the treatment, and thus the answer to Item 9 is No and a two-sample t test will be used for the sample size calculation. If the response variable is measured multiple times for each subject, then a different statistical method will be needed, and we need to clarify this with clients.

In Section 6.1.3.1, a one-sided two-sample t test will be used to calculate the required sample size.

5.5 Case Study: Impact of Dietary Supplements on Marathon Performance

Case 4: Impact of Dietary Supplements on Marathon Performance

This study examines the effects of three different dietary supplement regimens (BCAAs, Carb-loading, and Electrolyte-focused) on marathon completion times (in seconds) for recreational runners. The study tracked 90 runners with similar experience levels during a series of marathons over a 6-month period, and Table 5.4 shows the first two and the last two observations. Do these three dietary supplement regimens differ significantly and which one improves marathon performance the most?

TABLE 5.4: Supplement Data Table.

Supplement	Completion Time (seconds)
Carb-loading	14091
Carb-loading	15860
...	...
Electrolyte	15013
Electrolyte	14667

Initial Questions

1. What are the research objectives?

   Examines the effects of three different dietary supplement regimens on marathon performance for recreational runners.

2. What are the statistical questions?

   There are two statistical questions: (1) Do these three dietary supplement regimens differ significantly; (2) Which one improves marathon performance the most?

3. What is the response variable, and what is the data type of the response variable?

   The response variable is the marathon completion time (in seconds) and can be treated as continuous data.

4. What are the explanatory variables of interest?

   The dietary supplement regimen is the explanatory variable; it is categorical and has three different levels: BCAAs, Carb-loading, and Electrolyte-focused.

5. Are there covariates?

   No.

6. What is the population of interest?

   Unknown.

7. What is the subject, and what is the number of distinct subjects?

   The subject is each individual runner, and there are 30 distinct subjects for each of the 3 groups.

8. Are there subject-level data?

   Yes.

9. Are response variables dependent (repeated measures/clustered subjects)?

   No.

10. Are the subjects selected randomly?

    Unknown.

11. Are the subjects randomly assigned to different groups?

    Unknown.

Initial Thoughts

The answers to questions 6, 10, and 11 need to be clarified further for proper interpretation of the results. However, statistical analyses can still be performed. The response variable is continuous, and there are more than two groups, so we can first consider whether ANOVA is appropriate and whether there are significant differences between the means of the three groups. If so, then we can proceed with some post hoc analyses to identify which dietary supplement regimen leads to the best marathon performance.

A detailed analysis of this case study is reported in Section 6.1.4.

5.6 Case Study: Innovative Teaching on Posttest Improvement

Case 5: Innovative Teaching on Posttest Improvement

An educational research study examines the effectiveness of an innovative teaching approach. The study includes 60 students randomly sampled from a middle school and randomly assigned to two groups (30 students each). The same curriculum content is covered for both groups. The traditional approach (Method A: control) is used for one group and an innovative approach (Method B: experimental) is used for the other group. The pretest is administered at the beginning of an 8-week teaching intervention period, and the posttest is administered at the end of the intervention period. Standardized assessment of learning outcomes and standardized testing conditions are used for both the pretest and the posttest. Does the innovative approach significantly improve students' performance?

Initial Questions

1. What are the research objectives?

   Examines the effectiveness of an innovative teaching approach.
2. What are the statistical questions?

   Are the gain scores (posttest - pretest) for the innovative teaching approach significantly higher than those of the other group?
3. What is the response variable, and what is the data type of the response variable?

   The gain score, the difference between the posttest score and the pretest score is the response variable, and it can be treated as continuous data. Alternatively, the posttest score can also be used as the response variable.
4. What are the explanatory variables of interest?

   Teaching approach is the explanatory variable; it is categorical and has two different levels: innovative and traditional.
5. Are there covariates?

   Yes, the pretest score is a covariate.
6. What is the population of interest?

   The middle school involved in this study is the population of interest.
7. What is the subject, and what is the number of distinct subjects?

   The subject is the student, and there are 30 distinct subjects for each of the two groups.
8. Are there subject-level data?

   Yes.
9. Are response variables dependent (repeated measures/clustered subjects)?

   No.
10. Are the subjects selected randomly?

    Yes.
11. Are the subjects randomly assigned to different groups?

    Yes.

Initial Thoughts

Gain scores can be treated as a continuous variable. Since a covariate, namely the pretest score, is present, it is necessary to consider this covariate when comparing the gain scores between the two groups, as student performance may vary regardless of the teaching method implemented. An appropriate method for the analysis can involve a linear regression in which the gain score is the response variable, the teaching approach is the explanatory variable, and the pretest score serves as the covariate. Since the subjects are randomly selected and assigned to the two groups, we can use the study to interpret whether the different teaching approaches cause the different gain scores while accounting for the pretest scores; statistical inference can generalize to the entire school population as the subjects are randomly selected from the school.

A detailed analysis of this case study is reported in Section 6.1.5.

5.7 Case Study: Power Analysis for Innovative Teaching on Posttest Improvement

Case 6: Power Analysis for Innovative Teaching on Posttest Improvement

A researcher plans a randomized controlled trial to evaluate the effectiveness of an innovative teaching approach compared to a traditional method. Scores on a standardized final exam will be used to examine the effectiveness of the innovative teaching approach, and pretest scores will be collected as covariates. A previous, closely related study reported the effect size of the teaching method after adjusting for the pretest scores as is $\eta_p^2 = 0.06$. The researcher wants to conduct a power analysis to determine the required sample size for different target powers (0.7, 0.8, 0.9) at a significance level of 5%. Due to resource constraints, the researcher expects that the maximum feasible sample size is 100 students, so they also want to perform a sensitivity analysis to determine the minimum detectable effect size given a fixed sample size.

Initial Questions

1. What are the research objectives?

   Evaluate the effectiveness of an innovative teaching approach compared to a traditional method.

2. What are the statistical questions?

   (1) What is the required sample size to achieve target powers of 0.7, 0.8, and 0.9, respectively, to detect an effect size of $\eta_p^2 = 0.06$ at a significance level of 5%? (2) Given a fixed sample size of 100 students, what is the minimum detectable effect size at a significance level of 5% for target powers of 0.7, 0.8, and 0.9, respectively?

3. What is the response variable, and what is the data type of the response variable?

   The final exam score is the response variable and can be treated as continuous data.
4. What are the explanatory variables of interest?

   The teaching method is the explanatory variable; it is categorical and has two different levels: innovative and traditional.
5. Are there covariates?

   Yes, the pretest score is a covariate.
6. What is the population of interest?

   Unknown.
7. What is the subject, and what is the number of distinct subjects?

   Students is the subject, and the number of distinct subjects is to be determined.
8. Are there subject-level data?

   Subject-level data is expected, but not needed for power analysis.
9. Are response variables dependent (repeated measures/clustered subjects)?

   No.
10. Are the subjects selected randomly?

    Yes, the planned study is a randomized controlled trial.
11. Are the subjects randomly assigned to different groups?

    Yes, the planned study is a randomized controlled trial.

Initial Thoughts

This is a typical example of power analysis for ANCOVA. As the effect size is given in terms of η_p^2, the power analysis can be performed using the R function `pwr.f2.test()` from the `pwr` package for the power analysis for ANCOVA. The required sample size can be calculated for different target powers, and the minimum detectable effect size can also be calculated for a fixed sample size of 100 students.

A detailed analysis of this case study is reported in Section 6.1.6.

5.8 Case Study: Factors on Housing Prices

Case 7: Factors on Housing Prices

Real estate companies and homeowners need to accurately estimate house prices based on various characteristics. This case study analyzes residential properties in a midsize American city to determine the key factors that influence house prices and develop a predictive model. Research questions are the following:

1. Which housing characteristics significantly influence sale prices?
2. How well can we predict house prices using these characteristics?
3. What are the relative impacts of different features on house prices?

The dataset contains information on 200 house sales from 2024:

1. Sale_Price: Sale price in dollars (response variable)
2. Square_Feet: Total living area
3. Bedrooms: Number of bedrooms
4. Bathrooms: Number of bathrooms
5. Age: Age of house in years
6. Lot_Size: Lot size in square feet
7. Garage: Number of garage spaces
8. Distance_Downtown: Distance to downtown in miles
9. Pool: Presence of pool (binary: 0/1)
10. School_Rating: Nearby school rating (1-10)
11. Crime_Rate: Neighborhood crime rate per 1000 residents

Initial Questions

1. What are the research objectives?

   Estimate house prices based on various characteristics.

2. What are the statistical questions?

   (1) Which housing characteristics significantly influence sale prices? (2) How well can we predict house prices using these characteristics? (3) What are the relative impacts of different features on house prices?

3. What is the response variable, and what is the data type of the response variable?

   Housing price is the response variable and can be treated as a continuous variable.

4. What are the explanatory variables of interest?

   All the variables except the response variable could potentially be explanatory variables, and we need to identify those significant ones using statistical models.

5. Are there covariates?

   For this case, there is no need to distinguish between covariates and explanatory variables.

6. What is the population of interest?

   The midsize American city in the study.

7. What is the subject, and what is the number of distinct subjects?

   Each house is a subject and there are 200 distinct subjects.

8. Are there subject-level data?

   Yes.

9. Are response variables dependent (repeated measures/clustered subjects)?

   No.

10. Are the subjects selected randomly?

    Unknown.

11. Are the subjects randomly assigned to different groups?

    No.

Initial Thoughts

This is a typical example of using multiple linear regression to identify significant variables that can explain housing prices. This is an observational study, so there is no causality that can be established. Whether the study can lead to statistical inference for the population will depend on how the sample of 200 houses is sampled from the population, and it needs to be further clarified.

A detailed analysis of this case study is reported in Section 6.1.7.

5.9 Case Study: Evaluating Job Satisfaction in a Tech Company

Case 8: Evaluating Job Satisfaction in a Tech Company

A tech company surveys its employees to understand the factors that influence job satisfaction (measured on a scale of 0 to 100). The company collects data from 200 randomly sampled employees on several variables: annual salary (in thousands of dollars), years of experience, and employee ranks: Junior, Intermediate, Senior, and Lead. How do these factors, especially employee ranks, affect job satisfaction?

Initial Questions

1. What are the research objectives?

   Understand the factors that influence job satisfaction, with a particular focus on employee ranks.

2. What are the statistical questions?

   How do the factors, especially employee ranks, affect job satisfaction scores?

3. What is the response variable, and what is the data type of the response variable?

Job satisfaction score is the response variable, and it is a continuous variable.

4. What are the explanatory variables of interest?

   Annual salary, years of experience, and employee ranks are potential explanatory variables.

5. Are there covariates?

   For this case, there is no need to distinguish between covariates and explanatory variables.

6. What is the population of interest?

   The tech company.

7. What is the subject, and what is the number of distinct subjects?

   Each individual employee is a subject, and the number of distinct subject is 200.

8. Are there subject-level data?

   Yes.

9. Are response variables dependent (repeated measures/clustered subjects)?

   No.

10. Are the subjects selected randomly?

    Yes.

11. Are the subjects randomly assigned to different groups?

    No.

Initial Thoughts

The satisfaction scores on the scale of 0 to 100 should be treated as a continuous variable, so a multiple linear regression model can be used for this case. After looking at the data, both the salary and years of experience can be treated as continuous variables. However, the employee rank has four levels from lower to higher ranks: "Junior", "Intermediate", "Senior", and "Lead". If we simply treat it as a nominal categorical variable, then we will lose the ranking information among the data. If we code the four levels into rank numbers such as 1, 2, 3, and 4, and use it as a continuous variable in the regression model, then we not only assume that the distances between each consecutive rank are the same, but also that satisfaction scores are only linearly related to the rank numbers. A more appropriate way is to treat the employee ranks as ordinal data, and the model should also allow potentially nonlinear effects from the ranks.

A detailed analysis of this case study is reported in Section 6.1.8.

5.10 Case Study: Effectiveness of Study Techniques on Quiz Performance

Case 9: Evaluating the Effectiveness of Study Techniques on Quiz Performance

A university's educational research team is interested in assessing how different study techniques impact student quiz performance. 20 students are randomly selected from a pool of volunteers to ensure a diverse mix of academic abilities. Those participants vary in age, gender, and previous academic performance, although these factors are not explicitly analyzed in this study. Determine whether there are significant differences in quiz scores between the three techniques.

Initial Questions

1. What are the research objectives?

   Assessing how different study techniques impact student quiz performance.
2. What are the statistical questions?

   Determine whether there are significant differences in quiz scores between the three techniques.
3. What is the response variable, and what is the data type of the response variable?

   Quiz score is the response variable and can be treated as continuous data.
4. What are the explanatory variables of interest?

   Study technique is the explanatory variable, and there are three different techniques: A, B and C.
5. Are there covariates?

   No.
6. What is the population of interest?

   Unknown.
7. What is the subject, and what is the number of distinct subjects?

   Each volunteer is a subject and there 20 distinct subjects.
8. Are there subject-level data?

   Yes.
9. Are response variables dependent (repeated measures/clustered subjects)?

   Yes, each subject has been measured three times in terms of quiz scores.
10. Are the subjects selected randomly?

    No, based on volunteers.
11. Are the subjects randomly assigned to different groups?

    No.

Initial Thoughts

An important point of the case study is that 20 distinct subjects are measured three times, and the goal is to compare scores in three different categories. This is similar to ANOVA, but we need to consider repeated measurements, and therefore a repeated measure ANOVA may be appropriate for this case. Since the subjects are volunteers, they are not randomly selected from a larger population, and thus any statistical inference based on the data for a more general population than the sample is not applicable.

A detailed analysis of this case study is reported in Section 6.1.9.

5.11 Case Study: Chip Thickness by Machines

Case 10: Chip Thickness by Machines

A semiconductor manufacturing facility wants to understand the inherent variation in chip thickness produced by different wafer coating machines. They have randomly selected 20 identical machines from the facility and want to quantify the natural machine-to-machine variation in their production line. Each machine is maintained under identical conditions, uses the same materials, and follows the same protocols.

Initial Questions

1. What are the research objectives?

   Understand the inherent variation in chip thickness produced by different wafer coating machines.

2. What are the statistical questions?

   What are the natural machine-to-machine variations?

3. What is the response variable, and what is the data type of the response variable?

   Chip thickness is the response variable and can be treated as a continuous variable.

4. What are the explanatory variables of interest?

   No.

5. Are there covariates?

   No.

6. What is the population of interest?

   All the wafer coating machines in the facility.

7. What is the subject, and what is the number of distinct subjects?

   Each wafer coating machine is a subject, and there are 20 distinct subjects. Note that the chips are not the subjects, and the statistical question is to assess the variability of the machines in terms of the chip thickness they produce.

8. Are there subject-level data?

   Yes.

9. Are response variables dependent (repeated measures/clustered subjects)?

   Yes, each machine has been measured 10 times in terms of chip thickness.

10. Are the subjects selected randomly?

    Yes.

11. Are the subjects randomly assigned to different groups?

    No.

Initial Thoughts

It is critical to identify that the machines, not the chips, are the subjects, and there are repeated measures for each subject (machine). The statistical question is to assess the variability of the machines in terms of the chip thickness they produce, and therefore a random intercept model is appropriate. The 20 machines are randomly sampled, so the result can be generalized to all machines in the facility. There is no explanatory variable, and thus causality does not need to be studied.

A detailed analysis of this case study is reported in Section 6.1.10.

5.12 Case Study: Examining Student Math Performance

Case 11: Examining Student Math Performance

A researcher is studying math test scores of students. The study involves 200 randomly selected students from a large metropolitan area, with each student sampled independently. Each student is tested twice (Time 0 and Time 1) using standardized tests. The following variables are collected: Math score, Time, Student ID, and student socioeconomic status (SES). The researcher wants to study whether math scores improve over time and how SES affects math performance.

Initial Questions

1. What are the research objectives?

   Study math test scores of students in a large metropolitan area.

2. What are the statistical questions?

   Whether math scores improve over time, and how SES affects math performance.

3. What is the response variable, and what is the data type of the response variable?

   Math scores is the response variable, and it can be treated as continuous data.

4. What are the explanatory variables of interest?

   According to the statistical questions to be answered, Time and SES are the explanatory variables of interest.

5. Are there covariates?

   Except the above explanatory variables, there are no other covariates.

6. What is the population of interest?

   The students in the large metropolitan area.

7. What is the subject, and what is the number of distinct subjects?

   Each student is a subject and there are 200 distinct subjects.

8. Are there subject-level data?

   Yes.

9. Are response variables dependent (repeated measures/clustered subjects)?

   Yes, each student has been measured twice in terms of math scores.

10. Are the subjects selected randomly?

    Yes.

11. Are the subjects randomly assigned to different groups?

    No.

Initial Thoughts

Each subject has been measured twice, and they are potentially dependent, and we can assign a random effect to each subject to account for such dependency. Time can be treated as a binary variable, and SES is a continuous variable and both will be used as fixed effects in a linear mixed-effects model. As subjects are randomly selected, the result can be generalized to all students in the large metropolitan area. However, subjects are not randomly assigned to different schools, so causality cannot be established from the study.

If, instead, we treat the difference between the scores in Time 1 and Time 0 as the response variable and include the scores in Time 0, School, and SES as predictors, like that in Case 5, a multiple linear regression model without random effects can be used. However, such a model does not answer the question of how SES affects math performance, and it addresses the question of how SES affects the improvement in math performance over time, which is different from the original question.

A detailed analysis of this case study is reported in Section 6.1.11.

5.13 Case Study: Examining Student Math Performance with School Effects

Case 12: Examining Student Math Performance with School Effects

A researcher is studying math test scores of students in a large metropolitan area that has 20 schools. The researcher randomly selects 200 students from six different schools randomly selected from the metropolitan area. Each student is tested twice (Time 0 and Time 1) using standardized tests and the following variables are collected: Math score, Time, Student ID, School ID, and student socioeconomic status (SES). The researcher wants to study whether math scores improve over time, whether the rate of improvement varies between schools, and how SES affects math performance.

Initial Questions

1. What are the research objectives?

   Study math test scores of students in a large metropolitan area.

2. What are the statistical questions?

   Study whether math scores improve over time, whether the rate of improvement varies between schools, and how SES affects math performance.

3. What is the response variable, and what is the data type of the response variable?

   Math scores is the response variable, and it can be treated as continuous data.

4. What are the explanatory variables of interest?

   According to the statistical questions to be answered, Time, School ID, and SES are the explanatory variables of interest.

5. Are there covariates?

   Except the explanatory variables, there are no other covariates.

6. What is the population of interest?

   The students in the large metropolitan area.

7. What is the subject, and what is the number of distinct subjects?

   Each student is a subject and there are 200 distinct subjects.

8. Are there subject-level data?

   Yes.

9. Are response variables dependent (repeated measures/clustered subjects)?

   Yes, each student has been measured twice in terms of math scores.

10. Are the subjects selected randomly?

    Yes.

11. Are the subjects randomly assigned to different groups?

    No.

Initial Thoughts

Each subject has been measured two times and they are potentially dependent. We can assign a random effect to each subject to account for such dependency. The six schools are randomly selected, and we can treat `school_id` as a random effect. Time can be treated as a binary variable, and a random slope model can be used to answer the question "whether the rate of improvement over time varies between schools". Clearly, SES is a continuous variable and will be used as a fixed effect in a linear mixed-effects model.

A detailed analysis of this case study is reported in Section 6.1.11.2.

5.14 Case Study: Effectiveness of a Stress-Reduction Program

Case 13: Effectiveness of a Stress-Reduction Program

A health researcher evaluates a stress-reduction program designed to help individuals manage stress more effectively. The program includes mindfulness exercises, physical activity, and cognitive-behavioral techniques. The researcher collects data from 142 volunteers on stress levels before and after the program using a standardized stress assessment tool with stress scores range from 0-100, with higher scores indicating more stress. Due to incomplete participation, there are 55 participants completed both before and after measurements, 39 participants completed only before measurements, and 48 participants completed only after measurements. The objective is to assess whether the stress-reduction program is effective in reducing stress levels.

Initial Questions

1. What are the research objectives?

   Evaluate a stress-reduction program designed to help individuals manage stress more effectively.

2. What are the statistical questions?

   Assess whether the stress levels after the program are significantly lower than those before the program.

3. What is the response variable, and what is the data type of the response variable?

   Stress score is the response variable and can be treated as continuous data.

4. What are the explanatory variables of interest?

   Time (before vs. after) is the explanatory variable of interest.

5. Are there covariates?

   No.

6. What is the population of interest?

   Unknown.

7. What is the subject, and what is the number of distinct subjects?

   Each volunteer is a subject, and there are 142 distinct subjects.

8. Are there subject-level data?

   Yes.

9. Are response variables dependent (repeated measures/clustered subjects)?

   Yes, there are repeated measures for some subjects, but not all subjects have been measured at both time points.

10. Are the subjects selected randomly?

    No, based on volunteers.

11. Are the subjects randomly assigned to different groups?

    No.

Initial Thoughts

Not all subjects have been measured at the both time points, and thus the data are partially paired. We can use a linear mixed-effects model with a random intercept for each subject to account for the dependency of response variables, and such dependency is created by those 55 subjects who have been measured twice. Since subjects are not randomly selected, the result cannot be generalized to any population. This is an observational study, so causality cannot be established.

A detailed analysis of this case study is reported in Section 6.1.11.3.

5.15 Case Study: Environmental Factors on Antler Length

Case 14: Environmental Factors on Antler Length

In ecological research, it is often necessary to study a group of wild animals in a nonintrusive way, such as through camera traps or direct observation, where individual animals cannot be distinguished due to their similarity and elusive behavior. Researchers are investigating the antler length (in centimeters) of a small, isolated group of animals that belong to the Bovidae family over several months. The objective is to analyze the antler length data to assess whether environmental factors, such as habitat quality (e.g., forest density), influence antler length. The data includes repeated measures from multiple sightings, collected over time with an associated predictor: forest density (low or high).

Initial Questions

1. What are the research objectives?

   Assess whether environmental factors, such as habitat quality (e.g., forest density), influence antler length.

2. What are the statistical questions?

   Assess whether forest density significantly affect antler length.

3. What is the response variable, and what is the data type of the response variable?

   Antler length is the response variable and can be treated as continuous data.

4. What are the explanatory variables of interest?

   Forest density is the explanatory variable.

5. Are there covariates?

   No.

6. What is the population of interest?

   A small, isolated group of animals that belong to the Bovidae family.

7. What is the subject, and what is the number of distinct subjects?

   The group of animals is the subject as we cannot identify each individual and there is only one subject.

8. Are there subject-level data?

   Yes, but all the data are at the group level and group is the only subject.

9. Are response variables dependent (repeated measures/clustered subjects)?

   Yes, there is only one subject, and it has been measured many times.

10. Are the subjects selected randomly?

    No.

11. Are the subjects randomly assigned to different groups?

    No.

Initial Thoughts

At the first glance we may wrongly treat each individual animal as a subject. Although we have many measures for the group of animals, we cannot identify each of them. Therefore, this particular group of animals is the only one subject of this study, and we have repeated measures for this subject. Furthermore, if the group of animals is already the population of interest, and we are not trying to generalize beyond the group, then linear regression treating observations as independent can be used to describe relationships within the group. Although this may underestimate standard errors due to unaccounted within-subject correlation, this is less of a concern when inference is limited to the group itself.

A detailed analysis of this case study is reported in Section 6.1.12.

5.16 Case Study: Salary Comparisons Between Technology and Finance Sectors

Case 15: Salary Comparisons Between Technology and Finance Sectors

To analyze the salary differences between the technology and finance sectors, a random sample comprising 40 employees is chosen, as illustrated in Table 5.5. Is there a significant disparity in the salaries of these two industries?

TABLE 5.5: Salary by Industry.

Industry	Salary	Industry	Salary
Tech	105522	Finance	83877
Tech	196808	Finance	154135
Tech	127812	Finance	339442
Tech	151245	Finance	57313
Tech	131364	Finance	16052
Tech	130426	Finance	163410
Tech	264158	Finance	49931
Tech	119028	Finance	165694
Tech	284428	Finance	66215
Tech	61428	Finance	91758
Tech	178493	Finance	81148
Tech	141765	Finance	105841
Tech	148445	Finance	126888
Tech	161106	Finance	69263
Tech	103657	Finance	174170
Tech	112803	Finance	109604
Tech	80074	Finance	64334
Tech	77972	Finance	180623
Tech	155091	Finance	162016
Tech	166726	Finance	118643

Initial Questions

1. What are the research objectives?

   Analyze the salary differences between the technology and finance sectors.

2. What are the statistical questions?

   Assess whether there is a significant difference between the salaries of the technology and finance industries.

3. What is the response variable, and what is the data type of the response variable?

   Salary is the response variable and can be treated as continuous data.

4. What are the explanatory variables of interest?

   Industry is the explanatory variable, and it is binary data having two levels: Technology and Finance.

5. Are there covariates?

   No.

6. What is the population of interest?

   All employees in the technology and finance industries.

7. What is the subject, and what is the number of distinct subjects?

   Each employee is a subject and there are 40 distinct subjects.

8. Are there subject-level data?

   Yes.

9. Are response variables dependent (repeated measures/clustered subjects)?

   No.

10. Are the subjects selected randomly?

    Yes.

11. Are the subjects randomly assigned to different groups?

    No.

Initial Thoughts

To compare means of two independent groups, a two-sample t test could be relevant. However, salaries are often heavy-tailed and may not follow the normal distribution required for the two-sample t test. If the normal assumption does not hold, we need to consider using a nonparametric alternative of such a test or convert the salaries to be normally distributed.

A detailed analysis of this case study is reported in Section 6.2.1.

5.17 Case Study: Sleep Hygiene Practices on Sleep Quality

Case 16: Sleep Hygiene Practices on Sleep Quality

A sleep research institute develops a nonpharmacological intervention program focusing on sleep hygiene practices, and the study evaluates the effectiveness of this 4-week intervention program by measuring participants' sleep quality scores before and after the intervention. Twenty participants with mild to moderate sleep difficulties are recruited for the study. The Pittsburgh Sleep Quality Index (PSQI) is administered before and after the 4-week intervention. The PSQI provides a score ranging from 0 to 21, where higher scores indicate poorer sleep quality. Does the sleep hygiene intervention program significantly improve participants' sleep quality as measured by PSQI?

Initial Questions

1. What are the research objectives?

   Evaluate the effectiveness of a 4-week intervention program in terms of a sleep quality score.

2. What are the statistical questions?

   Assess whether the sleep hygiene intervention program significantly reduces the PSQI.

3. What is the response variable, and what is the data type of the response variable?

   The PSQI score is the response variable and can be treated as continuous data.

4. What are the explanatory variables of interest?

   The time of the measurement is the explanatory variable, and it has two levels: before and after the intervention.

5. Are there covariates?

   No.

6. What is the population of interest?

   Unknown, and it is potentially the people with mild to moderate sleep difficulties, but we need to further verify it.

7. What is the subject, and what is the number of distinct subjects?

   Each participant is a subject and there are 20 distinct subjects.

8. Are there subject-level data?

   Yes.

9. Are response variables dependent (repeated measures/clustered subjects)?

   Yes, the response variables are measured twice for each subject.

10. Are the subjects selected randomly?

    Unknown.

11. Are the subjects randomly assigned to different groups?

    No.

Initial Thoughts

This is a typical case study for comparing before and after effects for the same group of subjects. If the difference in scores before and after the intervention follows a normal distribution, then we may use a paired *t* test. If not, then we need to consider some nonparametric alternative to the paired *t* test, such as the Wilcoxon signed-rank test. There is no causality can be established, and we need to know the population of the study and how the participants are recruited to decide whether the study can be generalized to the population.

A detailed analysis of this case study is reported in Section 6.2.2.

5.18 Case Study: Impact of Different Training Methods on Employee Performance

Case 17: Impact of Different Training Methods on Employee Performance

A multinational company wants to evaluate the effectiveness of three different training methods (Traditional Classroom, Online Self-Paced, and Blended Learning) on employee performance improvement percentages. The HR department collects performance improvement data (percentage increase in performance after training) from randomly selected employees who are randomly assigned to carry out one of these different training approaches. Is there a significant difference in the percentages of improvement in employee performance between the three training methods, and which method is most effective?

Initial Questions

1. What are the research objectives?

   Evaluate the effectiveness of three different training methods in improving employee performance.

2. What are the statistical questions?

   Is there a significant difference in employee performance improvement percentages between the three training methods and which method is most effective?

3. What is the response variable, and what is the data type of the response variable?

   Improvement is the response variable and can be treated as continuous data.

4. What are the explanatory variables of interest?

   Training method is the explanatory variable and is categorical with three levels: Traditional Classroom, Online Self-Paced, and Blended Learning.

5. Are there covariates?

   No.

6. What is the population of interest?

   The employees in the multinational company.

7. What is the subject, and what is the number of distinct subjects?

   Each employee is a subject and there are 45 distinct subjects.

8. Are there subject-level data?

   Yes.

9. Are response variables dependent (repeated measures/clustered subjects)?

   No.

10. Are the subjects selected randomly?

    Yes.

11. Are the subjects randomly assigned to different groups?

    Yes.

Initial Thoughts

To compare means for multiple groups, we can use ANOVA if the assumptions for using ANOVA hold. Otherwise, we need to consider other methods, such as the Kruskal-Wallis test, a nonparametric alternative to parametric ANOVA. Subjects are randomly sampled and randomly assigned to different groups, so inference to the population and causality of effects can be feasible for this case study.

A detailed analysis of this case study is reported in Section 6.2.3.

5.19 Case Study: Teaching Methods and Students Engagement Scores

Case 18: Teaching Methods and Students Engagement Scores

A high school implemented three different teaching methods in their biology classes to improve student engagement. The study is carried out among 45 randomly selected students, who are then randomly divided into three groups of 15 each, each group receiving one of the following teaching methods: Traditional Lecture (Control), Project-Based Learning, and Flipped Classroom. The school measures student engagement through a standardized observation protocol that scores engagement on a scale from 0 to 100, as shown in Table 5.6. Engagement scores are collected over a six-week period, and trained observers rate student engagement using standardized criteria.

TABLE 5.6: Teaching Methods and Engagement Scores.

Method	Score	Method	Score	Method	Score
Traditional	40	Project_Based	60	Flipped	50
Traditional	41	Project_Based	61	Flipped	51
Traditional	42	Project_Based	62	Flipped	52
Traditional	42	Project_Based	63	Flipped	53
Traditional	43	Project_Based	64	Flipped	54
Traditional	44	Project_Based	90	Flipped	55
Traditional	45	Project_Based	91	Flipped	56
Traditional	46	Project_Based	92	Flipped	57
Traditional	47	Project_Based	93	Flipped	58
Traditional	48	Project_Based	94	Flipped	60
Traditional	50	Project_Based	95	Flipped	65
Traditional	55	Project_Based	96	Flipped	75
Traditional	70	Project_Based	97	Flipped	85
Traditional	85	Project_Based	98	Flipped	90
Traditional	95	Project_Based	99	Flipped	95

The following are the research questions:

- Is there a significant difference in student engagement levels across the three teaching methods?
- Which teaching methods, if any, show significantly different levels of engagement?

Initial Questions

1. What are the research objectives?

   Assessing effectiveness of three teaching methods for students engagement.

2. What are the statistical questions?

   There are two statistical questions: (1) Is there a significant difference in student engagement levels between the three teaching methods? (2) Which teaching methods, if any, show significantly different levels of engagement?

3. What is the response variable, and what is the data type of the response variable?

   The engagement score is the response variable and can be treated as continuous data.

4. What are the explanatory variables of interest?

   Teaching method is the explanatory variable; it is categorical and has three levels: Traditional, Project-based, and Flipped.

5. Are there covariates?

   No.

6. What is the population of interest?

   All the students in the biology classes at the high school.

7. What is the subject, and what is the number of distinct subjects?

   A student is a subject and there are 45 distinct subjects.

8. Are there subject-level data?

   Yes.

9. Are response variables dependent (repeated measures/clustered subjects)?

   No.

10. Are the subjects selected randomly?

    Yes.

11. Are the subjects randomly assigned to different groups?

    Yes.

Initial Thoughts

To compare means for multiple groups, we can use ANOVA if the assumptions for parametric ANOVA hold. Otherwise, we need to consider other methods, such as the Kruskal-Wallis test, a nonparametric alternative to parametric ANOVA. However, for using Kruskal-Wallis test to test group medians, we require that the shapes of the distributions of the three groups be similar. Otherwise, we need to consider some other methods that are more flexible and do not rely on too many assumptions. Subjects are randomly sampled and randomly assigned to different groups, so inference to the population and causality of effects can be feasible for this case study.

A detailed analysis of this case study is reported in Section 6.2.4.

5.20 Case Study: Student Psychological Well-Being Survey

Case 19: Student Psychological Well-Being Survey

A researcher is developing a survey to measure psychological well-being among university students. The survey consists of two subscales:

- Emotional Stability (ES): Measured by 5 items on a 5-point Likert scale (1 = Strongly Disagree, 5 = Strongly Agree).
- Life Satisfaction (LS): Measured by 5 items on a 10-point scale (1 = Not At All Satisfied, 10 = Completely Satisfied).

The researcher wants to evaluate whether the results are reliable and whether the survey measures what it is intended to measure.

Initial Questions

1. What are the research objectives?

   Measure psychological well-being among university students.
2. What are the statistical questions?

   Evaluate whether the results are reliable and whether the survey measures what it is intended to measure.
3. What is the response variable, and what is the data type of the response variable?

   Scores of the questionnaire is the response variable, and it is on the Likert scale with five questions ranging from 1 to 5, and the other five questions from 1 to 10. Likert scales can be treated as ordinal data or continuous data, and since there are relatively more numbers (at least five in this case), for practical reasons, we may proceed with treating it as continuous data unless there is a strong reason to be against it.
4. What are the explanatory variables of interest?

   No explanatory variables are provided, although some demographic information could be collected, but for the research questions, we do not need that information.
5. Are there covariates?

   No.
6. What is the population of interest?

   Unknown.
7. What is the subject, and what is the number of distinct subjects?

   Each student is a subject and there are 50 distinct subjects.
8. Are there subject-level data?

   Yes.
9. Are response variables dependent (repeated measures/clustered subjects)?

   Yes, each subject provides 10 scores associated with the 10 questions asked in the questionnaire.

10. Are the subjects selected randomly?

    Unknown.

11. Are the subjects randomly assigned to different groups?

    Unknown.

Initial Thoughts

There are two questions that are relevant to feedback from questionnaires: reliability and validity. Assessing the reliability and validity of a survey questionnaire means checking whether it consistently measures what it is supposed to measure and whether it actually measures the intended concept, respectively. We can use Cronbach's α to assess consistency between items measuring the same construct, and factor analysis to validate the underlying structures.

A detailed analysis of this case study is reported in Section 6.2.5.

5.21 Case Study: User Satisfaction with Three Software Interfaces

Case 20: Evaluating User Satisfaction with Three Software Interfaces

A tech company wants to evaluate user satisfaction with three different software interfaces (Interface A, Interface B, and Interface C) designed for a new application. They recruit 20 participants and ask each participant to use all three interfaces in randomized order. After using each interface, participants rate their satisfaction on a 5-point Likert scale (1 = Very Dissatisfied, 5 = Very Satisfied). The company wants to determine whether there are significant differences in satisfaction between the three interfaces and which interface has the highest level of satisfaction.

Initial Questions

1. What are the research objectives?

   Evaluate user satisfaction with three different software interfaces.

2. What are the statistical questions?

   Determine if there are significant differences in satisfaction between the three interfaces and which interface has the highest level of satisfaction.

3. What is the response variable, and what is the data type of the response variable?

   Rating of satisfaction is the response variable, and it is Likert scale data.

4. What are the explanatory variables of interest?

   Interface is the explanatory variable; it is categorical and has three levels: A, B, and C.

5. Are there covariates?

   No.

6. What is the population of interest?

   All customers of the tech company.

7. What is the subject, and what is the number of distinct subjects?

   A participant is a subject and there are 20 distinct subjects.

8. Are there subject-level data?

   Yes.

9. Are response variables dependent (repeated measures/clustered subjects)?

   Yes, each subject provides ratings for all three interfaces.

10. Are the subjects selected randomly?

    Unknown.

11. Are the subjects randomly assigned to different groups?

    No, all subjects are assigned to all groups/interfaces.

Initial Thoughts

This case seems to be related to one-way ANOVA for repeated measures; however, the Likert scales lead to concerns of treating the response variable as continuous data. There are five levels of responses, and it can be treated as continuous data if the distance between consecutive levels is the same or ordinal data if not. The way in which the participants are selected needs to be further verified, and no causality can be established from the case study.

A detailed analysis of this case study is reported in Section 6.2.6.

5.22 Case Study: Cognitive Training on Test Scores

Case 21: Cognitive Training on Test Scores

Table 5.7 summarizes three existing studies that investigated the effects of short cognitive training programs on standardized test scores. Each study reports the sample sizes, means, and standard deviations for both the treatment (cognitive training) and control (no training) groups, along with the mean age of participants and a quality rating (high/low) for each study.

TABLE 5.7: Existing Studies of Cognitive Training on Test Scores.

Study	Year	n_t	$\overline{X}_t$	SD_t	n_c	$\overline{X}_c$	SD_c	Mean Age
Wong_2017	2017	33	84.5	7.5	32	80.3	8.2	27.6
Patel_2018	2018	38	79.0	9.8	36	74.2	10.1	23.8
Lopez_2019	2019	60	81.4	8.9	58	78.0	9.2	33.9

Research question: Do short cognitive training programs improve standardized test scores compared to no training (control)?

Initial Questions

1. What are the research objectives?

   Evaluate the effects of short cognitive training programs on standardized test scores.

2. What are the statistical questions?

   Do short cognitive training programs improve standardized test scores compared to no training (control)?

3. What is the response variable, and what is the data type of the response variable?

   Test score is the response variable and can be treated as continuous data.

4. What are the explanatory variables of interest?

   Treatment group (cognitive training vs. control) is the explanatory variable.

5. Are there covariates?

   Yes, the mean age of each study can be considered as a covariate.

6. What is the population of interest?

   Unknown.

7. What is the subject, and what is the number of distinct subjects?

   Each participant in the studies is a subject. The total number of distinct subjects across all studies is 257.

8. Are there subject-level data?

   No, only summary statistics from each study are provided.

9. Are response variables dependent (repeated measures/clustered subjects)?

   Unknown.

10. Are the subjects selected randomly?

    Unknown.

11. Are the subjects randomly assigned to different groups?

    Unknown.

Initial Thoughts

This is a typical meta-analysis case study where only summary statistics from multiple studies are provided. As there are only three studies, between-study variability may be difficult to estimate accurately. Therefore, we prefer to use a fixed-effect meta-analysis model instead of a random-effects model unless there is strong evidence of heterogeneity among the studies. The mean age can be included as a covariates in a meta-regression model to assess its impact on the effect size. The way in which subjects are selected and assigned to groups in each study needs to be further verified to assess the validity of the meta-analysis, as well as the generalizability of the findings to the population of interest and causality.

A detailed analysis of this case study is reported in Section 6.3.

5.23 Case Study: Sport Brand Preference by Gender

Case 22: Sport Brand Preference by Gender

Two sports brands, Brand A and Brand B, compete in the market. To understand the brand preference by gender, a survey is conducted. Each consumer can choose only one of the two brands, and Table 5.8 shows the numbers of male and female consumers who prefer each brand. The total number of consumers surveyed is 750.

TABLE 5.8: Sport Brand Preference by Gender.

Gender	Brand A	Brand B
Male	200	150
Female	180	220

Determine if there is a significant association between gender and brand preference.

Initial Questions

1. What are the research objectives?

   Understand sport brand preference by gender.

2. What are the statistical questions?

   Determine if there is a significant association between gender and brand preference.

3. What is the response variable, and what is the data type of the response variable?

   Brand preference is the response variable; it is binary data and has levels: A and B.

4. What are the explanatory variables of interest?

   Gender is the explanatory variable.

5. Are there covariates?

   No.

6. What is the population of interest?

   All consumers of the two sports brands.

7. What is the subject, and what is the number of distinct subjects?

   Each consumer is a subject and there are 750 distinct subjects.

8. Are there subject-level data?

   No, but pseudo subject-level data can be constructed from the contingency table, and it is unique up to permutations of subject IDs.

9. Are response variables dependent (repeated measures/clustered subjects)?

   No.

10. Are the subjects selected randomly?

    Unknown.

11. Are the subjects randomly assigned to different groups?

    No groups to be assigned.

Initial Thoughts

This is a typical case study with a two-by-two contingency table. All 750 observations are independent, and to assess whether the distribution of the response variable (brand) is associated with the explanatory variable (gender), a Chi-square test of independence is appropriate. Whether the subjects are selected randomly needs to be further confirmed to decide whether the study can be used for inference to the population of interest, and causality cannot be established from the study.

A detailed analysis of this case study is reported in Section 7.1.1.

5.24 Case Study: Treatment Protocol and Adverse Reactions

Case 23: Treatment Protocol and Adverse Reactions

A research team at a medical center investigates the possible relationship between a new treatment protocol and adverse reactions in patients with chronic kidney disease. 40 patients are randomly selected and randomly assigned to the treatment and control groups. The research team organizes their findings in the following contingency table (Table 5.9):

TABLE 5.9: Contingency Table of Treatment Groups and Adverse Reactions.

	Adverse Reaction	No Adverse Reaction
Treatment Group	2	18
Control Group	8	12

Does the new treatment protocol significantly reduce the occurrence of adverse reactions in patients with chronic kidney disease?

Initial Questions

1. What are the research objectives?

   Investigates the potential relationship between a new treatment protocol and adverse reactions in patients with chronic kidney disease.

2. What are the statistical questions?

   Does the new treatment protocol significantly reduce the occurrence of adverse reactions in patients with chronic kidney disease?

3. What is the response variable, and what is the data type of the response variable?

   Adverse Reaction is the response variable, and it is binary and has levels: Yes and No.

4. What are the explanatory variables of interest?

   Group is the explanatory variable, and it is binary and has levels: Treatment and Control.

5. Are there covariates?

   No.

6. What is the population of interest?

   Patients with chronic kidney disease.

7. What is the subject, and what is the number of distinct subjects?

   Each patient is a subject and there are 40 distinct subjects.

8. Are there subject-level data?

   No, but pseudo subject-level data can be constructed from the contingency table, and it is unique up to permutations of subject IDs.

9. Are response variables dependent (repeated measures/clustered subjects)?

   No.

10. Are the subjects selected randomly?

    Yes.

11. Are the subjects randomly assigned to different groups?

    Yes.

Initial Thoughts

This is a case study for a two-by-two contingency table with all the observations being independent, and we need to assess whether the response variable (Adverse Reaction) is associated with the explanatory variable (Group). The sample size is very small, and, therefore, Fisher's exact test is more appropriate than the Chi-square test of independence. Subjects are randomly selected and assigned to different groups, so the results of the study can be generalized to the population of interest, and causality of the adverse reaction may be developed from the study.

A detailed analysis of this case study is reported in Section 7.1.2.

5.25 Case Study: Educational Intervention on Smoking Cessation

Case 24: Effectiveness of an Educational Intervention on Smoking Cessation

A health organization implements an educational intervention program to promote smoking cessation among a group of 100 participants. The program consists of educational sessions, counseling, and follow-up support over a three-month period. The effectiveness of the program is assessed by comparing the smoking status of the participants before and after the intervention. Before the intervention, the 100 participants are surveyed to determine their smoking status: Smoker or Nonsmoker. After the intervention, the same participants are surveyed again to determine their current smoking status. Table 5.10 summarizes the survey. Does the educational intervention affect smoking cessation?

TABLE 5.10: Smoking Status.

Before \ After	Smoker	Nonsmoker	Total
Smoker	58	12	70
Nonsmoker	2	28	30
Total	60	40	100

Initial Questions

1. What are the research objectives?

   Assess effectiveness of an educational intervention program on promoting smoking cessation.
2. What are the statistical questions?

   Does the educational intervention significantly affect smoking cessation?
3. What is the response variable, and what is the data type of the response variable?

   Smoking status is the response variable, and it is binary data: smoker or nonsmoker.
4. What are the explanatory variables of interest?

   Timing is the explanatory variable, and it is binary: before and after the intervention.
5. Are there covariates?

   No.
6. What is the population of interest?

   Unknown.
7. What is the subject, and what is the number of distinct subjects?

   Each participant is a subject and there are 100 distinct subjects.
8. Are there subject-level data?

   No, but pseudo subject-level data can be constructed from the contingency table, and it is unique up to permutations of subject IDs.
9. Are response variables dependent (repeated measures/clustered subjects)?

   Yes, each subject is measured twice.
10. Are the subjects selected randomly?

    Unknown.
11. Are the subjects randomly assigned to different groups?

    No, there are no groups to assign.

Initial Thoughts

For comparing paired binary data, we may use McNemar's test. How the subjects are selected and what the population is need to be further clarified to decide whether the study can be used for inference to the population of interest. There is no causality can be established from the study.

A detailed analysis of this case study is reported in Section 7.1.3.

5.26 Case Study: Advertisements and Purchase Intention

Case 25: Advertisements and Purchase Intention

In order to study whether advertising can increase consumers' purchase intention, a random sample of 100 people is asked to watch the ads. Before they see anything, they are asked if they intend to buy the product. Then after watching the ads, they are asked again to see if anyone has changed their mind. The following is the contingency table (Table 5.11):

TABLE 5.11: Does Watching Ads Increase Purchase Intention?

	Before	After	Total
Yes	30	10	40
No	70	90	160
Total	100	100	200

Does the ad have significant effects in changing the consumers' purchase intention?

Initial Questions

1. What are the research objectives?

   Study whether advertising can increase consumers' purchase intention.
2. What are the statistical questions?

   Does the ad have significant effects in changing the consumers' purchase intention?
3. What is the response variable, and what is the data type of the response variable?

   Purchase intention is the response variable and is binary: Yes and No.
4. What are the explanatory variables of interest?

   Timing of expressing the purchase intention is the explanatory variable, and it is binary: before and after watching the advertisement.
5. Are there covariates?

   No.
6. What is the population of interest?

   Unknown.
7. What is the subject, and what is the number of distinct subjects?

   Each consumer is a subject and there are 100 distinct subjects.
8. Are there subject-level data?

   No, from the provided tables we cannot recover the subject-level data.
9. Are response variables dependent (repeated measures/clustered subjects)?

   Yes, each subject has been measured twice.
10. Are the subjects selected randomly?

    Yes.
11. Are the subjects randomly assigned to different groups?

    No.

Initial Thoughts

To compare paired binary data, we may use McNemar's test if a cross-classified table like that in Case 24 is provided. However, the table only provides aggregated information, and there are multiple ways to recover tables like that in Case 24. Therefore, McNemar's test cannot be applied directly. To assess the effect of watching the ads, we can consider the possible contingency tables that can be constructed from the aggregated information, and then calculate the range of test statistics and thus the range of p-values of McNemar's test.

A detailed analysis of this case study is reported in Section 7.1.4.

5.27 Case Study: Drug Effects in Different Hospitals

Case 26: Drug Effects in Different Hospitals

A pharmaceutical company conducts a study in three hospitals to evaluate a new drug. Data is collected on whether patients received the drug and whether they recovered, with the following details:

- Hospital A: Smaller hospital, fewer patients.
- Hospital B: Medium-sized hospital, moderate volume of patients.
- Hospital C: Large hospital, higher patient volume.

TABLE 5.12: Treatment vs. Recovery Across Hospitals.

	Hospital A		Hospital B		Hospital C	
	Rec.	Unrec.	Rec.	Unrec.	Rec.	Unrec.
Treated	8	2	15	5	25	10
Untreated	3	5	7	10	12	18

Determine if there is a significant association between the treatment and recovery, adjusting for the confounding factor of the hospitals.

Initial Questions

1. What are the research objectives?

   Evaluate effectiveness of a new drug.

2. What are the statistical questions?

   Determine if there is a significant association between treatment and recovery, adjusting for the confounding factor of the hospitals.

3. What is the response variable, and what is the data type of the response variable?

   Recovery status is the response variable and is binary: Recovered, Unrecovered.

4. What are the explanatory variables of interest?

Treatment status is the explanatory variable and is binary: Treated, Untreated.

5. Are there covariates?

   Yes, hospital is the covariate.

6. What is the population of interest?

   Unknown.

7. What is the subject, and what is the number of distinct subjects?

   Each patient is a subject and there are 120 distinct subjects.

8. Are there subject-level data?

   No, but pseudo subject-level data can be constructed from the contingency tables, and it is unique up to permutations of subject IDs.

9. Are response variables dependent (repeated measures/clustered subjects)?

   No.

10. Are the subjects selected randomly?

    Unknown.

11. Are the subjects randomly assigned to different groups?

    No.

Initial Thoughts

If the odds ratios of recovery between treated and untreated are common among the three hospitals, we can use the Cochran-Mantel-Haenszel procedure to assess the association between treatment and recovery, adjusting for the confounding factor of the hospital. To assess whether there is a common odds ratio among hospitals, we can use the Breslow-Day test. If the odds ratios are not the same, we can use logistic regression to assess how treatment affects recovery while controlling for the effect of hospitals. The population of interest and how the subjects are selected need to be clarified to decide whether the study can be generalized to the population, and causality cannot be established from this study since this is an observational study.

A detailed analysis of this case study is reported in Section 7.1.5.

5.28 Case Study: Job Satisfaction Across Departments

Case 27: Job Satisfactory

A company wants to assess job satisfactions among different departments to improve employee satisfaction and retention. A random sample of 70 employees is selected to provide feedback on job satisfaction. In addition, employees' salaries are included. What conclusions can be drawn from the feedback?

Initial Questions

1. What are the research objectives?

   Assess job satisfactions among different departments to improve employee satisfaction and retention.
2. What are the statistical questions?

   Evaluating the effects from factors such as departments and salaries that can lead to satisfied or unsatisfied employees.
3. What is the response variable, and what is the data type of the response variable?

   Satisfaction feedback is the response variable, and it is a binary data: satisfied or unsatisfied.
4. What are the explanatory variables of interest?

   Department and salary are the explanatory variables of interest.
5. Are there covariates?

   For this case, there is no need to distinguish covariates and explanatory variables.
6. What is the population of interest?

   All employees of the departments surveyed in the company.
7. What is the subject, and what is the number of distinct subjects?

   Each individual employee is a subject, and there are 70 distinct subjects.
8. Are there subject-level data?

   Yes.
9. Are response variables dependent (repeated measures/clustered subjects)?

   No.
10. Are the subjects selected randomly?

    Yes.
11. Are the subjects randomly assigned to different groups?

    No.

Initial Thoughts

The response variable is binary, so a logistic regression would be appropriate to assess the effect of the department and the salary. The subjects are randomly sampled, so we can make a statistical inference on all the employees of the departments surveyed in this company. However, note that the inference will only be for the departments surveyed, not the whole company. This is an observational study, and therefore no causality can be established from this study.

A detailed analysis of this case study is reported in Section 7.1.6.

5.29 Case Study: Customer Satisfaction with Product Features

Case 28: Assessing Customer Satisfaction with Product Features

A tech company launches a new software product with various features designed to enhance user experience. To gather feedback and improve future iterations, they conduct a survey asking randomly selected 30 users to rate each feature as either "dislike" or "like", and Table 5.13 summarizes the feedback from the users.

TABLE 5.13: Product Features Feedback.

Feature	Dislike	Like
Feature 1	18	12
Feature 2	26	4
Feature 3	10	20

Determine if there is significant difference between these features according to the responses from the users.

Initial Questions

1. What are the research objectives?

   Evaluate three features designed to enhance the user experience.

2. What are the statistical questions?

   Determine if there is significant difference between these features according to the responses from the users.

3. What is the response variable, and what is the data type of the response variable?

   Rating of features is the response variable and is binary: like and dislike.

4. What are the explanatory variables of interest?

   Feature is the explanatory variable and is categorical with three levels: Features 1, 2, and 3.

5. Are there covariates?

   No.

6. What is the population of interest?

   All users of the new software product.

7. What is the subject, and what is the number of distinct subjects?

   Each user is a subject and there are 30 distinct subjects.

8. Are there subject-level data?

   No.

9. Are response variables dependent (repeated measures/clustered subjects)?

   Yes, each subject has been measured three times under three different conditions (features).

10. Are the subjects selected randomly?

    Yes.

11. Are the subjects randomly assigned to different groups?

    No.

Initial Thoughts

This case is similar to Case 24, where each subject has been measured twice under two different conditions (Before and After) with a binary response variable (Smoker/Nonsmoker). The differences between this case and Case 24 are two folds: (1) in this case, each subject has been measured three times under three different conditions (three features); (2) unlike Case 24, there are no cross-classified tables provided in this case. Therefore, McNemar's test cannot be employed directly. Cochran's Q test can be used to test whether there is a significant difference among these three features, and if so, then McNemar's test can be used as a post hoc test. However, for this case, neither subject-level data nor cross-classified tables are provided, similar to Case 25, we need to find the ranges of the *p*-values for both Cochran's Q test and McNemar's test. The result can be generalized to the population, and no causality can be established.

A detailed analysis of this case study is reported in Section 7.1.7.

5.30 Case Study: Customer Satisfaction with Product Features (with Subject-Level Data)

Case 29: Assessing Customer Satisfaction with Product Features (with Subject-Level Data)

A tech company launches a new software product with various features designed to enhance user experience. To gather feedback and improve future iterations, they conduct a survey asking randomly selected 30 users to rate each feature as either "dislike" or "like", and Table 5.14 summarizes the feedback from the users and subject-level data is provided.

TABLE 5.14: Product Features Feedback.

Feature	Dislike	Like
Feature 1	18	12
Feature 2	26	4
Feature 3	10	20

Determine if there is significant difference between these features according to the responses from the users.

Initial Thoughts

This case is the same as Case 28, except that here the subject-level data is provided. Each subject has been measured three times, and a logistic mixed-effects model is appropriate, with the user as the random intercept and the features as the fixed effects. The result can be generalized to the population, and no causality can be established.

A detailed analysis of this case study is reported in Section 7.1.8.

5.31 Case Study: Career Choices Among College Students

Case 30: Career Choices Among College Students

A university career center wants to understand how students choose their first job sector after graduation. Graduates typically enter one of three mutually exclusive job sectors: Industry, Academia, Government. The career center randomly collects 300 recently graduated and graduating students and records the following variables: JobSector, GPA, Internship (Yes/No), and Major (STEM/Non-STEM). The goal is to understand how these predictors influence the probability of choosing each job sector.

Initial Questions

1. What are the research objectives?

   Understand how students choose their first job sector after graduation.

2. What are the statistical questions?

   How GPA, internship experience, and major influence the probability of choosing each job sector.

3. What is the response variable, and what is the data type of the response variable?

   Job sector is the response variable, and it is nominal data with three categories: Industry, Academia, and Government.

4. What are the explanatory variables of interest?

   GPA, internship experience, and major are the explanatory variables of interest.

5. Are there covariates?

   For this case, there is no need to distinguish covariates and explanatory variables.

6. What is the population of interest?

   All recently graduated and graduating students of the university.

7. What is the subject, and what is the number of distinct subjects?

   Each student is a subject, and there are 300 distinct subjects.

8. Are there subject-level data?

   Yes.

9. Are response variables dependent (repeated measures/clustered subjects)?

   No.

10. Are the subjects selected randomly?

    Yes.

11. Are the subjects randomly assigned to different groups?

    No.

Initial Thoughts

The response variable is nominal with more than two categories, making multinomial logistic regression an appropriate model. Since the subjects are randomly sampled, we can make statistical inferences on all recently graduated and graduating students of the university. However, this is an observational study, so causality cannot be established from the study.

A detailed analysis of this case study is reported in Section 7.2.1.

5.32 Case Study: Customer Satisfaction Levels

Case 31: Customer Satisfaction Levels

A company collects survey data from 200 customers about their satisfaction with a service. The goal is to understand how service speed (in minutes) and staff friendliness (rated 1-10) influence satisfaction levels.

Initial Questions

1. What are the research objectives?

   Understand factors that influence customer satisfaction.

2. What are the statistical questions?

   How service speed and staff friendliness affect satisfaction levels.

3. What is the response variable, and what is the data type of the response variable?

   Satisfaction level is the response variable, and it is ordinal data.

4. What are the explanatory variables of interest?

   Service speed and staff friendliness are the explanatory variables.

5. Are there covariates?

   For this case, there is no need to distinguish covariates and explanatory variables.

6. What is the population of interest?

   The customers of the company.

7. What is the subject, and what is the number of distinct subjects?

   Each customer is a subject of the study, and the number of distinct subjects is 200.

8. Are there subject-level data?

   Yes.

9. Are response variables dependent (repeated measures/clustered subjects)?

   No.

10. Are the subjects selected randomly?

    Unknown.

11. Are the subjects randomly assigned to different groups?

    No.

Initial Thoughts

The response variable is ordinal with three ordered categories: Dissatisfied, Neutral, and Satisfied. The predictors are continuous data. A regression model with an ordinal response variable, such as the proportional odds model, may be useful in this case. To decide whether the study can be used to make inferences on all customers of the company, we need to further verify whether the 200 customers are randomly sampled. This is an observational study and the causality will not be established from the study.

A detailed analysis of this case study is reported in Section 7.2.2.

5.33 Case Study: Patient Satisfaction with Healthcare Services

Case 32: Analyzing Patient Satisfaction with Healthcare Services

A hospital conducted a survey to assess patient satisfaction with its services. The response variable has three levels: Low Satisfaction, Moderate Satisfaction, High Satisfaction. The hospital collected data on the following predictors: Age: Patient age (years), Wait Time: Time spent waiting before being attended (minutes), and insurance: Whether the patient has private insurance (0 = No, 1 = Yes). The goal is to model how these predictors influence satisfaction levels.

Initial Questions

1. What are the research objectives?

   Assess patient satisfaction with services provided by a hospital.

2. What are the statistical questions?

   How age, waiting time, and whether having private insurance influence patient satisfaction levels.

3. What is the response variable, and what is the data type of the response variable?

   Patients' satisfaction levels, and this is ordinal data with three levels in order: Low, Moderate, and High.

4. What are the explanatory variables of interest?

   Age, waiting time, whether having private insurance are the explanatory variables of interest.

5. Are there covariates?

   For this case, there is no need to distinguish covariates and explanatory variables.

6. What is the population of interest?

   All patients of the hospital.

7. What is the subject, and what is the number of distinct subjects?

   Each patient is a subject and there are 200 distinct subjects.

8. Are there subject-level data?

   Yes.

9. Are response variables dependent (repeated measures/clustered subjects)?

   No.

10. Are the subjects selected randomly?

    Unknown.

11. Are the subjects randomly assigned to different groups?

    No.

Initial Thoughts

The response variable is ordinal with three ordered categories: Low, Moderate, and High. The predictors have continuous and binary data. A regression model with an ordinal response variable, such as the proportional odds model, can be attempted. To decide whether the study can be used to make inferences on all hospital patients, we need to further verify whether the 200 patients are randomly sampled. This is an observational study, and the causality of these predictors on satisfaction levels cannot be established.

A detailed analysis of this case study is reported in Section 7.2.3.

5.34 Case Study: Dietary Intervention on Cholesterol Levels

Case 33: Effect of a Dietary Intervention on Cholesterol Levels

A nutritionist assesses the impact of a dietary intervention on cholesterol levels. The study involves measuring cholesterol levels in patients before and after a dietary intervention. The cholesterol level of each patient is classified into three levels: Low, Medium, and High. The following contingency table (Table 5.15) summarizes the cholesterol levels of 100 patients before and after the dietary intervention:

TABLE 5.15: Cholesterol Levels.

Before \ After	Low	Medium	High
Low	30	10	5
Medium	8	20	10
High	2	5	10

Determine if there is significant difference in cholesterol levels before and after the dietary intervention.

Initial Questions

1. What are the research objectives?

   Evaluating the impact of a dietary intervention on cholesterol levels.

2. What are the statistical questions?

   Determine if there is significant difference in cholesterol levels before and after the dietary intervention.

3. What is the response variable, and what is the data type of the response variable?

   Cholesterol level is the response variable and is ordinal and has three levels: Low < Medium < High.

4. What are the explanatory variables of interest?

   Time of the measurement is the explanatory variable, and it is binary: Before and After.

5. Are there covariates?

   No.

6. What is the population of interest?

   Unknown.

7. What is the subject, and what is the number of distinct subjects?

   Each patient is a subject and there are 100 distinct subjects.

8. Are there subject-level data?

   No, but pseudo subject-level data can be constructed from the contingency table, and it is unique up to permutations of subject IDs.

9. Are response variables dependent (repeated measures/clustered subjects)?

   Yes, each subject has been measured twice.

10. Are the subjects selected randomly?

    Unknown.

11. Are the subjects randomly assigned to different groups?

    No.

Initial Thoughts

This case is about comparing before-and-after effects for categorical and ordinal responses. For 2 by 2 contingency table, we can use McNemar's test, and here we have a similar situation but with more than 2 categories. Some extensions to McNemar's tests such as the Stuart-Maxwell test, the Bhapkar test, and the McNemar-Bowker test can be employed. The population of interest and how the patients are selected need to be clarified to decide whether the study can generalize to the population, and no causality can be established.

A detailed analysis of this case study is reported in Section 7.2.4.

5.35 Case Study: Environment Noise Level and Exam Confidence

Case 34: Environment Noise Level and Exam Confidence

An education researcher investigates whether the level of noise in a student's study environment is associated with their self-reported confidence before an exam. While excessive noise may be distracting, students may adapt differently, so only a weak association is expected. The researcher collects data from 30 student volunteers, recording their study environment noise level (1 = Very Quiet, 2 = Quiet, 3 = Moderate, 4 = Noisy, 5 = Very Noisy) and their self-reported exam confidence (1 = Very Low, 2 = Low, 3 = Moderate, 4 = High, 5 = Very High) before a major exam. The goal is to determine if there is a significant association between study environment noise level and exam confidence.

Initial Questions

1. What are the research objectives?

   To determine whether the level of noise in a student's study environment is associated with their self-reported confidence before an exam.
2. What are the statistical questions?

   Is there a significant association between study environment noise level and exam confidence?
3. What is the response variable, and what is the data type of the response variable?

   Exam confidence is the response variable, and it is ordinal data with five levels: Very Low < Low < Moderate < High < Very High.
4. What are the explanatory variables of interest?

   Environment noise level is the explanatory variable, and it is ordinal data with five levels: Very Quiet < Quiet < Moderate < Noisy < Very Noisy.
5. Are there covariates?

   No.
6. What is the population of interest?

   Unknown.
7. What is the subject, and what is the number of distinct subjects?

   Each student volunteer is a subject, and there are 30 distinct subjects.
8. Are there subject-level data?

   Yes.
9. Are response variables dependent (repeated measures/clustered subjects)?

   No.
10. Are the subjects selected randomly?

    No.
11. Are the subjects randomly assigned to different groups?

    No.

Initial Thoughts

Both the response variable and the explanatory variable are ordinal data. We can use statistical methods designed for assessing associations between two ordinal variables, such as Spearman's rank correlation or Kendall's tau coefficient. To decide whether the study can be used to make inferences on a larger population, we need to further verify how the 30 student volunteers are selected. This is an observational study, and causality cannot be established from the study.

A detailed analysis of this case study is reported in Section 7.2.5.

5.36 Case Study: Modeling Property Crimes with Multiple Predictors

Case 35: Modeling Property Crimes with Multiple Predictors

Local government officials and community planners want to evaluate how socioeconomic conditions are associated with property crimes across counties. They have collected data from 30 counties over a single calendar year. For each county, the following variables are recorded: the total number of reported property crimes, total population, the poverty percentage, the unemployment rate, population density (persons per square mile), and median household income (in thousands of dollars). The goal is to understand what factors are associated with the number of property crimes across counties.

Initial Questions

1. What are the research objectives?

   The research objective is to understand what socioeconomic factors are associated with the number of property crimes across counties.

2. What are the statistical questions?

   The statistical question is to evaluate what those provided factors are associated with the number of property crimes and how they affect it.

3. What is the response variable, and what is the data type of the response variable?

   The response variable is the number of reported property crimes, and it is count data.

4. What are the explanatory variables of interest?

   The explanatory variables of interest are poverty percentage, unemployment rate, population density, and median household income. As larger population expects to have more crimes, total population should be included in the model as an offset term so that essentially the crime rate (number of crimes per person) is modeled instead of the total number of crimes.

5. Are there covariates?

   For this case, there is no need to distinguish covariates and explanatory variables.

6. What is the population of interest?

   Unknown.

7. What is the subject, and what is the number of distinct subjects?

   The subject is the county, and there are 30 distinct subjects.
8. Are there subject-level data?

   Yes, each county has its own set of measurements for the variables.
9. Are response variables dependent (repeated measures/clustered subjects)?

   No.
10. Are the subjects selected randomly?

    Unknown.
11. Are the subjects randomly assigned to different groups?

    No.

Initial Thoughts

The response variable is count data, and therefore a count regression model such as Poisson regression would be appropriate. The key for modeling it correctly is to treat the total population as an offset term in the model, so that we are essentially modeling the crime rate instead of the total number of crimes which expects to be higher in counties with larger population. To decide whether the study can be used to make inferences on a larger population, we need to further verify how the 30 counties are selected. This is an observational study, and causality cannot be established from the study.

A detailed analysis of this case study is reported in Section 7.3.1.

5.37 Case Study: Assessing Factors Affecting Customer Complaints

Case 36: Assessing Factors Affecting Customer Complaints

A telecommunications company serves many customers through phone calls, and customers have the option to leave a complaint about the service they received from the service representative right after their service interaction. The company wants to understand factors associated with the number of complaints received by service representatives during a workday, which can help the company to improve customer satisfaction and service quality. The company collects data from 40 service representatives randomly selected from multiple call centers for a single day. For each representative, the following variables are recorded: number of complaints, number of customers served, and years of experience of the service representative. The goal is to understand what factors influence the number of complaints about the service provided by service representatives.

Initial Questions

1. What are the research objectives?

   The research objective is to understand what factors are associated with the number of complaints received by service representatives during a workday. Because the number of complaints is expected to be higher for representatives who serve more customers, so the actual research objective is to understand what factors are associated with the complaint rate (number of complaints per customer served) of service representatives during a workday.

2. What are the statistical questions?

   The statistical question is to evaluate what those provided factors are associated with the complaint rate and how they affect it.

3. What is the response variable, and what is the data type of the response variable?

   The response variable is the number of complaints out of the number of customers served, and it is count data.

4. What are the explanatory variables of interest?

   The explanatory variables of interest are years of experience of the service representative, and potentially the number of customers served.

5. Are there covariates?

   For this case, there is no need to distinguish covariates and explanatory variables.

6. What is the population of interest?

   All service representatives of the company.

7. What is the subject, and what is the number of distinct subjects?

   The subject is a service representative, and there are 40 distinct subjects.

8. Are there subject-level data?

   Yes, the data are collected at the subject level.

9. Are response variables dependent (repeated measures/clustered subjects)?

   No, each subject is observed only once.

10. Are the subjects selected randomly?

    Yes, the subjects are randomly selected from multiple call centers.

11. Are the subjects randomly assigned to different groups?

    No, this is an observational study.

Initial Thoughts

The response variable is the number of complaining customers out of the number of customers served, and therefore a binomial regression would be appropriate. Since the subjects are randomly sampled, we can make statistical inferences on all service representatives of the company. However, this is an observational study, so causality cannot be established from the study.

A detailed analysis of this case study is reported in Section 7.3.2.

6

Formal Analysis – Continuous Data

This chapter covers detailed statistical analyses of case studies that have continuous response variables. Normal and nonnormal response variables lead to different procedures, so they are discussed separately. For each case study, we discuss the background and the data, then present the statistical analysis, and finally conclude with a summary of the findings. When appropriate, discussions on relevant methodologies are included, and references are provided for further reading.

6.1 Normal Data

When the response variable is approximately normally distributed, we can use standard statistical methods such as *t* tests, linear regression, analysis of variance (ANOVA), and linear mixed-effects models to analyze the data. In this section, we present several case studies that all have response variables that are continuous and approximately normally distributed.

6.1.1 Two-Sample *t* Test

Two-sample *t* tests are commonly used to compare the means of two independent groups. We use the following case study to illustrate how to conduct a two-sample *t* test in R. Note that, when we say statistical tests, we emphasize that these are inferential procedures and not just computational recipes. Understanding the underlying assumptions, interpretations, and limitations of these tests is crucial for their proper application in real-world scenarios.

DOI: 10.1201/9781003683827-6

Case 1: Effectiveness of Learning Formats

This study aims to compare the effectiveness of two distinct learning formats: in-person and online. A random sample of 42 students is drawn from a large class consisting of two groups based on their choice of learning format. Both groups undergo the same final exam after one semester. The scores are listed in Table 5.2. Evaluate whether the final exam scores differ significantly between students in the two learning formats and whether the effectiveness of the two learning formats is significantly different.

TABLE 5.2: Learning Format and Final Score Comparison.

Format	Final Score	Format	Final Score
Online	70	In-Person	77.5
Online	74	In-Person	72
Online	68.5	In-Person	72.5
Online	69.5	In-Person	68.5
Online	74.5	In-Person	76
Online	68	In-Person	73.5
Online	72.5	In-Person	76.5
Online	69	In-Person	71
Online	73.5	In-Person	75
Online	71.5	In-Person	73
Online	71	In-Person	78
Online	72.5	In-Person	73.5
Online	68.5	In-Person	75.5
Online	71	In-Person	74.5
Online	70	In-Person	76.5
Online	73	In-Person	74
Online	71.5	In-Person	72
Online	70	In-Person	71.5
Online	69.5	In-Person	74.5
Online	68	In-Person	72.5
Online	70.5		
Online	69.5		

For answers to the initial questions and some initial thoughts about this case study, refer to Section 5.2.

To assess whether the final exam scores differ significantly between students in the two learning formats, we can carry out a two-sample t test with the following hypothesis where μ_1 and μ_2 are the respective group means:

$$H_0 : \mu_1 = \mu_2$$
$$H_1 : \mu_1 \neq \mu_2.$$

This is a two-sided test, as we are interested in whether they are different without concerns about which one is larger. Prior to conducting a formal analysis, an exploratory data analysis is presented below.

```
library(dplyr)
library(ggplot2)
```

```
# Read the data
data <- read.csv("data/learning-format.csv")
data$Format <- as.factor(data$Format)

# Calculate descriptive statistics by groups
desc_stats <- data %>% group_by(Format) %>%
  summarise(n = n(), mean = mean(Final_Score), sd = sd(Final_Score),
    se = sd/sqrt(n), min = min(Final_Score), max = max(Final_Score))
print(desc_stats)
: # A tibble: 2 x 7
:   Format        n  mean    sd    se   min   max
:   <fct>     <int> <dbl> <dbl> <dbl> <dbl> <dbl>
: 1 In-Person    20  73.9  2.38 0.533  68.5  78
: 2 Online       22  70.7  1.94 0.414  68    74.5
```

The mean score for the in-person group is higher than that for the online group. The following box plot further indicates that there is probably a difference between the two groups.

```
# Create box plot
ggplot(data, aes(y = Format, x = Final_Score)) +
  geom_boxplot() + theme_bw() +
  stat_summary(fun = "mean", geom = "point", shape = 23, size = 2) +
  labs(x = "Final Score", y = "Learning Format")
```

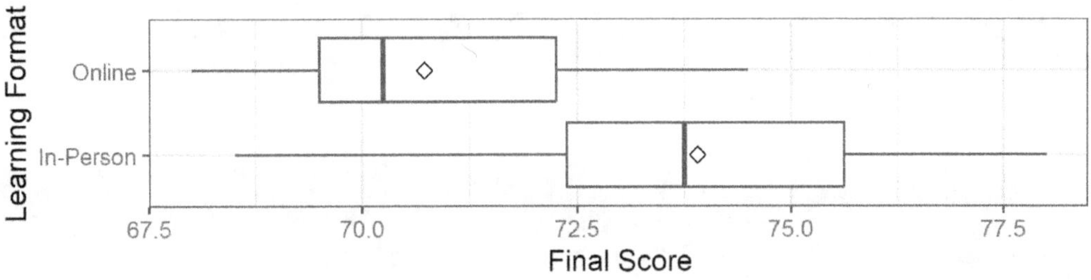

FIGURE 6.1: Box Plots for Different Learning Formats.

Density plots can also help visualize the distribution of final scores for the two learning formats.

```
# Create density plot
ggplot(data, aes(x=Final_Score, fill=Format)) +
  geom_density(alpha=0.5) + theme_bw() +
  labs(title = "Distribution of Final Scores by Learning Format",
       x = "Final Score", y = "Density") +
  scale_fill_grey(start = 0.3, end = 0.8)
```

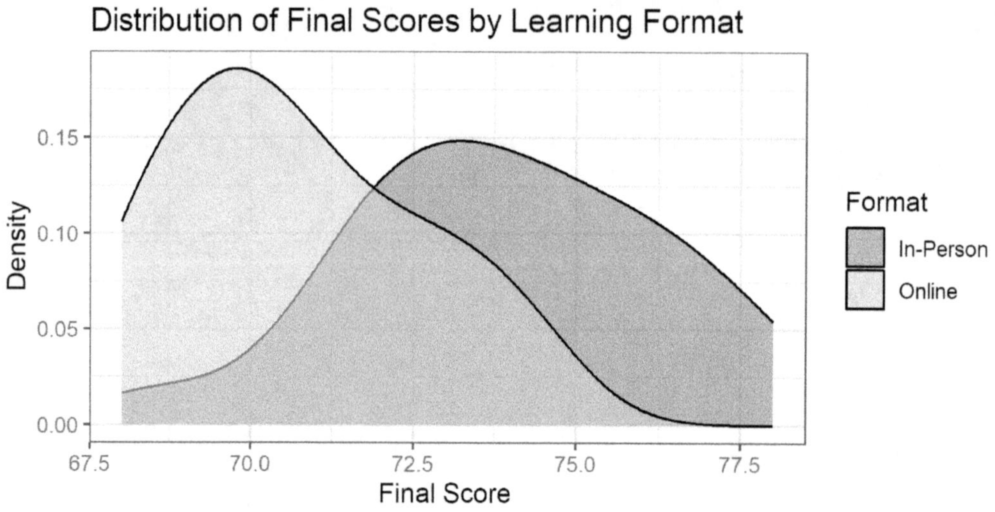

FIGURE 6.2: Density Plots for Different Learning Formats.

Although it is visually clear that the group means are very different, a formal analysis is needed to assess to what extent the difference is. Here we perform an independent two-sample t test. For more theoretical details, see Note 6.4.

First, we check the required assumptions for using the two-sample t test, namely, the normality and the equality of variances between the groups.

```
# Normality tests: Shapiro-Wilk for each group
shapiro.test(data$Final_Score[data$Format=="In-Person"])
:
:   Shapiro-Wilk normality test
:
: data:  data$Final_Score[data$Format == "In-Person"]
: W = 0.98209, p-value = 0.9581
shapiro.test(data$Final_Score[data$Format=="Online"])
:
:   Shapiro-Wilk normality test
:
: data:  data$Final_Score[data$Format == "Online"]
: W = 0.9492, p-value = 0.3044
```

Then we can use QQ plots to visually check the normality assumption for each group, as shown below.

```
# Create QQ plots for each group
ggplot(data, aes(sample = Final_Score)) + geom_qq() + geom_qq_line() +
  facet_wrap(~Format) + theme_bw() +
  labs(title = "QQ Plots by Learning Format",
       y = "Sample quantiles of final scores", x = "Theoretical quantiles")
```

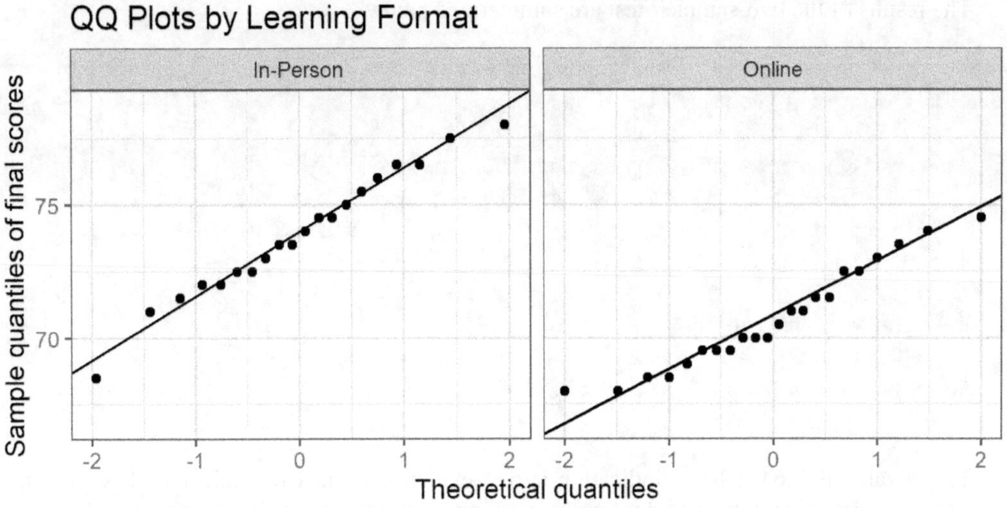

The QQ plots also suggest that the normality assumption is reasonable for both groups. Next, we check the homogeneity of variance assumption using several tests.

```
# Homogeneity of variance (Levene's test)
car::leveneTest(Final_Score ~ Format, data = data)

: Levene's Test for Homogeneity of Variance (center = median)
:       Df F value Pr(>F)
: group  1  0.6184 0.4363
:       40
```

```
# F test (assuming normality, two groups only)
var.test(Final_Score ~ Format, data = data)
:
:  F test to compare two variances
:
: data:  Final_Score by Format
: F = 1.5013, num df = 19, denom df = 21, p-value = 0.3662
: alternative hypothesis: true ratio of variances is not equal to 1
: 95 percent confidence interval:
:  0.6146748 3.7426292
: sample estimates:
: ratio of variances
:          1.501284
```

Both the normality and the equal variances assumptions seem reasonable based on the above tests and plots, so we can proceed with the standard two-sample t test.

```
# Perform two-sample t test
t_test_result <- t.test(Final_Score ~ Format, data = data,
                        alternative = "two.sided",  var.equal = TRUE)
```

The results of the two-sample t test are summarized below.

```
cat(t_test_result$method, ":", t_test_result$alternative, "\n",
    "t =", t_test_result$statistic, "\n",
    "p-value =", t_test_result$p.value, "\n",
    "Mean difference =", t_test_result$estimate[1] -
                        t_test_result$estimate[2], "\n",
    "95 percent confidence interval:", t_test_result$conf.int, "\n")
:  Two Sample t-test : two.sided
:  t = 4.747377
:  p-value = 2.63873e-05
:  Mean difference = 3.172727
:  95 percent confidence interval: 1.822019 4.523436
```

The p-value is 2.64×10^{-5}, indicating strong evidence against the null hypothesis of equal means. As statistical significance does not necessarily imply practical significance, we should calculate effect size and uncertainty, and let the client and / or domain experts to decide whether the effect is practically meaningful.

The following gives a calculation of the effect size using Cohen's d (1.47), which is considered a large effect size. For more details on Cohen's d and it's interpretation, see Note 6.6.

```
# Calculate effect size (Cohen's d)
cohens_d <- function(x, y){ nx <- length(x); ny <- length(y)
  pooled_sd <- sqrt(((nx-1)*var(x) + (ny-1)*var(y))/(nx+ny-2))
  (mean(x) - mean(y))/pooled_sd}

effect_size <- cohens_d(data$Final_Score[data$Format=="In-Person"],
  data$Final_Score[data$Format=="Online"])
cat("Cohen's d:", round(effect_size, 2))
: Cohen's d: 1.47
```

Cohen's d can also be calculated using an R package called `effectsize`, as shown below.

```
# Alternatively, use an R package effectsize
effectsize::cohens_d(Final_Score ~ Format, data=data)
: Cohen's d |         95% CI
: -------------------------
: 1.47      | [0.77, 2.14]
:
: - Estimated using pooled SD.
```

Conclusions

In this case study, we conduct an independent two-sample t test to assess whether the final exam scores differ significantly between students in two different learning formats: in-person and online. The two-sample t test indicates strong evidence of a difference between groups ($t = 4.75, p = 2.64 \times 10^{-5}$). The estimated mean difference is 3.2 units (95% CI: 1.82 to 4.52). The standardized effect size is Cohen's $d = 1.47$, indicating that the difference between groups is large on the scale of within-group variability. This is an observational study, and therefore, potential confounding factors and the lack of randomization limit causal interpretation. A way to address such an issue is to randomly assign students to the two different groups. Furthermore, the statistical analysis does not assess the substantive importance of the difference of 3.2 units, and it is up to the client and / or domain experts to determine whether this difference is practically meaningful.

Note 6.1 (The Shapiro-Wilk Test)**.** The *Shapiro-Wilk test* ([99]) is one of the most powerful tests for checking normality, particularly effective for small sample sizes ($n < 50$). The null hypothesis is that the sample comes from a normally distributed population. The test statistic is

$$W = \frac{(\sum_{i=1}^{n} a_i X_{(i)})^2}{\sum_{i=1}^{n} (X_i - \bar{X})^2},$$

where $X_{(i)}$ are the ordered sample values, a_i are constants derived from the means and covariance matrix of the order statistics, and $\bar{X}$ is the sample mean. The test statistic W ranges between 0 and 1, values close to 1 indicate normality, and small values lead to rejection of normality. Critical values depend on sample size and significance levels. For small samples, exact critical values can be used, while for larger samples, approximations or simulations are often employed. The Shapiro-Wilk test is sensitive to outliers and may reject normality if outliers are present, even if the underlying distribution is approximately normal. Therefore, it is advisable to use the Shapiro-Wilk test together with QQ plots to evaluate normality comprehensively. For more details on the Shapiro-Wilk test and other normality tests, see [108].

Note 6.2 (*F* Test for Equality of Variances)**.** The *F test for equality of variances* is a classical parametric procedure used to assess whether two independent normal populations have equal variances. Let $X_1, \ldots, X_{n_1} \sim \mathcal{N}(\mu_1, \sigma_1^2)$ and $Y_1, \ldots, Y_{n_2} \sim \mathcal{N}(\mu_2, \sigma_2^2)$ be two independent samples. The null and alternative hypotheses are respectively

$$H_0 : \sigma_1^2 = \sigma_2^2, \qquad H_1 : \sigma_1^2 \neq \sigma_2^2.$$

Let S_1^2 and S_2^2 denote the sample variances of the two groups. The test statistic is defined as

$$F = \frac{S_1^2}{S_2^2},$$

where, without loss of generality, $S_1^2 \geq S_2^2$ is often chosen to ensure $F \geq 1$. Under the null hypothesis and the assumption of normality, the statistic follows an F distribution with $(n_1 - 1, n_2 - 1)$ degrees of freedom,

$$F \sim F_{n_1-1, n_2-1}.$$

The null hypothesis is rejected at significance level α if F lies outside the acceptance region determined by the corresponding critical values of the F distribution. Although the F test is optimal under normality, it is highly sensitive to departures from this assumption, and it can only compare two groups at a time. Therefore, it is often recommended to use more robust alternatives, such as Levene's test, when the normality assumption is in doubt or when comparing more than two groups. A reference for the F test and its properties is [102].

Note 6.3 (Levene's Test for Homogeneity of Variance). *Levene's Test for Homogeneity of Variance* ([67]) is a widely used method for testing the quality of variances across multiple groups. The hypotheses are

$$H_0: \text{The population variances are equal among all groups,}$$
$$H_1: \text{At least one variance differs.}$$

The mathematical formulation of the test statistic W is defined as

$$W = \frac{N-k}{k-1} \times \frac{\sum_{i=1}^{k} n_i(\bar{Z}_{i\cdot} - \bar{Z}_{\cdot\cdot})^2}{\sum_{i=1}^{k}\sum_{j=1}^{n_i}(Z_{ij} - \bar{Z}_{i\cdot})^2},$$

where N is the sample size, k is the number of groups, n_i is the size of ith group, and $Z_{ij} = |Y_{ij} - \bar{Y}_{i\cdot}|$, $\bar{Z}_{i\cdot}$ is the mean of Z_{ij} for group i, $\bar{Z}_{\cdot\cdot}$ is the overall mean of Z_{ij}. Compare W to $F_{k-1,N-k,\alpha}$ distribution, and reject H_0 if $W > F_{k-1,N-k,\alpha}$, suggesting that variances are not equal. Levene's test is more robust to non-normality than the F test but less powerful for normal data, for which the standard F test is preferred. For more robust versions of Levene's test when the normality assumption is violated, see [17], where median or trimmed mean instead of mean is used to calculate Z_{ij}.

Note 6.4 (Two-Sample t Test with Equal Variances). *Two-sample t tests* are used to compare the means of two independent groups. The key assumptions for the standard two-sample t test are that the two samples are independent, the response variable is approximately normally distributed in each group, and the population variances of the two groups are equal. Assuming the equal population variances ($\sigma_1^2 = \sigma_2^2 = \sigma^2$), the hypothesis for a two-sided test, where μ_1 and μ_2 are the respective group means, is

$$H_0: \mu_1 = \mu_2,$$
$$H_1: \mu_1 \neq \mu_2.$$

For a one-sided test change the equality sign in H_1 by an inequality sign accordingly. The test statistic is

$$t = \frac{\bar{X}_1 - \bar{X}_2}{s_p\sqrt{\frac{1}{n_1} + \frac{1}{n_2}}},$$

where $\bar{X}_1, \bar{X}_2$ are the sample means, n_1, n_2 are the sample sizes, and s_p is the pooled standard deviation. The pooled standard deviation s_p is calculated as

$$s_p = \sqrt{\frac{(n_1-1)s_1^2 + (n_2-1)s_2^2}{n_1+n_2-2}},$$

where s_1^2, s_2^2 are the sample variances and $n_1 + n_2 - 2$ is the degrees of freedom. Under H_0, the test statistic follows a t distribution with $n_1 + n_2 - 2$ degrees of freedom. The pooled variance estimator s_p^2 is the weighted average of the two sample variances. This test is more powerful than Welch's t test when the equal variance assumption holds. If the variances are not equal, the pooled variance estimator can be misleading and lead to incorrect conclusions, and thus it is recommended to use Welch's t test. We refer to [102] for more details on the two-sample t test and its properties.

Note 6.5 (Two-Sample t Test without Equal Variances). When the assumption of equal variances is violated, a more robust alternative to the standard two-sample t test using the pooled variance estimator is Welch's t test [114]. Compared to the standard two-sample t test, the differences are how we calculate the variance to standardize the group mean difference and how to calculate the degree of freedoms. The following is the test statistic of Welch's t test:

$$t = \frac{\bar{X}_1 - \bar{X}_2}{\sqrt{\frac{s_1^2}{n_1} + \frac{s_2^2}{n_2}}},$$

where $\bar{X}_1, \bar{X}_2$ are sample means, s_1^2, s_2^2 are sample variances, and n_1, n_2 are sample sizes. Under the null hypothesis, the test statistic approximately follows a t distribution with degrees of freedom calculated using the Welch-Satterthwaite equation [95].

Note 6.6 (Cohen's d, Effect Size for Comparing Two Group Means)**.** *Cohen's d* is one of the widely used standardized mean difference effect sizes for comparing two groups, and it essentially measures the standardized difference between group means as follows:

$$d = \frac{\bar{X}_1 - \bar{X}_2}{s_p},$$

where

$$s_p = \sqrt{\frac{(n_1 - 1)s_1^2 + (n_2 - 1)s_2^2}{n_1 + n_2 - 2}}.$$

Cohen [27] proposed benchmarks for interpreting d: values of 0.2, 0.5, and 0.8 are often described as "small", "medium", and "large" effects, respectively. However, these thresholds should not be treated as rigid cutoffs but rather as rough guidelines that may vary across disciplines and research contexts [66]. For smaller sample sizes, Hedges' g is a less biased effect size measure; see Note 6.39 for more details. Effect sizes such as Cohen's d are not commonly taught in statistic classes but can be useful in research field such as psychology and social sciences. For more practical details on how to calculate and report effect sizes for t tests and ANOVA, [66] is a good reference.

Note 6.7 (Relationship Between Cohen's d and the t Statistic)**.** Cohen's d and the t statistic in a two-sample t test with equal variances are related as

$$d = t\sqrt{1/n_1 + 1/n_2},$$

where d is Cohen's d, t is the t statistic, and n_1 and n_2 are the sample sizes of the two groups. The t statistic in Note 6.4 assesses the magnitude of the group mean difference on the scale of variability in group means. The variability in the group mean difference decreases in sample size, that is, a larger sample size more likely leads to a larger t statistic, and thus statistical significance can be achieved simply by increasing the sample size. However, Cohen's d is sample-size independent, and it assesses the magnitude of the group mean difference on the scale of within-group variability that is not affected by sample size. Although given a fixed sample size, Cohen's d increases as the t statistic increases, as both measure the separation between groups, they are not interchangeable. Cohen's d is more informative about the practical significance of the group mean difference, while the t statistic and the corresponding p-value mainly indicate how incompatible the observed sample data are with the null hypothesis of no difference.

6.1.2 Paired t Test

Paired t tests are used to compare the means of two related groups. This is typically applied in before-and-after studies, where the same subjects are measured at two different time points or under two different conditions. The key feature of paired t tests is that they account for the correlation between the paired observations, which can lead to increased statistical power compared to independent two-sample t tests. The following case study illustrates how to conduct a paired t test in R.

Case 2: Exercises on Blood Pressures

This case study examines the effect of a 12-week exercise program on blood pressure. A group of 25 volunteers participate in the study. Table 5.3 shows the blood pressures measured before and after the program. Does the 12-week exercise program significantly reduce blood pressure?

TABLE 5.3: Blood Pressure Before and After the Exercise Program.

Patient ID	BP Before	BP After
1	139	120
2	143	143
3	161	136
4	146	125
5	146	146
6	162	139
7	150	132
8	132	143
9	138	143
10	141	142
11	157	141
12	149	140
13	149	134
14	146	132
15	139	132
16	163	129
17	150	133
18	125	124
19	152	155
20	140	146
21	134	125
22	143	131
23	135	131
24	138	142
25	139	134

For answers to the initial questions and some initial thoughts about this case study, refer to Section 5.3. We first load the blood pressure measurement data and do some basic descriptive statistics.

```
library(dplyr)
library(ggplot2)
data <- read.csv("data/bp-measurements.csv")

# Add the difference column
data$Difference <- data$BP_Before - data$BP_After

# Basic descriptive statistics
summary_stats <- data.frame(
  Time = c("Before", "After"),
  Mean = c(mean(data$BP_Before), mean(data$BP_After)),
  SD = c(sd(data$BP_Before), sd(data$BP_After)),
```

```
  Min = c(min(data$BP_Before), min(data$BP_After)),
  Max = c(max(data$BP_Before), max(data$BP_After))
)
print(summary_stats)
:     Time   Mean        SD Min Max
: 1 Before 144.68 9.551265 125 163
: 2  After 135.92 8.255907 120 155
```

After creating a new column for the difference of blood pressure measures, we draw a spaghetti plot to visualize the changes of blood pressures for each individual.

```
# Spaghetti plot for individual changes
data_long <- data %>%
  tidyr::gather(key = "TimePoint", value = "BP", BP_Before, BP_After)

data_long$TimePoint <- factor(data_long$TimePoint,
                              levels = c("BP_Before", "BP_After"))
ggplot(data_long, aes(x = TimePoint, y = BP, group = PatientID)) +
  geom_line(alpha = 0.3) + geom_point() + theme_bw() +
  labs(title = "Individual Blood Pressure Changes",
       y = "Blood Pressure (mmHg)", x = "Time Point")
```

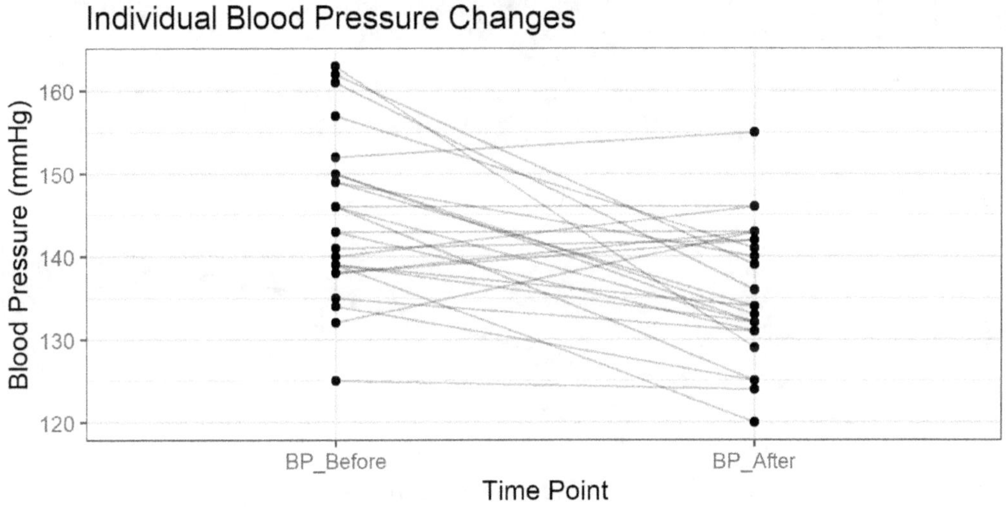

We then generate a density plot to gain an initial idea of how the differences in blood pressures appear.

```
ggplot(data, aes(x=Difference)) + geom_density() + theme_bw() +
  labs(title = "Distribution of Differences (Before - After)",
       x = "Difference in Blood Pressure (mmHg)", y = "Density")
```

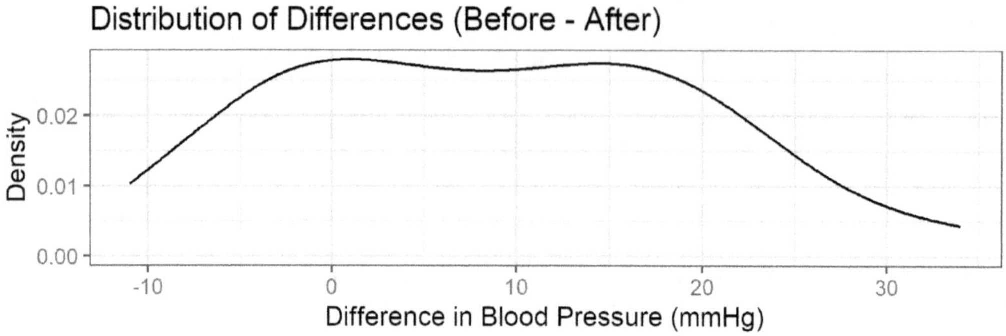

We now formally check the main assumption for the paired t test as follows:

```
# Check normality of differences
shapiro.test(data$Difference)
:
:  Shapiro-Wilk normality test
:
: data:  data$Difference
: W = 0.97642, p-value = 0.8062
```

```
# QQ plot of differences
ggplot(data, aes(sample = Difference)) + geom_qq() + geom_qq_line() +
  theme_bw() + labs(title = "QQ Plots",
                    y = "Sample quantiles of difference",
                    x = "Theoretical quantiles")
```

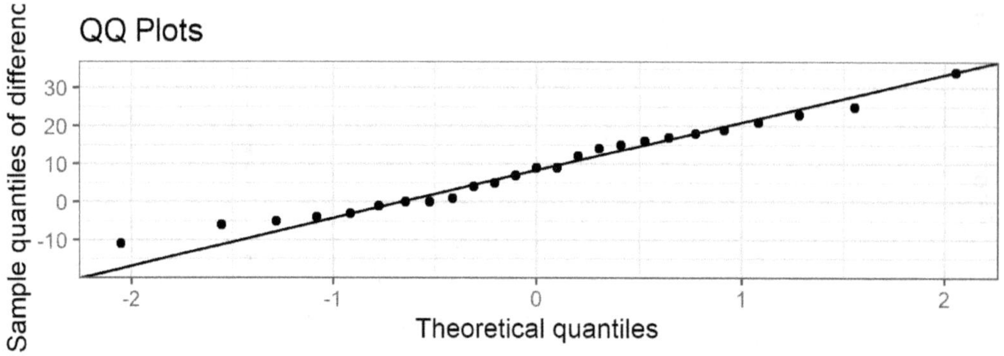

The p-value of the normality test is 0.81, which, along with the QQ plot, indicates that the normality assumption is valid. Now we are ready to do the paired t test. Since we are interested in whether blood pressure is reduced, we perform a one-sided test.

```
# Perform paired t test
t.test(data$BP_Before, data$BP_After, paired = TRUE,
       alternative = "greater")
:
:  Paired t-test
:
: data:  data$BP_Before and data$BP_After
: t = 3.861, df = 24, p-value = 0.000374
```

```
: alternative hypothesis: true mean difference is greater than 0
: 95 percent confidence interval:
:  4.878295      Inf
: sample estimates:
: mean difference
:            8.76
```

The p-value < 0.001, suggesting a strong evidence that the blood pressure is reduced. The effect size in terms of Cohen's d can be calculated as follows:

```
# Calculate Cohen's d for paired data
mean_diff <- mean(data$Difference)
sd_diff <- sd(data$Difference)
cohens_d <- mean_diff / sd_diff
print(cohens_d)
: [1] 0.7722034
```

```
# Alternatively, use an R package effectsize
effectsize::cohens_d(x=data$BP_Before, y=data$BP_After, paired = TRUE)
: Cohen's d |       95% CI
: -------------------------
: 0.77      | [0.32, 1.21]
```

Conclusions

In this case study, we use a paired t test to evaluate whether a new drug effectively reduces blood pressure. The paired t test indicates strong evidence of reduced blood pressure ($t = 3.86, p < 0.001$). The estimated mean difference is 8.8 units (95% CI: $(4.88, +\infty)$). The size of the standardized effect is Cohen'$d = 0.77$, indicating a medium effect on the scale of variability of the before-and-after differences. Although these results provide clear evidence against the null hypothesis of no difference, the result is based on an observational study, and therefore potential confounding factors and lack of randomization limit causal interpretation. In addition, subjects are not randomly selected; therefore, the study has limitations in generalizing to the population. Moreover, statistical analysis cannot be used directly to assess the substantive importance of reducing 8.76 units of blood pressure, and this should be evaluated in the context of the reduction of blood pressure.

Note 6.8 (Paired t Test). *Paired t Tests* are used to compare the means of two related groups. The key assumptions are that differences between before and after measurements are normally distributed and that the differences are independent of each other. The test statistic is

$$t = \frac{\bar{X_1} - \bar{X_2}}{s_D/\sqrt{n}},$$

where s_D is the estimated population standard deviation of the differences and n is the number of pairs, and thus $s_D/\sqrt{n}$ is the standard error of the mean difference. Under H_0, the test statistic follows the Student's t distribution with the degree of freedom $n-1$. The paired t test is the same as one-sample t test when treating the n pair differences as a sample of size n and then comparing them with 0. See (page 766 of [100]) for more details on the paired t test.

Note 6.9 (What If a Two-Sample t Test is Wrongly Used for Paired Data)**.** If a two-sample t test is incorrectly used for paired data, then it ignores the correlation between pairs, and when the two pairs are positively correlated (most likely in repeated measurements), the misuse typically reduces statistical power and may fail to detect genuine differences.

The following mathematical expression helps us to understand the impact. For paired data (X_1, X_2), let $n_1 = n_2 = n$, σ_1^2 and σ_2^2 be the variances of X_1 and X_2, respectively, and the pairs be positively correlated, i.e., $\text{Corr}(X_1, X_2) = \rho \geq 0$. If X_1 and X_2 were independent, then $\text{Var}(\bar{X}_1 - \bar{X}_2) = (\sigma_1^2 + \sigma_2^2)/n$. However, when X_1 and X_2 are positively correlated, the variance of the difference $\bar{D} = \bar{X}_1 - \bar{X}_2$ is smaller than that under independence. Specifically, $\text{Var}(\bar{D}) = \text{Var}(\bar{X}_1) + \text{Var}(\bar{X}_2) - 2\text{Cov}(\bar{X}_1, \bar{X}_2) = \sigma_1^2/n + \sigma_2^2/n - 2\rho\sigma_1\sigma_2/n$. Therefore, the test statistic of the paired t test is larger than that of the independent two-sample t test, leading to a higher chance of rejecting the null hypothesis when it is false. The degrees of freedom of the test statistics are $2(n-1)$ and $n-1$, respectively, for the independent two-sample t test and the paired t test. Although increasing the degree of freedom will typically make the critical value relatively smaller (Student's t converges to Normal as $\nu \to \infty$ and Normal has lighter right tail), making it relatively easier to reject the null hypothesis, when the sample size and effect size δ are not too small, the gain of power from a lower variance in the paired t test substantially outweighs the advantage of higher degrees of freedom in the two-sample t test. Therefore, using an independent two-sample t test incorrectly for paired data will typically reduce the statistical power. We refer to Section 6.13 of [102] for more discussions on this topic.

To better understand how power of a test is affected by effect size and sample size, refer to Section 6.1.3, where some commonly used cases are discussed to demonstrate how to perform power analysis and calculate sample sizes to achieve a certain level of power.

6.1.3 Sample Size Calculation and Power Analysis

A larger sample size generally leads to higher statistical power for detecting a true effect. However, larger sample sizes also incur higher costs and longer study durations. Therefore, it is crucial to determine an appropriate sample size that balances the need for sufficient power with practical constraints. Power analysis is a statistical technique used to calculate the minimum sample size required to achieve a desired level of statistical power for a given effect size and significance level. The following case study illustrates how to perform power analysis and calculate the required sample size for a clinical trial.

Case 3: Evaluating a New Blood Pressure Medication (Sample Size Calculation)

A pharmaceutical company is developing a new blood pressure (BP) drug. They want to design a clinical trial to compare the effectiveness of their new drug with current standard treatment. The researchers hypothesize that the new drug will result in a greater reduction in systolic BP compared to standard treatment. The expected mean reduction in systolic BP with the new treatment is 12 mmHg, and the expected mean reduction in systolic BP with the standard treatment is 8 mmHg. The expected standard deviation of the reduction in systolic BP is 10 mmHg, and the minimum clinically significant difference is 4 mmHg. How many volunteers must be recruited to achieve at least 90% power of the study at a significance level of 2.5%?

For answers to the initial questions and some initial thoughts about this case study, refer to Section 5.4. In what follows, we will use a one-sided two-sample t test to calculate the required sample size.

6.1.3.1 One-Sided Two-Sample *t* Test (Normal Approximation)

The following discussion is based on a one-sided two-sample t test, assuming that the response variable follows a normal distribution with a known variance parameter σ. The following is the hypothesis, and Figure 6.3 shows how the power is calculated. As we assume that the variance is known, the illustrative plot is simply based on a z test.

$$\begin{cases} H_0: & \mu_1 - \mu_2 = 0 \quad \text{(The new drug does not differ from the standard drug)} \\ H_1: & \mu_1 - \mu_2 = \delta > 0 \quad \text{(The new drug is more effective than the standard drug)} \end{cases}$$

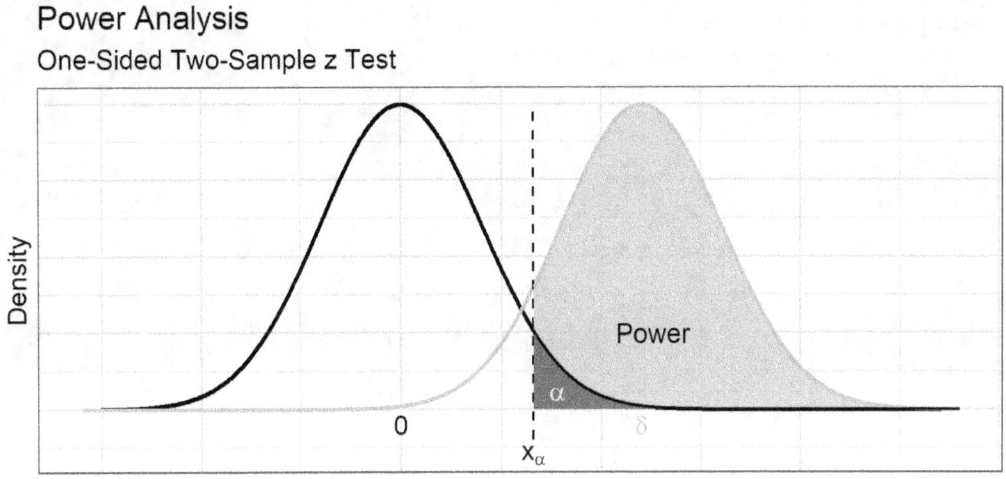

FIGURE 6.3: Illustration of Power Analysis for a One-Sided Two-Sample t Test. (Normal Approximation)

Mathematical formulars for calculating the required sample size can be derived as follows. Clearly, $\text{Var}(\bar{x}_1 - \bar{x}_2) = \frac{\sigma_1^2}{n_1} + \frac{\sigma_2^2}{n_2}$. If $n_1 = n_2 = n$ and $\sigma_1 = \sigma_2 = \sigma$ then $\text{sd}(\bar{x}_1 - \bar{x}_2) = \sqrt{\frac{2}{n}} \cdot \sigma$. Denote $x_\alpha = z_{1-\alpha}\sqrt{\frac{2}{n}} \cdot \sigma$, then

$$\begin{aligned}
\text{Power} &= 1 - \beta = \mathbb{P}[\bar{x}_1 - \bar{x}_2 > x_\alpha | H_1] \\
&= \mathbb{P}\left[\frac{\bar{x}_1 - \bar{x}_2 - \delta}{\sqrt{\frac{2}{n}} \cdot \sigma} > \frac{z_{1-\alpha}\sqrt{\frac{2}{n}}\sigma - \delta}{\sqrt{\frac{2}{n}} \cdot \sigma}\right] \\
&= \mathbb{P}\left[Z > z_{1-\alpha} - \frac{\sqrt{\frac{n}{2}} \cdot \delta}{\sigma}\right] \\
&= \mathbb{P}\left[Z \leq \frac{\sqrt{\frac{n}{2}} \cdot \delta}{\sigma} + z_\alpha\right] \\
&= \Phi\left(z_\alpha + \frac{\sqrt{\frac{n}{2}} \cdot \delta}{\sigma}\right).
\end{aligned}$$

Let m be the minimum n such that the power $\geq 1-\beta$, then $z_\alpha + \frac{\sqrt{\frac{n}{2}}\cdot\delta}{\sigma} = z_{1-\beta} \Rightarrow m = \frac{2\cdot\left(z_{1-\beta}-z_\alpha\right)^2\cdot\sigma^2}{\delta^2}$, and $n_1 = n_2 = m$.

Implementing the above formula into the following R function, and then the minimum sample size can be calculated.

```
# Function to calculate sample size for one-sided two-sample t test
calculate_sample_size <- function(delta, sd, alpha, power) {
  # Z-scores for alpha and beta
  z_alpha <- qnorm(alpha)
  z_beta <- qnorm(power)
  # Calculate sample size per group
  n <- ceiling(2 * (sd^2) * (z_beta-z_alpha)^2 / delta^2)
  return(n)
}

# Set parameters
delta <- 4          # Minimum clinically significant difference
sd <- 10            # Standard deviation
alpha <- 0.025      # One-sided significance level
power <- 0.90       # Desired power

# Calculate sample size
n <- calculate_sample_size(delta, sd, alpha, power)

# Print results
cat("Required sample size per group:", n, "\n")
: Required sample size per group: 132
cat("Total sample size:", 2*n, "\n")
: Total sample size: 264
```

The step-by-step calculation shown above helps us to understand the power analysis. However, a much easier way is to use the R function `power.t.test()` directly as follows:

```
# Sample size can be calculated directly
power.t.test(delta = delta, sd = sd, sig.level = alpha, power = 0.9,
                          type = "two.sample", alternative = "one.sided")
:
:      Two-sample t test power calculation
:
:               n = 132.3106
:           delta = 4
:              sd = 10
:       sig.level = 0.025
:           power = 0.9
:     alternative = one.sided
:
: NOTE: n is number in *each* group
```

After the required sample size is calculated, we can verify the power using the calculated sample size as follows:

```
# Verify power using power.t.test
power_verify <- power.t.test(n = n, delta = delta, sd = sd,
                             sig.level = alpha,  power = NULL,
                             type = "two.sample", alternative = "one.sided")
cat("\nVerified power:", round(power_verify$power, 3))
:
: Verified power: 0.899
```

Both methods lead to the same value of sample sizes, and this is a fairly large number, and thus the normal approximation is reasonable. However, if the calculated sample size is too small, say, less than 5 for each group, then the normal approximation is not applicable and a direct use of the R function `power.t.test()` is preferred. In practice, `power.t.test()` can be used regardless of whether the resulting sample size is small or large.

Conclusions

In this case study, assuming that the minimum clinically significant difference is 4 mmHg, the standard deviation of blood pressure reduction is 10 mmHg, the one-sided significance level is 2.5%, and the desired statistical power is 90%, then the required sample size for each of the two groups is 132.

The subsequent sections are not directly tied to the case study, and they are included here to illustrate how to perform power analysis and calculate sample sizes for some commonly used cases.

6.1.3.2 One-Sided One-Sample z Test (Population σ is Known)

In what follows, we illustrate how to calculate the required sample size to obtain a certain statistical power for a one-sided one-sample z test. The following are the null and alternative hypotheses, and Figure 6.4 shows how the statistical power is calculated.

$$\begin{cases} H_0: & \mu_1 = \mu_0 \\ H_1: & \mu_1 > \mu_0 \end{cases}$$

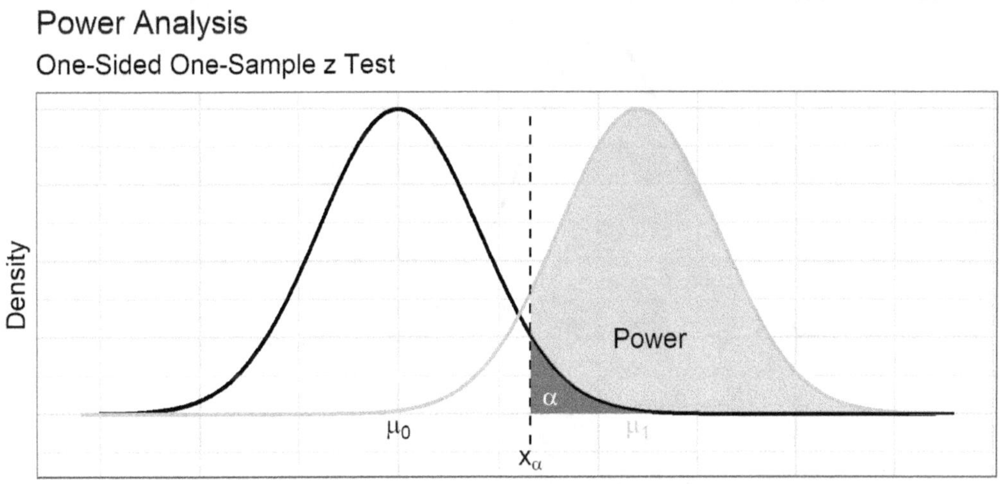

FIGURE 6.4: Illustration of Power Analysis for a One-Sided One-Sample z Test.

The following are mathematical formulas for calculating the statistical power and the required sample size. Letting

$$x_\alpha = \mu_0 + z_{1-\alpha} \cdot \frac{\sigma}{\sqrt{n}},$$

$$\begin{aligned}
\text{Power} = 1-\beta &= \mathbb{P}[\bar{x} > x_\alpha | H_1] \\
&= \mathbb{P}\left[\frac{\bar{x}-\mu_1}{\frac{\sigma}{\sqrt{n}}} > \frac{\mu_0 + z_{1-\alpha}\frac{\sigma}{\sqrt{n}} - \mu_1}{\frac{\sigma}{\sqrt{n}}}\right] \\
&= \mathbb{P}\left[Z > \frac{\sqrt{n}(\mu_0-\mu_1) + \sigma \cdot z_{1-\alpha}}{\sigma}\right] \\
&= \mathbb{P}\left[Z \le \frac{\sqrt{n}(\mu_1-\mu_0)}{\sigma} + z_\alpha\right] \\
&= \Phi\left(z_\alpha + \frac{\sqrt{n}(\mu_1-\mu_0)}{\sigma}\right).
\end{aligned}$$

Letting m be the minimum n such that the power $\ge 1-\beta$, then

$$z_\alpha + \frac{\sqrt{m}(\mu_1-\mu_0)}{\sigma} = z_{1-\beta} \Rightarrow m = \frac{(z_{1-\beta}-z_\alpha)^2 \cdot \sigma^2}{(\mu_1-\mu_0)^2}.$$

6.1.3.3 Two-Sided One-Sample *z* Test

The following are the null and alternative hypotheses of a two-sided one-sample z test, and Figure 6.5 shows how the statistical power is calculated.

$$\begin{cases} H_0: & \mu = \mu_0 \\ H_1: & \mu = \mu_1 \ne \mu_0 \end{cases}$$

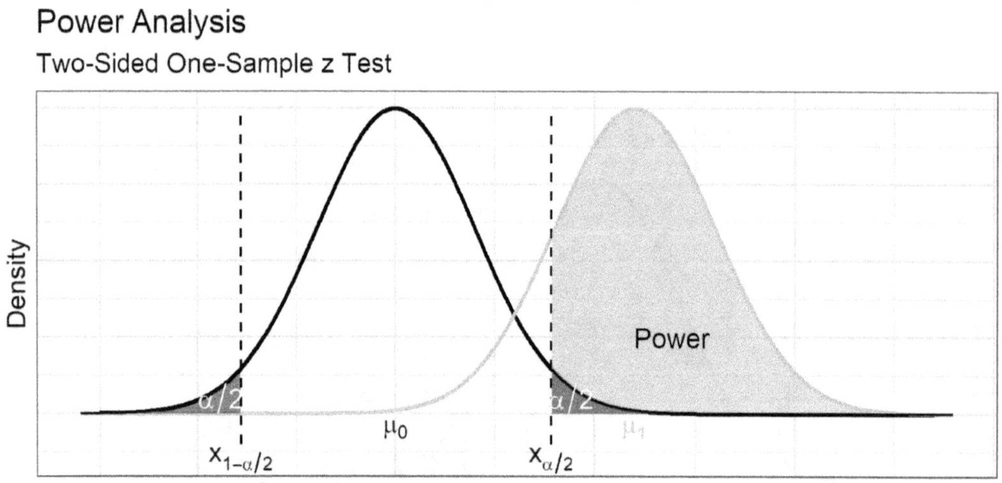

FIGURE 6.5: Illustration of Power Analysis for a Two-Sided One-Sample z Test.

The following are mathematical formulas for calculating the statistical power and the required sample size. Letting

$$x_{\alpha/2} = \mu_0 + z_{1-\alpha/2} \cdot \frac{\sigma}{\sqrt{n}},$$

$$\begin{aligned}
\text{Power} = 1-\beta &= \mathbb{P}\left[\bar{x} > x_{\alpha/2} \middle| H_1\right] + \mathbb{P}\left[\bar{x} < x_{1-\alpha/2} \middle| H_1\right] \\
&\approx \mathbb{P}\left[\frac{\bar{x}-\mu_1}{\frac{\sigma}{\sqrt{n}}} > \frac{\mu_0 + z_{1-\alpha/2}\frac{\sigma}{\sqrt{n}} - \mu_1}{\frac{\sigma}{\sqrt{n}}}\right] + 0 \\
&= \mathbb{P}\left[Z > \frac{\sqrt{n}(\mu_0-\mu_1) + \sigma \cdot z_{1-\alpha/2}}{\sigma}\right] \\
&= \mathbb{P}\left[Z \leq \frac{\sqrt{n}(\mu_1-\mu_0)}{\sigma} + z_{\alpha/2}\right] \\
&= \Phi\left(z_{\alpha/2} + \frac{\sqrt{n}(\mu_1-\mu_0)}{\sigma}\right).
\end{aligned}$$

Letting m be the minimum n such that the power $\geq 1-\beta$, then

$$z_{\alpha/2} + \frac{\sqrt{m}(\mu_1-\mu_0)}{\sigma} = z_{1-\beta} \Rightarrow m = \frac{\left(z_{1-\beta} - z_{\alpha/2}\right)^2 \cdot \sigma^2}{(\mu_1-\mu_0)^2}.$$

6.1.3.4 One-Sided One-Sample *t* Test (Small Sample Size)

The following are the null and alternative hypotheses of a one-sided one-sample t test, and Figure 6.6 shows how the statistical power is calculated.

$$\begin{cases} H_0: & \mu = \mu_0 \\ H_1: & \mu = \mu_1 > \mu_0 \end{cases}$$

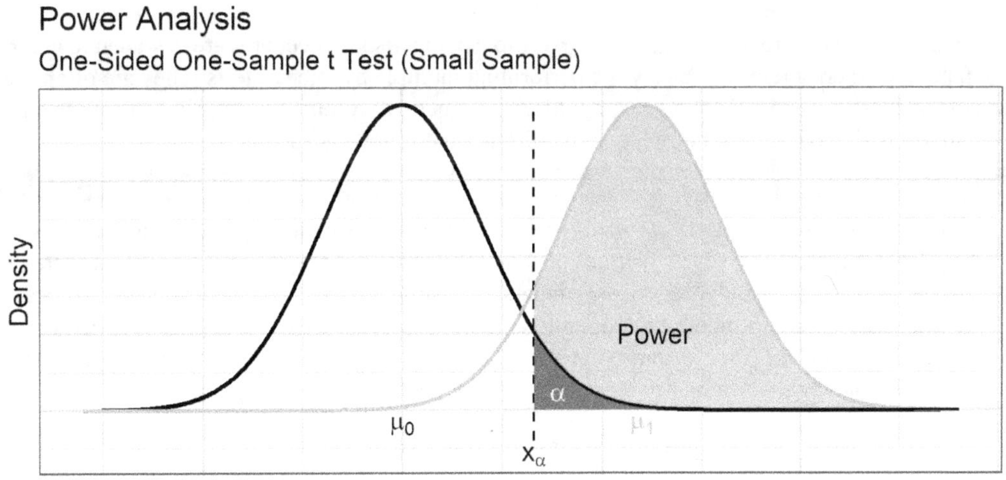

FIGURE 6.6: Illustration of Power Analysis for a One-Sided One-Sample t Test (Small Sample).

The following are mathematical formulas for calculating the statistical power and the required sample size. Write $\delta = \mu_1 - \mu_0$ and $x_\alpha = \mu_0 + t_{1-\alpha,n-1}\frac{s}{\sqrt{n}}$, where $n-1$ is the degree of freedom of Student's t distribution and the sample standard deviation $s = \sqrt{\sum_{i=1}^{n}(x_i-\bar{x})^2/(n-1)}$.

$$\begin{aligned}
\text{Power } &= 1-\beta = \mathbb{P}[\bar{x} > x_\alpha | H_1] \\
&= \mathbb{P}\left[\frac{\bar{x}-\mu_1+\delta}{\frac{s}{\sqrt{n}}} > \frac{t_{1-\alpha,n-1}\frac{s}{\sqrt{n}}}{\frac{s}{\sqrt{n}}} \middle| H_1\right] \\
&= \mathbb{P}\left[\frac{\bar{x}-\mu_1+\delta}{\frac{\sigma}{\sqrt{n}}\cdot\frac{s}{\sigma}} > t_{1-\alpha,n-1} \middle| H_1\right] \\
&= \mathbb{P}\left[\frac{Z+\sqrt{n}\cdot\delta/\sigma}{\frac{s}{\sigma}} > t_{1-\alpha,n-1} \middle| H_1\right] \\
&= \mathbb{P}\left[\frac{Z+\sqrt{n}\cdot\delta/\sigma}{\sqrt{\frac{V}{n-1}}} > t_{1-\alpha,n-1}\right],
\end{aligned}$$

where

$$V = \frac{(n-1)s^2}{\sigma^2} \sim \chi(n-1); \quad \frac{Z+\sqrt{n}\delta/\sigma}{\sqrt{\frac{V}{n-1}}} \sim t\left(n-1, \sqrt{n}\delta/\sigma\right),$$

and $\sqrt{n}\delta/\sigma$ is the noncentrality parameter of the noncentral Student's t distribution.

Note that in order to conduct power analysis and calculate the required sample size, we need to specify the statistical model in advance. Due to the limited space, we will not go into further details here. The freeware `G*Power` can be used to perform power analysis and sample size calculation for many commonly used statistical tests, including the one-sided one-sample t test.

6.1.4 ANOVA and Multiple Comparisons

When multiple groups are involved, we often want to compare the means of different groups and find out which group has the largest or smallest group means. ANOVA is a procedure to detect overall whether such a difference exists at all, and if that is detected, one is often interested in knowing which groups are different from the others. Multiple comparison procedures are designed to control the family-wise error rate (FWER) while performing multiple hypothesis tests. They are often used as post hoc tests after ANOVA suggesting differences between groups' means. Before we dive into formal discussions of the methods, let us first discuss the following case study.

Case 4: Impact of Dietary Supplements on Marathon Performance

This study examines the effects of three different dietary supplement regimens (BCAAs, Carb-loading, and Electrolyte-focused) on marathon completion times (in seconds) for recreational runners. The study tracked 90 runners with similar experience levels during a series of marathons over a 6-month period, and Table 5.4 shows the first two and the last two observations. Do these three dietary supplement regimens differ significantly and which one improves marathon performance the most?

TABLE 5.4: Supplement Data Table.

Supplement	Completion Time (seconds)
Carb-loading	14091
Carb-loading	15860
...	...
Electrolyte	15013
Electrolyte	14667

For answers to the initial questions and some initial thoughts about this case study, refer to Section 5.5. Some preliminary analyses can be conducted as follows:

```
library(dplyr)
library(ggplot2)
supp_data <- read.csv("data/supplement-marathon.csv")
supp_data$supplement <- as.factor(supp_data$supplement)

# Summary statistics
summary_stats <- supp_data %>% group_by(supplement) %>%
  summarise(n = n(),  mean_time = mean(completion_time),
            sd = sd(completion_time), se = sd/sqrt(n),
            min = min(completion_time), max = max(completion_time))
print(summary_stats)
: # A tibble: 3 x 7
:   supplement        n mean_time    sd    se   min   max
:   <fct>         <int>     <dbl> <dbl> <dbl> <int> <int>
: 1 BCAA             30    15861.  758.  138. 13921 17591
: 2 Carb-loading     30    15509. 1041.  190. 13754 17352
: 3 Electrolyte      30    14800.  939.  171. 12851 16699
```

The following boxplot shows that there might be an overall difference among the three groups, as the group "Electrolyte" seems to be lower than the other two.

```
# Create box plots
ggplot(supp_data, aes(y = supplement, x = completion_time)) +
  geom_boxplot() + theme_bw() +
  stat_summary(fun = "mean", geom = "point", shape = 23, size = 2) +
  labs(y = "Supplement Regimen", x = "Completion Time (minutes)")
```

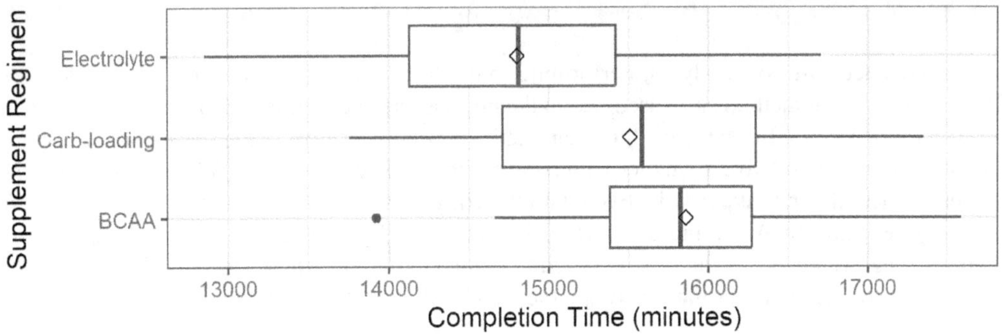

FIGURE 6.7: Marathon Completion Times by Supplement Type.

Now we conduct a formal analysis with ANOVA. First, we check the assumptions for ANOVA. The p-values for the normality tests of the three groups are $0.60, 0.29$ and 0.91, respectively; the p-value for the test of homogeneity of variances is 0.08. There is no strong evidence to reject either the normality assumption or the assumption of equal variances, and therefore we proceed with ANOVA.

```
# Normality test for each group
supp_data %>% group_by(supplement) %>%
  summarise(statistic = shapiro.test(completion_time)$statistic,
            p_value = shapiro.test(completion_time)$p.value)
: # A tibble: 3 x 3
:   supplement    statistic p_value
:   <fct>             <dbl>   <dbl>
: 1 BCAA              0.972   0.599
: 2 Carb-loading      0.959   0.286
: 3 Electrolyte       0.983   0.906
```

```
# Test of homogeneity of variance
car::leveneTest(completion_time ~ supplement, data = supp_data)
: Levene's Test for Homogeneity of Variance (center = median)
:       Df F value  Pr(>F)
: group  2  2.6331 0.07758 .
:       87
: ---
: Signif. codes:  0 '***' 0.001 '**' 0.01 '*' 0.05 '.' 0.1 ' ' 1
```

The following ANOVA leads to a p-value < 0.0001, strong evidence that there is a difference between the group means.

```
# Perform one-way ANOVA
anova_result <- aov(completion_time ~ supplement, data = supp_data)
summary(anova_result)
:             Df   Sum Sq Mean Sq F value   Pr(>F)
: supplement   2 17499944 8749972   10.34 9.39e-05 ***
: Residuals   87 73653921  846597
: ---
: Signif. codes:  0 '***' 0.001 '**' 0.01 '*' 0.05 '.' 0.1 ' ' 1
```

We conduct post hoc analyses, performing multiple comparison procedures as shown below. The details of these methods are discussed in the notes at the end of this section.

The following is a Bonferroni adjustment that is very conservative in the sense that the required significance level is relatively smaller with a given family-wise error rate (FWER), and it becomes harder to reject the null hypothesis. Even with the conservative criterion, the p-value < 0.0001 for "Electrolyte" and "BCAA", and the p-value is 0.006 for "Electrolyte" and "Carb-loading", providing strong evidence that the time required to complete the marathon course is shorter for those taking "Electrolyte" than for those taking either "BCAA" or "Carb-loading".

```
# Bonferroni corrections
pairwise.t.test(supp_data$completion_time, supp_data$supplement,
                alternative = "less", p.adjust.method="bonferroni")
:
:  Pairwise comparisons using t tests with pooled SD
:
: data:  supp_data$completion_time and supp_data$supplement
:
:              BCAA    Carb-loading
```

```
: Carb-loading 0.2132  -
: Electrolyte  3.6e-05 0.0056
:
: P value adjustment method: bonferroni
```

The following Tukey-Kramer procedure leads to a consistent result as the Bonferroni adjustment, with the p-value < 0.001 for "Electrolyte" and "BCAA", and the p-value being 0.005 for "Electrolyte" and "Carb-loading".

```
# Tukey-Kramer procedure - one-sided
library(multcomp)
summary(glht(anova_result, linfct = mcp(supplement = "Tukey"),
             alternative = "less"))
:
:   Simultaneous Tests for General Linear Hypotheses
:
: Multiple Comparisons of Means: Tukey Contrasts
:
:
: Fit: aov(formula = completion_time ~ supplement, data = supp_data)
:
: Linear Hypotheses:
:                                Estimate Std. Error t value  Pr(<t)
: Carb-loading - BCAA >= 0          -351.9      237.6  -1.481 0.17269
: Electrolyte - BCAA >= 0          -1060.3      237.6  -4.463 < 0.001 ***
: Electrolyte - Carb-loading >= 0   -708.4      237.6  -2.982 0.00527 **
: ---
: Signif. codes:  0 '***' 0.001 '**' 0.01 '*' 0.05 '.' 0.1 ' ' 1
: (Adjusted p values reported -- single-step method)
```

If we are interested in comparisons of one particular supplement to the others, we can use Dunnett's procedure. Making "Electrolyte" the reference level of the explanatory variable, and then the following Dunnett's procedure shows that "Electrolyte" leads to less time than that of those who take either "BCAA" or "Card-loading" (p-values: < 0.0001, 0.004, respectively).

```
# Dunnett's Procedure
supp_data$supplement <- relevel(supp_data$supplement, ref="Electrolyte")
anova_relevel <- aov(completion_time ~ supplement, data = supp_data)
summary(glht(anova_relevel, linfct = mcp(supplement = "Dunnett"),
        alternative="greater"))
:
:   Simultaneous Tests for General Linear Hypotheses
:
: Multiple Comparisons of Means: Dunnett Contrasts
:
:
: Fit: aov(formula = completion_time ~ supplement, data = supp_data)
:
: Linear Hypotheses:
:                             Estimate Std. Error t value   Pr(>t)
: BCAA - Electrolyte <= 0        1060.3      237.6   4.463 2.39e-05 ***
```

```
: Carb-loading - Electrolyte <= 0    708.4       237.6   2.982  0.00357 **
: ---
: Signif. codes:  0 '***' 0.001 '**' 0.01 '*' 0.05 '.' 0.1 ' ' 1
: (Adjusted p values reported -- single-step method)
```

The following is used to calculate the effect size in terms of η^2, and the result suggests that about 19% of the variance in marathon performance is explained by the supplement types.

```
# Calculate effect size eta squared (manually)
eta_squared <- function(aov_result) {
  ss_effect <- sum((summary(aov_result)[[1]]$"Sum Sq")[1])
  ss_total <- sum(summary(aov_result)[[1]]$"Sum Sq")
  return(ss_effect/ss_total)
}
effect_size <- eta_squared(anova_result)
print(paste("Effect Size (Eta-squared):", round(effect_size, 3)))
: [1] "Effect Size (Eta-squared): 0.192"
```

```
# Alternatively, use an R package directly
effectsize::eta_squared(anova_result, partial = FALSE)
: # Effect Size for ANOVA (Type I)
:
: Parameter  | Eta2 |       95% CI
: ---------------------------------
: supplement | 0.19 | [0.07, 1.00]
:
: - One-sided CIs: upper bound fixed at [1.00].
```

```
# Cohen's d between Electrolyte and BCAA
effectsize::cohens_d(completion_time ~ supplement, data =
    supp_data[supp_data$supplement %in% c("Electrolyte","BCAA"), ])
: Cohen's d |         95% CI
: --------------------------
: -1.24     | [-1.79, -0.68]
:
: - Estimated using pooled SD.
```

```
# Cohen's d between Electrolyte and Carb-loading
effectsize::cohens_d(completion_time ~ supplement, data =
    supp_data[supp_data$supplement %in% c("Electrolyte","Carb-loading"), ])
: Cohen's d |         95% CI
: --------------------------
: -0.71     | [-1.23, -0.19]
:
: - Estimated using pooled SD.
```

Conclusions

In this case study, we use ANOVA to compare the marathon completion times among three different supplement regimens: Electrolyte, BCAA, and Carb-loading. The ANOVA indicates strong evidence of the three supplements associated with marathon completion time ($p < 0.0001$), with effect size $\eta^2 = 0.19$. Post hoc analyses, which encompasses various multiple comparison methods, indicates that "Electrolyte" supplement is associated with less completion time than "BCAA" (p-values: < 0.0001 Bonferroni; < 0.001 Tukey-Kramer; < 0.0001 Dunnett) and "Carb-loading" (p-values: 0.006 Bonferroni; 0.005 Tukey-Kramer; < 0.004 Dunnett), but completion time is not different between "BCAA" and "Carb-loading" (p-values: 0.21 Bonferroni; 0.17 Tukey-Kramer). Cohen's d is -1.24 and -0.71, respectively, for the pair of Electrolyte and BCAA and the pair of Electrolyte and Carb-loading. Additional information regarding the selection of subjects and their assignment to various groups is necessary to properly interpret the aforementioned results in terms of causality and generalization to the target population.

Note 6.10 (Analysis of Variance (ANOVA)). *ANOVA* is a statistical method used to compare the means of three or more groups to determine if there are statistically significant differences among them. The key idea is to compare the variance between groups to the variance within groups, and if the between-group variance is significantly larger than the within-group variance, we can conclude that there are differences among the group means. For one-way ANOVA, the hypothesis is

$$\begin{aligned} H_0 &: \mu_1 = \cdots = \mu_k, \\ H_1 &: \text{At least one mean differs.} \end{aligned}$$

The model is written as

$$Y_{ij} = \mu_i + \varepsilon_{ij}, \quad i = 1, \ldots, k, j = 1, \ldots, n_i,$$

where Y_{ij} is the jth observation in group i, $\mu_i = \mu + \alpha_i$ with μ the grand mean and α_i the fixed effect of treatment i, and ε_{ij} is the random error. The sum of squares decomposition is

$$\text{Total SS (SST)} = \text{Between SS (SSB)} + \text{Within SS (SSE)},$$

where $\text{SST} = \sum_{i=1}^{k} \sum_{j=1}^{n_i} (Y_{ij} - \bar{Y}_{..})^2$, $\text{SSB} = \sum_{i=1}^{k} n_i (\bar{Y}_{i.} - \bar{Y}_{..})^2$, and within-group sum of squares is also referred to as sum of squares due to error and thus $\text{SSE} = \sum_{i=1}^{k} \sum_{j=1}^{n_i} (Y_{ij} - \bar{Y}_{i.})^2$. The F statistic is

$$F = \frac{\text{MS}_{\text{Between}}}{\text{MS}_{\text{Within}}} = \frac{\text{SSB}/(k-1)}{\text{SSE}/(N-k)},$$

which under H_0 follows an F distribution with degrees of freedom $k-1$ and $N-k$, with $N = \sum_{i=1}^{k} n_i$. The key assumptions for ANOVA are Y_{ij}'s are independent and $\varepsilon_{ij} \overset{\text{iid}}{\sim} N(0, \sigma^2)$. One-way ANOVA can be extended to two-way ANOVA, which includes two factors and their interaction, and to more complex designs such as repeated measures ANOVA (Note 6.20). Many standard textbooks cover the details of ANOVA, and we refer readers to [52] and [89] for more information.

Note 6.11 (Eta-Squared (η^2) as an Effect Size Measure). *Eta-squared (η^2)* is a commonly used effect size measure in ANOVA, and it represents the proportion of variance in the response variable explained by the explanatory variable(s). Eta-squared (η^2) is defined as

$$\eta^2 = \frac{SS_{\text{effect}}}{SS_{\text{total}}},$$

where SS_{effect} is the sum of squares for the effect (e.g. between groups) and SS_{total} is the total sum of squares. η^2 ranges from 0 to 1, and a value closer to 1 indicates a stronger effect. For example, suppose we have the following ANOVA results:

Source	Sum of Squares (SS)	Degrees of Freedom (df)
Between Groups	120	2
Within Groups	80	27
Total	200	29

Then $\eta^2 = SS_{\text{effect}}/SS_{\text{total}} = 120/200 = 0.6$, and 60% of the variance in the response variable is explained by the explanatory variable. A similar effect size measure is partial eta-squared, which will be discussed in Note 6.14. Refer to [27] and [86] for more information on effect size measures and their interpretation.

Note 6.12 (Multiple Comparison Procedures)**.** *Multiple comparison procedures* are designed to control the family-wise error rate (FWER), that is, the probability of making at least one Type I error (false positive) across all comparisons, while performing multiple hypothesis tests. They are often used as post hoc tests after ANOVA suggesting differences between groups' means. To motivate the need for multiple comparison procedures, consider the following scenario. Suppose we have m independent hypothesis tests, each with a significance level α. The probability of making at least one Type I error (false positive) across all m tests is given by

$$\text{FWER} = \text{P}(\text{At least one Type I error}) = 1-(1-\alpha)^m \geq \alpha, \quad m \geq 1.$$

For example, let $m = 10, \alpha = 0.05$, then $\text{P}(\text{at least one Type I error}) > 40\%$. That is, without multiple comparison procedures, the probability of Type I error may increase substantially as the number of comparisons increases, leading to a higher chance of false positives. Multiple comparison procedures adjust the significance level or the test statistics to control the FWER at a desired level α, ensuring that the probability of making at least one Type I error across all comparisons does not exceed α. In the following, we discuss several commonly used multiple comparison procedures. For more details on multiple comparison procedures, refer to [56] for a comprehensive review.

Bonferroni's procedure [12] is the simplest multiple comparison procedure and it controls FWER at level α: for m tests, use α/m as a threshold for each individual test, and then the FWER is controlled at level α by the union bound:

$$\text{FWER} \leq \sum_{i=1}^{m} \text{P}(\text{Type I Error for Test } i) \leq m \cdot \frac{\alpha}{m} = \alpha.$$

This procedure is simple to use and easy to explain, but can be too conservative. Some less conservative procedures are discussed next.

Sidak's procedure [101] is a method for controlling the FWER when performing multiple comparisons and it is best used when the tests are independent. It is similar to the Bonferroni method but slightly less conservative, especially when the number of comparisons is large. When testing m hypotheses, the Sidak's method adjusts each individual test's significance level to: $\alpha' = 1-(1-\alpha)^{1/m}$, where α is the desired family-wise error rate, m is the number of comparisons, α' is the per-comparison error rate. The adjusted p-value for a test is $p_{\text{Sidak}} = 1-(1-p)^m$. A p-value must be smaller than α' to be considered significant under the Sidak's correction.

The Tukey-Kramer procedure [110, 111, 64] is designed for all pairwise comparisons, while controlling FWER at level α for balanced and unbalanced designs, and it is more powerful than the Bonferroni procedure for pairwise comparisons. For groups i and j, the test statistic is

$$q_{ij} = \frac{|\bar{X}_i - \bar{X}_j|}{\sqrt{\frac{\text{MSE}}{2}(n_i^{-1} + n_j^{-1})}},$$

where $\bar{X}_i, \bar{X}_j$ are group means, MSE is mean square error, and n_i, n_j are group sample sizes. Then compare q_{ij} to the critical value $q_{\alpha,k,\nu}$ according to a studentized range distribution, where k is the number of groups, ν is the degree of freedom of errors, and α is the significance level. The main idea of the studentized range distribution is to consider the maximum difference between any two group means, standardized by the estimated standard error; see [82] for details.

Dunnett's procedure [37] is specifically designed for comparing treatments to a control group and is more powerful than Bonferroni or Tukey-Kramer when only comparisons with the control group are needed. This procedure controls FWER for comparisons with the control group. The test statistic for comparing treatment i to control c is

$$t_i = \frac{\bar{X}_i - \bar{X}_c}{\sqrt{\text{MSE}(n_i^{-1} + n_c^{-1})}},$$

where $\bar{X}_i$ is the mean of the treatment group i, $\bar{X}_c$ is the mean of the control group, MSE is the mean square error of ANOVA and n_i, n_c are respective sample sizes. The test statistics follow a multivariate t distribution:

$$(t_1, t_2, \ldots, t_k) \sim t_\nu(0, R),$$

where the correlation matrix R has the elements:

$$r_{ij} = \frac{n_c^{-1}}{\sqrt{(n_i^{-1} + n_c^{-1})(n_j^{-1} + n_c^{-1})}}.$$

Scheffé's procedure [96] controls FWER for all possible contrasts and is the most conservative and most flexible procedure. For a contrast

$$\psi = \sum_{i=1}^{k} c_i \bar{Y}_i, \quad \sum_{i=1}^{k} c_i = 0,$$

the standard error $SE_\psi = \sqrt{\text{MSE}\sum_{i=1}^{k}(c_i^2/n_i)}$, where MSE is mean square error, k is the number of groups, n_i is the sample size of group i, $\bar{Y}_i$ is the mean of group i, and c_i's are contrast coefficients. The Scheffé's test statistic is $F_\psi = \psi^2/SE_\psi^2$. Then the contrast ψ is significant at level α if

$$F_\psi > (k-1)F_{\alpha,k-1,N-k},$$

where $F_{\alpha,k-1,N-k}$ is the critical value from an F distribution with N the total sample size.

Holm's step-down procedure [54] is designed to control FWER in a less conservative manner than the Bonferroni procedure. For ordered p-values $p_{(1)} \leq p_{(2)} \leq \cdots \leq p_{(m)}$ of m tests, the procedure is as follows: start with $i = 1$, and then compare $p_{(i)}$ with $\alpha/(m-i+1)$; stop at the first nonsignificant result and all remaining hypotheses are not rejected. This procedure controls FWER at level α and is more powerful than the Bonferroni procedure, as it allows for a less stringent threshold for significance as you move through the ordered p-values. The key idea is that as you reject hypotheses, the number of remaining tests decreases, which allows for a less stringent threshold for the remaining tests. This procedure ensures that the overall FWER is controlled at level α, meaning that the probability of making at least one Type I error across all comparisons does not exceed α. It can be applied to any set of p-values, regardless of the dependence structure between tests.

The Benjamini-Hochberg procedure [10] compares ordered p-values with $\frac{i}{m}\alpha$, and it controls the false discovery rate (FDR), the proportion of false discoveries among all rejected hypotheses, while FWER controls the probability of making any Type I errors. The BH procedure is less conservative than those controlling FWER, and more powerful for high-dimensional settings. Suppose we have m independent hypothesis tests with ordered p-values $p_{(1)} \leq p_{(2)} \leq \cdots \leq p_{(m)}$. The BH procedure is as follows: find the largest i such that $p_{(i)} \leq \frac{i}{m}\alpha$, and reject all hypotheses corresponding to $p_{(1)}, p_{(2)}, \ldots, p_{(i)}$. This procedure controls the FDR at level α, meaning that the expected proportion of false discoveries among all rejected hypotheses does not exceed α.

A brief summary of multiple comparison procedures is in Table 6.1.

TABLE 6.1: Comparison of Multiple Comparison Procedures.

Procedure	Control	Best Use Case
Bonferroni	FWER	Few tests
Sidak	FWER	Independent tests
Tukey-Kramer	FWER	All pairwise
Dunnett	FWER	vs. control
Scheffé	FWER	All contrasts
Holm	FWER	General use
BH	FDR	Many tests

The power is typically ranked as

$$\text{BH} > \text{Holm} > \text{Dunnett} > \text{Tukey} > \text{Sidak} > \text{Bonferroni} > \text{Scheffé}.$$

However, this is a general ranking and the actual power of each procedure can depend on the specific context, such as the number of comparisons, the sample size, the effect size, and the correlation between tests. Therefore, it is important to consider these factors when choosing a multiple comparison procedure for a specific case.

6.1.5 ANCOVA

In many practical situations, researchers are interested in comparing group means while accounting for the influence of continuous covariates that may affect the response variable. When continuous covariates are present, analysis of covariance (ANCOVA) can be performed to assess the difference between group means while controlling for the covariates. We use the following case study as an example to illustrate how to conduct ANCOVA.

Case 5: Innovative Teaching on Posttest Improvement

An educational research study examines the effectiveness of an innovative teaching approach. The study includes 60 students randomly sampled from a middle school and randomly assigned to two groups (30 students each). The same curriculum content is covered for both groups. The traditional approach (Method A: control) is used for one group and an innovative approach (Method B: experimental) is used for the other group. The pretest is administered at the beginning of an 8-week teaching intervention period, and the posttest is administered at the end of the intervention period. Standardized assessment of learning outcomes and standardized testing conditions are used for both the pretest and the posttest. Does the innovative approach significantly improve students' performance?

For answers to the initial questions and some initial thoughts about this case study, refer to Section 5.6. We begin by constructing the response variable for gain scores and calculating their basic statistics as shown below.

```
library(ggplot2)
library(dplyr)
teach_data <- read.csv("data/teaching-Pretest-Posttest.csv")
teach_data$Method <- as.factor(teach_data$Method)
teach_data$Gain <- teach_data$Posttest - teach_data$Pretest

# Summary statistics
summary_stats <- teach_data %>% group_by(Method) %>%
  summarise(n = n(), mean = mean(Gain), sd = sd(Gain),
    se = sd/sqrt(n), min = min(Gain), max = max(Gain))
print(summary_stats)
: # A tibble: 2 x 7
:   Method      n  mean    sd    se   min   max
:   <fct>  <int> <dbl> <dbl> <dbl> <dbl> <dbl>
: 1 A         30  5.77  4.62 0.844    -7  17
: 2 B         30  6.92  4.79 0.874     0  17.5
```

The following plot shows the relationship between the gain scores and pretest scores for two different teaching approaches. A smoothing curve is added to illustrate the trend of gain scores against pretest scores. It seems that the innovative approach (Group B) improves the scores more than the traditional group (Group A).

```
# First look at the relationship between pretest and gain scores
pretest_gain_plot <- ggplot(teach_data,
    aes(x = Pretest, y = Gain, shape=Method, color = Method)) +
  geom_point() + theme_bw() +
  geom_smooth(method="lm", se=TRUE, linewidth=0.5, formula="y ~ x") +
  labs(title = "Relationship Between Pretest Scores and Gains",
      x = "Pretest Score", y = "Gain Score (Posttest - Pretest)") +
  scale_color_manual(values = c("grey30", "grey80"))
print(pretest_gain_plot)
```

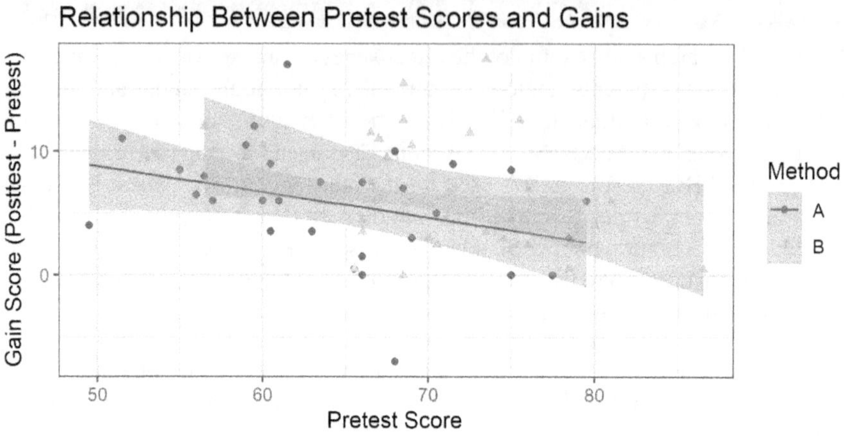

FIGURE 6.8: Relationship Between Pretest Scores and Gain Scores.

A Spaghetti plot is useful for displaying individual score changes, and it can be created by the following codes.

```
library(tidyr)
teach_data$ID <- seq(nrow(teach_data))
teach_long <- teach_data %>% pivot_longer(cols = Pretest:Posttest,
                                          names_to = "Test", values_to = "Score")

# Spaghetti plot for individual changes
teach_long$Test <- factor(teach_long$Test, levels=c("Pretest", "Posttest"))
ggplot(teach_long, aes(x = Test, y = Score, group = ID)) +
  geom_line(alpha = 0.3) + geom_point() + theme_bw() +
  labs(title = "Individual Test Score Changes", y = "Score", x = "Test")
```

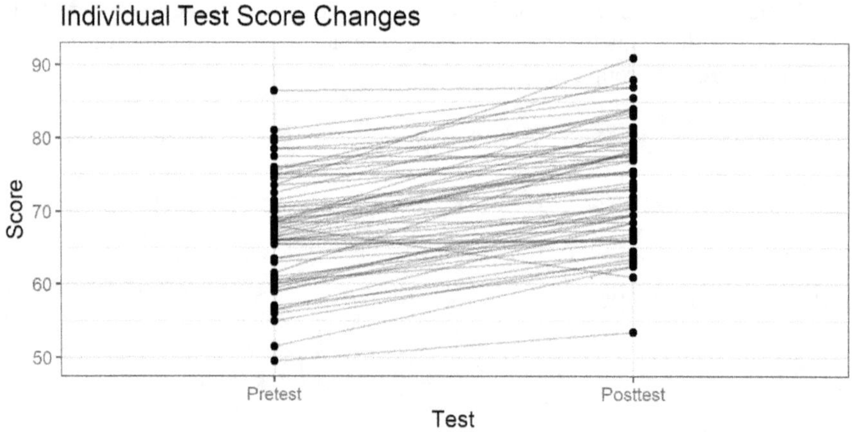

FIGURE 6.9: Spaghetti Plot for Individual Score Changes.

ANCOVA can be performed using a linear regression model. Then assumptions such as normality and homogeneity of variance need to be checked. We continue with ANCOVA because the p-values for the normality test and the test of equal variance are 0.68 and 0.40, respectively, indicating that there is no strong evidence to reject these two assumptions.

```
# Linear regression for ANCOVA
gain_lm <- lm(Gain ~ Pretest + Method, data = teach_data)
```

```
# Normality assumption check
shapiro.test(residuals(gain_lm))
:
:  Shapiro-Wilk normality test
:
: data:  residuals(gain_lm)
: W = 0.98525, p-value = 0.6832
```

```
# Homogeneity of variance assumption check
car::leveneTest(Gain ~ Method, data = teach_data)
: Levene's Test for Homogeneity of Variance (center = median)
:       Df F value Pr(>F)
: group  1  0.7114 0.4024
:       58
```

The following is used to check whether the regression slopes are the same for the two different groups. That is, roughly speaking, we test whether the two lines in Figure 6.8 are parallel. The ANOVA table shows that the p-value is 0.805 for the term “Method:Pretest”, suggesting that there is no strong evidence to reject the null hypothesis that the slopes are the same.

```
# Homogeneity of regression slopes
interaction_model <- lm(Gain ~ Method * Pretest, data = teach_data)
anova(interaction_model)
: Analysis of Variance Table
:
: Response: Gain
:                Df  Sum Sq Mean Sq F value   Pr(>F)
: Method          1   19.84  19.838  0.9842 0.325422
: Pretest         1  153.98 153.985  7.6400 0.007713 **
: Method:Pretest  1    1.23   1.234  0.0612 0.805483
: Residuals      56 1128.69  20.155
: ---
: Signif. codes:  0 '***' 0.001 '**' 0.01 '*' 0.05 '.' 0.1 ' ' 1
```

Therefore, we carry out the ANCOVA without the interaction term, and then the p-value for “Method” is 0.053 after controlling for the pretest scores. This is a good case to demonstrate that a binary decision should not be made simply on the basis of whether the p-value is greater or less than 5%. Instead of saying whether “Method” is significant or not, we should report the p-value and calculate effect sizes, and leave it to domain experts to assess whether such an effect is practically meaningful.

```
# ANCOVA results
anova(gain_lm)
```

```
: Analysis of Variance Table
:
: Response: Gain
:           Df  Sum Sq Mean Sq F value  Pr(>F)
: Pretest    1   96.44  96.436  4.8648 0.03146 *
: Method     1   77.39  77.386  3.9038 0.05302 .
: Residuals 57 1129.92  19.823
: ---
: Signif. codes:  0 '***' 0.001 '**' 0.01 '*' 0.05 '.' 0.1 ' ' 1
```

To obtain an intuitive way of understanding the magnitude of the gain scores for different groups, we can derive the adjusted mean using the following codes from the R package `emmeans`. For an average pretest score, Method A results in a gain score of 5.12, while Method B results in a gain score of 7.57.

```
# Calculate adjusted means (given average pretest scores of both groups)
emmeans::emmeans(gain_lm, "Method")
:  Method emmean    SE df lower.CL upper.CL
:  A        5.12 0.846 57     3.42     6.81
:  B        7.57 0.846 57     5.87     9.26
:
: Confidence level used: 0.95
```

```
# Same as:
predict(gain_lm, newdata=data.frame(Method=c("A", "B"),
                                    Pretest=mean(teach_data$Pretest)))
:        1        2
: 5.115845 7.567488
```

The effect sizes in terms of partial η^2 (denoted as η_p^2) and Cohen's f^2 can be derived. Both the teaching method and pretest scores made moderate contributions to explaining the gain scores. After controlling for the other variable, the teaching method accounted for about 6% of the variance, while the pretest scores accounted for about 8%. This indicates that the pretest scores explain a slightly greater proportion of variance than the teaching method does, although their contributions are fairly comparable in size.

```
# effect sizes: partial eta-squared and Cohen's f
effectsize::eta_squared(gain_lm, partial = TRUE)
: # Effect Size for ANOVA (Type I)
:
: Parameter | Eta2 (partial) |       95% CI
: -----------------------------------------
: Pretest   |           0.08 | [0.00, 1.00]
: Method    |           0.06 | [0.00, 1.00]
:
: - One-sided CIs: upper bound fixed at [1.00].
effectsize::cohens_f_squared(gain_lm, partial = TRUE)
: # Effect Size for ANOVA (Type I)
:
: Parameter | Cohen's f2 (partial) |       95% CI
```

```
: ----------------------------------------
: Pretest   |              0.09 | [0.00, Inf]
: Method    |              0.07 | [0.00, Inf]
:
: - One-sided CIs: upper bound fixed at [Inf].
```

Conclusions

In this case study, ANCOVA is used to compare the effectiveness of two different teaching methods on students' gain scores while controlling for their pretest scores. The ANCOVA provides evidence that the two teaching methods are associated with different gain scores after controlling for the pretest scores ($p = 0.053$). The adjusted means of gain scores given the average pretest score of all subjects are $5.12, 95\%$ CI : $(3.42, 6.81)$ for Method A and $7.57, 95\%$ CI : $(5.87, 9.26)$ for Method B. The effect size of the teaching methods $\eta_p^2 = 6\%$ after controlling for the pretest scores. Since the subjects are randomly selected and assigned to the two groups, there is evidence that different teaching methods caused a difference in gain scores while controlling for the pretest scores. Such results can be generalized to the entire school population as subjects are randomly selected from the school. However, whether the effective size is practically meaningful needs to be further assessed by domain experts in the context of teaching method effectiveness evaluations.

Note that it is important to include the pretest score as a covariate. Let us imagine that we have a case study without the pretest scores and that we may proceed with a two-sample t test. The following result leads to p-value $= 0.348$, and thus there is no strong evidence to claim that the innovative approach is effective. Clearly, it has missed important information about the pretest scores, and thus the result can be misleading. In practice, the rule of thumb is to collect as many covariates as possible, especially those that could have an effect on the response variable of interest.

```
# Simple t test on gains (without covariate)
gain_ttest <- t.test(Gain ~ Method, data = teach_data)
print(gain_ttest$p.value)
: [1] 0.3477481
```

An alternative way to deal with this case study is to treat the posttest score as the response variable, and the explanatory variable and the covariate are the same as before. As $E[Y - X|X = x] = E[Y|X = x] - x$, $E[Y - X|X = x]$ corresponds to linear regression with gain score as response variable, and $E[Y|X = x]$ corresponds to linear regression with posttest score as response variable, a similar result can be obtained, and the conditional linear dependencies between these two models are identical up to the constant x. The following ANCOVA table using posttest score as the response variable results in the same p-value for "Method".

```
# ANCOVA (posttest as response variable)
posttest_lm <- lm(Posttest ~ Pretest + Method, data = teach_data)
anova(posttest_lm)
: Analysis of Variance Table
:
: Response: Posttest
:           Df  Sum Sq Mean Sq  F value    Pr(>F)
: Pretest    1 2486.24 2486.24 125.4204 5.035e-16 ***
```

```
: Method      1   77.39   77.39   3.9038   0.05302 .
: Residuals 57 1129.92   19.82
: ---
: Signif. codes:  0 '***' 0.001 '**' 0.01 '*' 0.05 '.' 0.1 ' ' 1
```

Note 6.13 (Analysis of Covariance (ANCOVA)). *ANCOVA* is a statistical procedure that extends analysis of variance (ANOVA) by incorporating one or more continuous covariates to control their effects on the response variable. The assumptions of ANCOVA include normality of residuals, homogeneity of variances, and homogeneity of regression slopes. The homogeneity of regression slopes assumption requires that the relationship between the covariate and the response variable is the same across all groups. An ANCOVA model can be fitted using linear regression, and the adjusted means can be calculated to compare group differences while controlling for covariates. The following is the general form of a one-way ANCOVA model:

$$Y_{ij} = \mu + \tau_i + \beta(X_{ij} - \bar{X}_{..}) + \varepsilon_{ij},$$

where Y_{ij} is the response variable for the jth subject in the ith group, μ is the overall mean, τ_i is the effect of the ith group, β is the regression coefficient for the covariate, X_{ij} is the covariate, $\bar{X}_{..}$ is the grand mean of the covariate and $\varepsilon_{ij} \overset{\text{iid}}{\sim} N(0, \sigma^2)$. The adjusted mean of the ith group is

$$\bar{Y}_{i(\text{adj})} = \bar{Y}_i - \hat{\beta}(\bar{X}_i - \bar{X}_{..}).$$

It removes the effect of covariates, allowing for a more accurate comparison between groups. We refer to [75, 57] for more details on ANCOVA.

Note 6.14 (Effect Sizes for ANCOVA). Effect sizes of treatment effects in ANCOVA can be quantified using various measures, such as partial eta squared (η_p^2) and Cohen's f^2. Let η_p^2 denote the partial eta squared associated with the treatment effect, then it can be calculated as

$$\eta_p^2 = \frac{SS_{\text{treatment}}}{SS_{\text{treatment}} + SS_{\text{error}}},$$

where $SS_{\text{treatment}}$ and SS_{error} are the sums of squares for treatment and residual error after adjustment for covariate(s), respectively. Cohen's f^2 is then defined as

$$f^2 = \frac{\eta_p^2}{1 - \eta_p^2}.$$

Unlike η^2 in ANOVA (Note 6.11), η_p^2 in ANCOVA represents the proportion of variance explained by the treatment effect after controlling for the covariate(s), and that is why in the calculation of η_p^2, the denominator includes only the sum of squares for treatment and error, excluding the sum of squares for covariate(s). Cohen's f^2 and η_p^2 can be solely determined by each other, and in Section 6.1.6, we will use Cohen's f^2 to conduct power analysis for ANCOVA. We refer to [51] for more discussions on various effect sizes for ANCOVA.

6.1.6 Power Analysis for ANCOVA

In the previous case study, we use ANCOVA to evaluate the effect of an intervention while controlling for a continuous covariate (pretest scores). This is a common scenario in research, and in this section, we will use a case study to illustrate how to conduct power analysis for ANCOVA and determine the required sample size to achieve a desired statistical power.

Case 6: Power Analysis for Innovative Teaching on Posttest Improvement

A researcher plans a randomized controlled trial to evaluate the effectiveness of an innovative teaching approach compared to a traditional method. Scores on a standardized final exam will be used to examine the effectiveness of the innovative teaching approach, and pretest scores will be collected as covariates. A previous, closely related study reported the effect size of the teaching method after adjusting for the pretest scores as is $\eta_p^2 = 0.06$. The researcher wants to conduct a power analysis to determine the required sample size for different target powers $(0.7, 0.8, 0.9)$ at a significance level of 5%. Due to resource constraints, the researcher expects that the maximum feasible sample size is 100 students, so they also want to perform a sensitivity analysis to determine the minimum detectable effect size given a fixed sample size.

For answers to the initial questions and some initial thoughts about this case study, refer to Section 5.7. We can use the R function `pwr.f2.test()` from the `pwr` package to conduct both the sample size determination and sensitivity analysis for ANCOVA. The R function uses Cohen's f^2 as input, so we need to convert the given effect size measure η_p^2 to Cohen's f^2 using the formula in Section 4.5.

```
library(pwr)
f2 <- 0.06 / (1 - 0.06)
f2
: [1] 0.06382979
```

Next, we can determine the required total sample size for different target powers (0.7, 0.8, 0.9) using the following code. Here we use $u = 1$ for the numerator degrees of freedom which is the number of groups minus 1; there are two groups (innovative vs. traditional). We use the commonly used $\alpha = 0.05$ for the significance level. The required total sample sizes are 99, 125, 167, respectively, for target powers of 0.7, 0.8, and 0.9.

```
u <- 1        # numerator df (groups - 1)
alpha <- 0.05
powers <- c(0.7, 0.8, 0.9)
sample_sizes <- sapply(powers, function(p) {
  result <- pwr.f2.test(u = u, f2 = f2, sig.level = alpha, power = p)
  ceiling(result$v + u + 1)})
sample_sizes
: [1]  99 125 167
```

If we have a fixed total sample size of 100, we can conduct a sensitivity analysis to determine the minimum detectable effect size with target powers. The following code shows that with a total sample size of 100, we can detect an effect size of approximately $\eta_p^2 = 0.059, 0.074, 0.097$, respectively, for target powers of 0.7, 0.8, and 0.9; that is, the test becomes less powerful for smaller effect sizes.

```
# sensitivity analysis
N_total <- 100
v <- N_total - u - 1
f2 <- sapply(powers, function(p) {
  result <- pwr.f2.test(u = u, v = v, sig.level = 0.05, power = p)
  result$f2})
f2
: [1] 0.06294703 0.08005759 0.10717665
```

```
# Convert f2 back to eta-squared
eta2 <- f2 / (1 + f2)
eta2
: [1] 0.05921935 0.07412344 0.09680176
```

Conclusions

In this case study, we conduct a power analysis for ANCOVA to determine the required sample size to detect an effect size ($\eta_p^2 = 0.06$) with different target powers. We find that the required total sample sizes are 99, 125, 167, respectively, for target powers of 0.7, 0.8, and 0.9. Additionally, we perform a sensitivity analysis to determine the minimum detectable effect size with a total sample size of 100 and find that we can detect an effect size of approximately $\eta_p^2 = 0.059, 0.074, 0.097$, respectively, for target powers of 0.7, 0.8, and 0.9.

6.1.7 Multiple Linear Regression

Multiple linear regression is widely used in various fields to model the relationship between a continuous response variable and multiple predictors. It helps to understand how each predictor affects the response while controlling for the effects of other predictors. We use the following case study to illustrate how to conduct multiple linear regression analysis.

Case 7: Factors on Housing Prices

Real estate companies and homeowners need to accurately estimate house prices based on various characteristics. This case study analyzes residential properties in a midsize American city to determine the key factors that influence house prices and develop a predictive model. Research questions are the following:

1. Which housing characteristics significantly influence sale prices?
2. How well can we predict house prices using these characteristics?
3. What are the relative impacts of different features on house prices?

The dataset contains information on 200 house sales from 2024:

1. Sale_Price: Sale price in dollars (response variable)
2. Square_Feet: Total living area
3. Bedrooms: Number of bedrooms
4. Bathrooms: Number of bathrooms
5. Age: Age of house in years
6. Lot_Size: Lot size in square feet
7. Garage: Number of garage spaces
8. Distance_Downtown: Distance to downtown in miles
9. Pool: Presence of pool (binary: 0/1)
10. School_Rating: Nearby school rating (1-10)
11. Crime_Rate: Neighborhood crime rate per 1000 residents

For answers to the initial questions and some initial thoughts about this case study, refer to Section 5.8. First, we need to prepare the data for further analysis. We center the continuous predictors and define categorical variables as factors in R. Centering continuous predictors is recommended, and it leads to better interpretation, as any effect of a predictor will then be conditioning on the other predictors taking their average values, and it is especially useful when there are interaction terms and when variance inflation factors (VIF) are to be checked; see [7]. Note that we usually do not center the response variable to keep interpretation and predictions straightforward.

```
library(ggplot2)
library(corrplot)
library(car)
data <- read.csv("data/housing-price.csv")

data$Square_Feet <- scale(data$Square_Feet, scale = FALSE)
data$Bedrooms <- scale(data$Bedrooms, scale = FALSE)
data$Bathrooms <- scale(data$Bathrooms, scale = FALSE)
data$Age <- scale(data$Age, scale = FALSE)
data$Lot_Size <- scale(data$Lot_Size, scale = FALSE)
data$Garage <- scale(data$Garage, scale = FALSE)
data$Distance_Downtown <- scale(data$Distance_Downtown, scale = FALSE)
data$Pool <- as.factor(data$Pool)
data$School_Rating <- scale(data$School_Rating, scale = FALSE)
data$Crime_Rate <- scale(data$Crime_Rate, scale = FALSE)
```

Then a pairwise scatter plot can be useful to get the first impression about the data and the relationships between the predictors and the response variable.

```
# Scatter plots of Sale_Price vs. predictors
pairs(data)
```

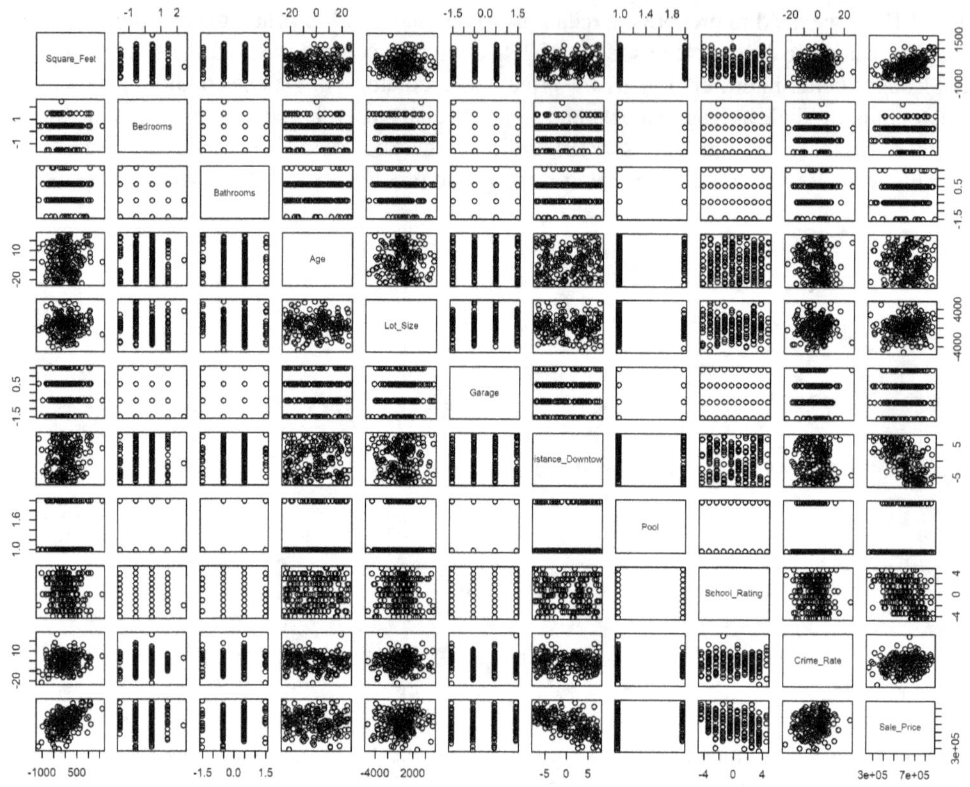

FIGURE 6.10: Scatter Plots of the Variables.

A correlation matrix can be plotted as follows, and it indicates that sale prices are positively associated with the square feet, etc., and are negatively associated with distance to downtown, etc.

```
cor_matrix <- cor(subset(data, select=-Pool)) # Correlation matrix
grey_colors <- colorRampPalette(c("grey90", "grey20"))(nrow(cor_matrix))
corrplot(cor_matrix, col = grey_colors, tl.col = "black", method = "number")
```

We start with a linear regression model that includes all main and two-way interactions. Note that including all two-way interactions can lead to a very large model, especially when there are many predictors. In practice, we may need to use domain knowledge to pre-select some important interactions to include in the initial model. The below R code `(.)^2` indicates that all main effects and two-way interactions are included.

```
full_model <- lm(Sale_Price ~ (.)^2, data = data)
```

To decide which predictors are significant, we can use either `drop1()` or `step()` functions, and they use different criteria. The `drop1()` function uses the F test to compare the full model with a reduced model that excludes one predictor at a time. We can repeat this process until all remaining predictors are significant. The following R code implements this backward elimination process. When the task is for statistical inference, this method is preferred. When the task is for prediction, stepwise selection based on AIC or BIC is more commonly used, and it is implemented in the `step()` function as shown later.

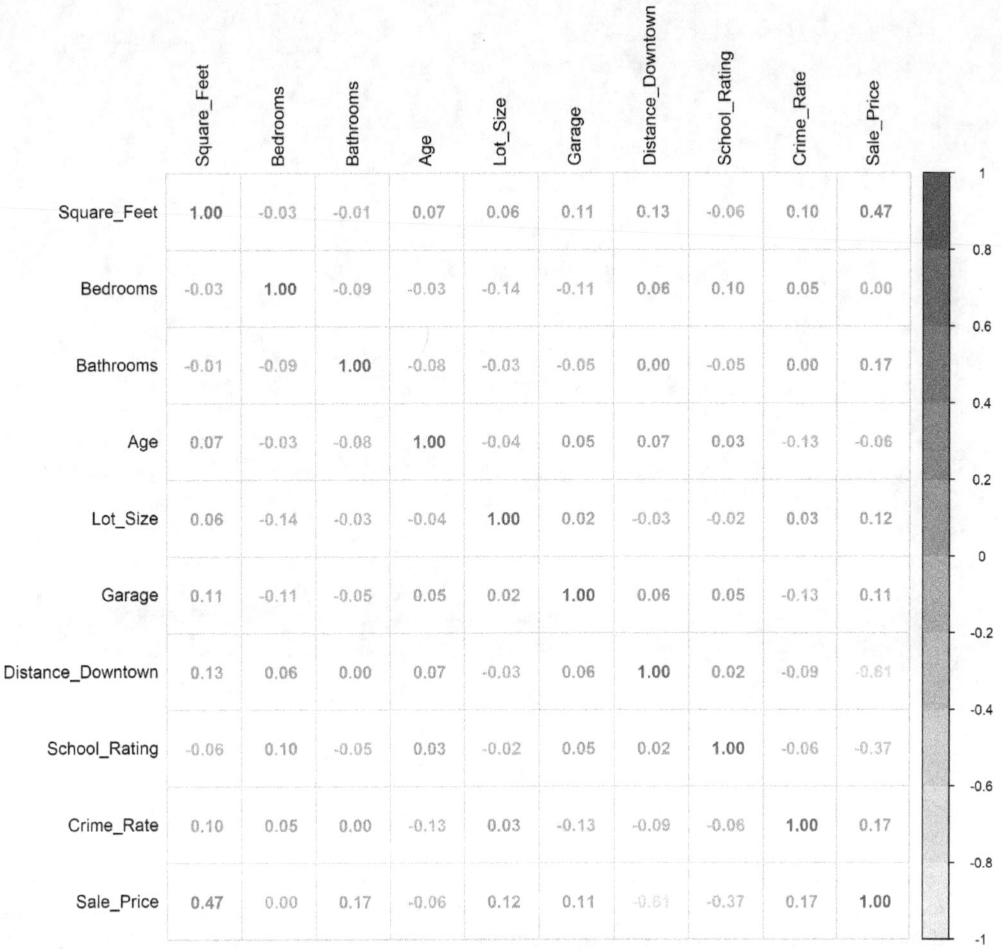

FIGURE 6.11: Plot for Correlation Matrix.

```
repeat {test <- drop1(full_model, test="F")
        worst <- which.max(test$`Pr(>F)`[-1])
        if(test$`Pr(>F)`[worst+1] < 0.05) break
        term <- rownames(test)[worst+1]
        full_model <- update(full_model, paste(". ~ . -", term))}
```

Here we use stepwise selection to select a reduced model based on a criterion such as AIC or BIC. The R codes are shown below, where `k=log(nrow(data))` is to indicate that BIC is used as the criterion to encourage a simpler model than AIC does, although AIC can also be used.

```
step_model <- step(full_model, direction = "both",trace = FALSE,
                   k=log(nrow(data))) # BIC for a simpler model
summary(step_model)
:
: Call:
```

```
: lm(formula = Sale_Price ~ Square_Feet + Bedrooms + Bathrooms +
:     Age + Garage + Distance_Downtown + Pool + School_Rating +
:     Age:School_Rating + Distance_Downtown:School_Rating + Pool:School_Rating +
:     Square_Feet:Age, data = data)
:
: Residuals:
:     Min      1Q  Median      3Q     Max
: -109768  -31041   -2564   32816  164769
:
: Coefficients:
:                                   Estimate Std. Error t value Pr(>|t|)
: (Intercept)                      5.897e+05  3.807e+03 154.912  < 2e-16 ***
: Square_Feet                      1.478e+02  7.492e+00  19.731  < 2e-16 ***
: Bedrooms                         1.580e+04  4.126e+03   3.829 0.000175 ***
: Bathrooms                        3.378e+04  4.848e+03   6.966 5.38e-11 ***
: Age                             -5.312e+02  2.437e+02  -2.180 0.030505 *
: Garage                           1.886e+04  3.566e+03   5.289 3.41e-07 ***
: Distance_Downtown               -2.008e+04  7.730e+02 -25.982  < 2e-16 ***
: Pool1                            2.002e+04  8.605e+03   2.327 0.021048 *
: School_Rating                   -1.822e+04  1.489e+03 -12.232  < 2e-16 ***
: Age:School_Rating               -2.592e+02  9.572e+01  -2.707 0.007412 **
: Distance_Downtown:School_Rating -1.755e+03  2.862e+02  -6.132 5.04e-09 ***
: Pool1:School_Rating              8.756e+03  3.395e+03   2.579 0.010666 *
: Square_Feet:Age                  1.396e+00  5.316e-01   2.626 0.009362 **
: ---
: Signif. codes:  0 '***' 0.001 '**' 0.01 '*' 0.05 '.' 0.1 ' ' 1
:
: Residual standard error: 47660 on 187 degrees of freedom
: Multiple R-squared:  0.8729, Adjusted R-squared:  0.8648
: F-statistic: 107.1 on 12 and 187 DF,  p-value: < 2.2e-16
```

We can compare the reduced model (after model selection) with the full model, and the following F test leads to a p-value (0.59) much greater than commonly used significance levels. We keep the reduced model, as it is not significantly different from the full model to account for the variability of the response variable, but it is much smaller and easier to interpret. We can also further explore the possibility of involving three-way interactions and beyond. To keep the model and interpretation simple, we choose to use the reduced model with the selected two-way interactions and the main effects only. Note that, when interactions are included in the model, the main effects involved in the interactions should also be included in the model, even if they are not significant, to keep the hierarchy of the model.

```
# Compare models
anova(step_model, full_model)
: Analysis of Variance Table
:
: Model 1: Sale_Price ~ Square_Feet + Bedrooms + Bathrooms + Age + Garage +
:     Distance_Downtown + Pool + School_Rating + Age:School_Rating +
:     Distance_Downtown:School_Rating + Pool:School_Rating + Square_Feet:Age
: Model 2: Sale_Price ~ (Square_Feet + Bedrooms + Bathrooms + Age + Lot_Size +
:     Garage + Distance_Downtown + Pool + School_Rating + Crime_Rate)^2
:   Res.Df        RSS Df  Sum of Sq      F Pr(>F)
: 1    187 4.2469e+11
: 2    144 3.3188e+11 43 9.2802e+10 0.9364 0.5873
```

In what follows, we conduct model diagnostics. First, we check whether there are serious multicollinearity issues. Because there are categorical predictors that have multiple levels and interaction terms in the selected model, we use the option `type = "predictor"` in the R function `car::vif` and the adjusted GVIF "`GVIF^(1/(2*Df))`" to decide whether there is multicollinearity. Since all the values are less than 2, we decide that there are no concerns of multicollinearity, and thus we keep the model as before.

```
# VIF check for multicollinearity
# use type = "predictor" as there are interactions in the model
vif_step <- car::vif(step_model, type = "predictor")
vif_step[,c("GVIF", "Df", "GVIF^(1/(2*Df))")]
:                        GVIF Df GVIF^(1/(2*Df))
: Square_Feet        1.161781  3        1.025307
: Bedrooms           1.078241  1        1.038384
: Bathrooms          1.075182  1        1.036910
: Age                1.500256  5        1.041397
: Garage             1.041012  1        1.020300
: Distance_Downtown  1.420080  3        1.060194
: Pool               1.187628  3        1.029074
: School_Rating      1.210118  7        1.013716
```

The next step for model diagnostic is to check the normality assumption, the equal variance assumption, etc. The following contains four diagnostic plots. The top-left plot has a level smoothing line and residuals are evenly distributed above and below the line, indicating that there are no severe violations of the equal variance assumption. The bottom-left is used to check the equal variance from another angel. The top-right is used to check the normality assumption. The button-right is used to identify influential points: vertical dimension for outliers and horizontal dimension for high-leverage points, and points with the largest 3 Cook's distances are labeled: 6, 9, 54. All plots suggest that there are no serious violations of the assumptions, and a few influential data points are highlighted for attention.

```
# Model Diagnostics: Residual plots
par(mfrow = c(2, 2))
plot(step_model)
```

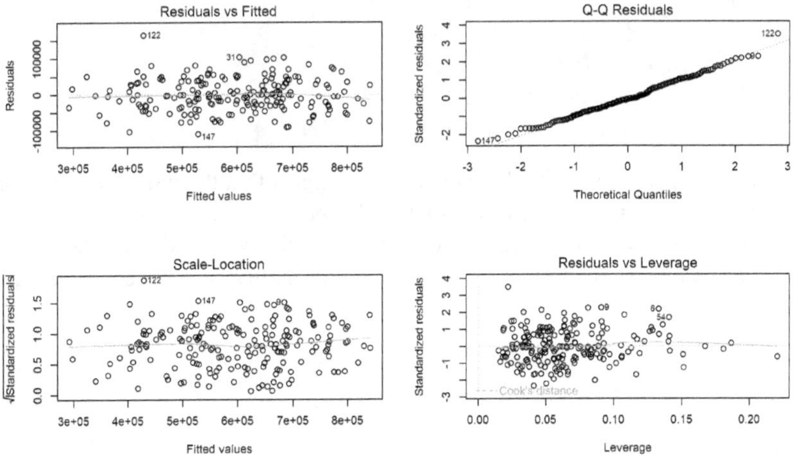

FIGURE 6.12: Diagnostic Plots for Multiple Linear Regression.

```
par(mfrow = c(1, 1))
```

Alternatively, we can use the following R function to obtain the influence plot. The labeled data points either have large residuals, leverages, or Cook's distance, and a larger circle corresponds to a larger Cook's distance. We refer to Note 6.17 and 6.18 for details on the concepts of outliers, leverage, and Cook's distance in linear regression.

```
car::influencePlot(step_model, fill.col="black")
```

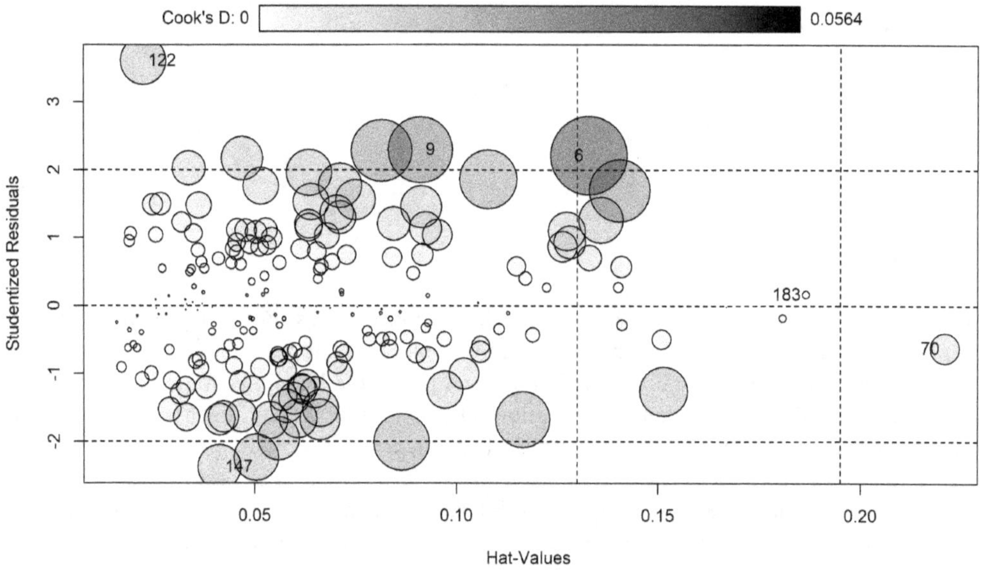

FIGURE 6.13: Influence Plot for Multiple Linear Regression.

```
:          StudRes        Hat       CookD
: 6    2.2097654 0.13287363 0.056387080
: 9    2.2943911 0.09103054 0.039649451
: 70  -0.6222846 0.22107361 0.008482046
: 122  3.6069904 0.02215121 0.021302879
: 147 -2.3814762 0.04115939 0.018270730
: 183  0.1707820 0.18642074 0.000516768
```

Although visual inspections of the diagnostic plots suggest that the model assumptions are reasonably satisfied, some formal tests can also be performed to confirm this as follows, and they suggest that there is no significant evidence to reject the null hypothesis that the model assumptions hold (p-values are 0.30, 0.48, 0.08 for the tests on normality, heteroscedasticity, and outliers, respectively).

```
# Normality test for residuals
shapiro.test(residuals(step_model))
:
:  Shapiro-Wilk normality test
:
: data:  residuals(step_model)
: W = 0.9916, p-value = 0.3014
```

```
# Breusch-Pagan test for heteroscedasticity
lmtest::bptest(step_model)
:
:  studentized Breusch-Pagan test
:
: data:  step_model
: BP = 11.534, df = 12, p-value = 0.4838
```

```
# Outlier detection
outlierTest(step_model)
: No Studentized residuals with Bonferroni p < 0.05
: Largest |rstudent|:
:     rstudent unadjusted p-value Bonferroni p
: 122  3.60699         0.00039792      0.079583
```

The following plot shows how close what the fitted model predicts is to the actual values.

```
# Predicted vs. actual plot
library(ggplot2)
predicted <- predict(step_model)
actual <- data$Sale_Price
plot_data <- data.frame(Predicted = predicted, Actual = actual)
ggplot(plot_data, aes(x = Predicted, y = Actual)) +
  geom_point(color = "grey") +
  geom_abline(intercept = 0, slope = 1, linetype = "dashed") +
  labs(x = "Predicted Price", y = "Actual Price", title = "") + theme_bw()
```

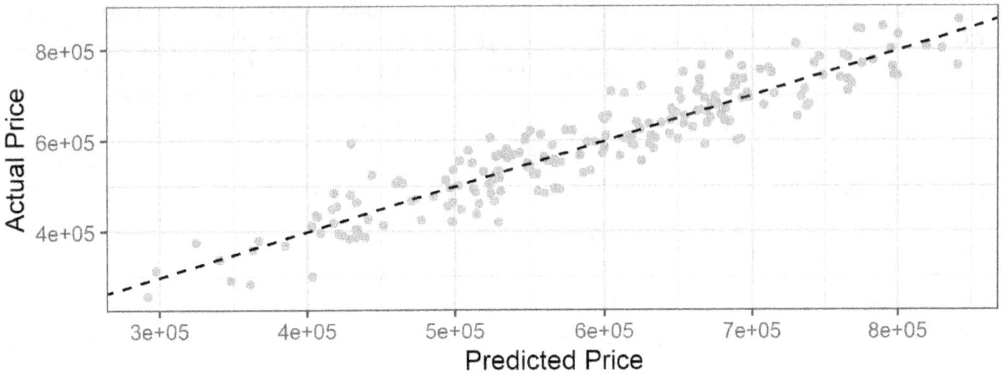

FIGURE 6.14: Predicted vs. Actual House Prices.

Conclusions

In this case study, a multiple linear regression analysis is conducted to model housing prices based on various predictors. The multiple linear regression analysis identifies several significant predictors and interactions that explain housing prices, including "square feet", "bedrooms", "bathrooms", etc. However, some predictors, such as "lot size" and "crime rate", are not significant in the presence of other predictors. The final model passes various diagnostic checks, indicating that the assumptions of linear regression are reasonably satisfied. This is an observational study, so there is no causality that can be derived. Whether the study can lead to statistical inference for the population will depend on how the sample of 200 houses is sampled from the population, and it needs to be further clarified.

Note 6.15 (Variance Inflation Factors (VIF))**.** *Variance Inflation Factor (VIF)* measures multicollinearity among predictors in regression models, and it applies only to single, continuous predictors. As VIF is applied to predictors, it can be used for various types of regression models, including linear regression, generalized linear regression, etc. For a predictor X_j, VIF is defined as

$$\text{VIF}_j = \frac{1}{1-R_j^2},$$

where R_j^2 is the coefficient of determination from regressing X_j on all other predictors. A VIF value of 1 indicates no multicollinearity. VIF > 5 or 10 suggests significant multicollinearity. For dealing with multicollinearity, we can remove highly correlated predictors, combine predictors into a single variable, use regularization techniques such as ridge regression, or use principal components. For the original paper on VIF please refer to [73], and a useful resource for practical caution of applying VIF is [85].

Note 6.16 (Generalized Variance Inflation Factors (GVIF))**.** *GVIF* is an extension of the traditional VIF, and it tells us how much a group of variables together (like a categorical variable with multiple dummy variables, or an interaction term) is correlated with the other predictors in a regression model. The more overlap or redundancy there is, the higher the GVIF. GVIF uses determinants of matrices to summarize how spread out or independent the predictors are in a multidimensional space. For a given predictor with d degrees of freedom, GVIF is commonly adjusted as $\text{GVIF}^{1/(2d)}$ to allow comparison across predictors with different numbers of parameters, as multi-column predictors introduce structural multicollinearity and naturally inflate GVIF. There is no strict cutoff for GVIF, but a $\text{GVIF}^{1/(2d)}$ value less than 2 brings little concerns of multicollinearity, and a value greater than 5 or 10 may indicate significant multicollinearity. For more details, we refer to [44].

Note 6.17 (Leverage in Linear Regression)**.** In regression analysis, *leverage* quantifies the influence of an individual data point on the fitted model. High-leverage points have potential to significantly affect the estimated regression coefficients. Consider the standard multiple linear regression model:

$$\mathbf{y} = \mathbf{X}\beta + \varepsilon, \quad \varepsilon \sim \mathcal{N}(0, \sigma^2\mathbf{I}),$$

where $\mathbf{y}$ is an $n \times 1$ response vector, $\mathbf{X}$ is an $n \times p$ matrix of predictors (including intercept) with $\text{rank}(\mathbf{X}) = p$, β is a $p \times 1$ vector of coefficients, ε is an $n \times 1$ vector of errors. The fitted values are

$$\hat{\mathbf{y}} = \mathbf{X}(\mathbf{X}^\top\mathbf{X})^{-1}\mathbf{X}^\top\mathbf{y} = \mathbf{H}\mathbf{y},$$

where $\mathbf{H}$ is the hat matrix, given by $\mathbf{H} = \mathbf{X}(\mathbf{X}^\top\mathbf{X})^{-1}\mathbf{X}^\top$. The diagonal elements of $\mathbf{H}$, denoted by $h_i = H_{ii}$, are the leverages. Each h_i represents the leverage of the ith observation. As $\text{Var}(\hat{\varepsilon}_i) = \sigma^2(1-h_i)$ and $0 \le h_i \le 1$, a point with high leverage leads to a small $\text{Var}(\hat{\varepsilon}_i)$ and thus pulls the regression line toward itself, potentially distorting the overall fit. For more details, we refer to [29, 9].

Note 6.18 (Cook's Distance in Linear Regression)**.** *Cook's Distance* is a valuable diagnostic tool for identifying influential observations in regression analysis, and it combines both residual information and leverage. Influential observations can be outliers or high-leverage points: the former have large residuals, that is, their actual values differ significantly from the predicted values, and the latter have unusual values of the predictors. Cook's Distance for the ith observation is defined as

$$D_i = \frac{(\hat{\mathbf{y}} - \hat{\mathbf{y}}_{(i)})^\top (\hat{\mathbf{y}} - \hat{\mathbf{y}}_{(i)})}{p \cdot \hat{\sigma}^2} = \frac{1}{p} r_i^2 \frac{h_i}{1 - h_i},$$

where $\hat{\mathbf{y}}$ is estimated $\mathbf{y}$ using all data, $\hat{\mathbf{y}}_{(i)}$ is estimated $\mathbf{y}$ without the ith observation, p is the number of parameters, $\hat{\sigma}^2$ is estimated residual variance, r_i is the standardized residual, and h_i is the leverage for the ith observation. Because r_i reflects residual information for identifying outliers and h_i identifies high-leverage points, Cook's distance summarizes both into one quantity. We refer to [29] for more details on Cook's distance.

6.1.8 Ordinal Variables as Predictors

When an ordinal variable is used as a predictor in a regression model, we can use polynomial contrasts to capture the trends across ordered levels. In what follows, we illustrate this approach using a case study on job satisfaction based on employee ranks.

Case 8: Evaluating Job Satisfaction in a Tech Company

A tech company surveys its employees to understand the factors that influence job satisfaction (measured on a scale of 0 to 100). The company collects data from 200 randomly sampled employees on several variables: annual salary (in thousands of dollars), years of experience, and employee ranks: Junior, Intermediate, Senior, and Lead. How do these factors, especially employee ranks, affect job satisfaction?

For answers to the initial questions and some initial thoughts about this case study, refer to Section 5.9. In what follows, we will treat the employee rank as ordinal data and use it as a predictor in a multiple linear regression model.

Ordinal data are categorical data, and in R we need to first convert them into factors. Note that in the following R codes, the levels are specified in a particular sequence from lower rank to higher rank. We cannot skip this because by default R assigns levels in alphabetical order which leads to "Intermediate", "Junior", "Lead", "Senior", not what we needed. Use `ordered = TRUE` to define ordinal data.

```
library(MASS)
library(ggplot2)
data <- read.csv("data/job-satisfactory-ranks.csv")

data$Rank <- factor(data$Rank, ordered = TRUE,
                    levels = c("Junior", "Intermediate", "Senior", "Lead"))
```

Figure 6.15 suggests that there may be a nonlinear relationship between satisfaction scores and employee ranks.

```
# Visualize the relationship between Rank and Job_Satisfaction
ggplot(data, aes(y = Rank, x = Job_Satisfaction)) + geom_boxplot() +
  stat_summary(fun = "mean", geom = "point", shape = 23, size = 2) +
  labs(title = "", y = "Rank (1 = Junior, 4 = Lead)",
       x = "Job Satisfaction (0-100)") + theme_bw()
```

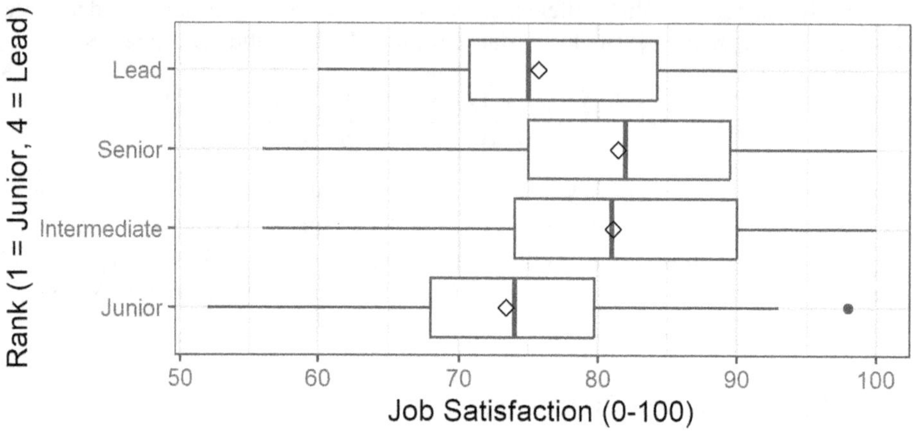

FIGURE 6.15: Job Satisfaction by Employee Rank.

In R, the default contrasts for ordinal data should be "contr.poly" the polynomial contrast; this can be verified by using `options("contrasts")`. Otherwise, use the following to assign the polynomial contrasts to the ordinal variable. Here, ".L", ".Q" and ".C" correspond to "Linear", "Quadratic" and "Cubic".

```
# Set polynomial contrasts for Rank
contrasts(data$Rank) <- contr.poly(4)  # 4 levels
contrasts(data$Rank)
:                        .L   .Q         .C
: Junior       -0.6708204  0.5 -0.2236068
: Intermediate -0.2236068 -0.5  0.6708204
: Senior        0.2236068 -0.5 -0.6708204
: Lead          0.6708204  0.5  0.2236068
```

The polynomial contrast converts the one column of the variable "employee ranks" into three columns according to the numbers in the contrasts; for example, "Junior" occupying one column is converted to the dummy variables −0.6708204, 0.5, −0.2236068 that occupy three columns, and then the three columns of dummy variables will be used for regression analysis. Figure 6.16 illustrates the three patterns.

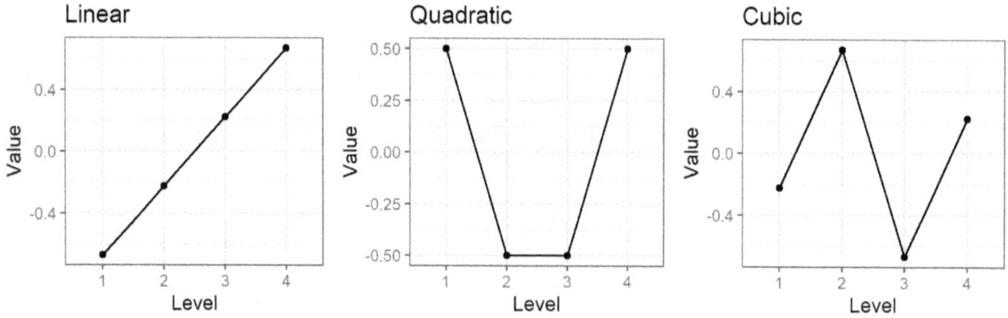

FIGURE 6.16: Three Patterns Created by the Polynomial Contrasts.

The specific values of the dummy variables are calculated to make them orthonormal, that is, letting X be the 4 by 3 matrix created by `contr.poly(4)`, then $X^\top X = I$, the 3 by 3 identity matrix:

```
X <- contr.poly(4)
t(X) %*% X
:              .L            .Q            .C
: .L 1.000000e+00 0.000000e+00 2.775558e-17
: .Q 0.000000e+00 1.000000e+00 8.326673e-17
: .C 2.775558e-17 8.326673e-17 1.000000e+00
# The above 3 by 3 matrix is very close to I
```

Then a multiple linear regression can be carried out as follows. The result tells us that the cubic pattern for the rank variable is not significant.

```
# Fit the multiple linear regression model
model <- lm(Job_Satisfaction ~ Salary + Experience + Rank, data = data)
summary(model)
:
: Call:
: lm(formula = Job_Satisfaction ~ Salary + Experience + Rank, data = data)
:
: Residuals:
:      Min       1Q   Median       3Q      Max
: -23.4444  -6.0142   0.6212   6.0233  22.6051
:
: Coefficients:
:             Estimate Std. Error t value Pr(>|t|)
: (Intercept) 51.72824    3.06194  16.894  < 2e-16 ***
: Salary       0.25731    0.03191   8.064 7.49e-14 ***
: Experience   1.14571    0.30159   3.799 0.000194 ***
: Rank.L       2.83486    1.57391   1.801 0.073231 .
: Rank.Q      -6.41152    1.36817  -4.686 5.23e-06 ***
: Rank.C       0.28753    1.14115   0.252 0.801337
: ---
: Signif. codes:  0 '***' 0.001 '**' 0.01 '*' 0.05 '.' 0.1 ' ' 1
:
: Residual standard error: 8.439 on 194 degrees of freedom
: Multiple R-squared:  0.3715, Adjusted R-squared:  0.3553
: F-statistic: 22.94 on 5 and 194 DF,  p-value: < 2.2e-16
```

So, we now only use the linear and quadratic patterns together with the other predictors to fit the data. Note that we need to use `how.many=2` in the `contrasts()` function to indicate the number of patterns. The p-value of "Rank.L" is about 7%, although insignificant at a significance level of 5%, suggests that there is an increasing trend of satisfaction scores over ranks.

```
# only linear and quadratic
contrast_L_Q <- contr.poly(4)[,1:2]  # 2 levels, linear, quadratic
contrasts(data$Rank, how.many=2) <- contrast_L_Q
model_L_Q <- lm(Job_Satisfaction ~ Salary + Experience + Rank, data = data)
summary(model_L_Q)
```

```
:
: Call:
: lm(formula = Job_Satisfaction ~ Salary + Experience + Rank, data = data)
:
: Residuals:
:      Min       1Q   Median       3Q      Max
: -23.6192  -6.1113   0.5656   5.8838  22.5555
:
: Coefficients:
:             Estimate Std. Error t value Pr(>|t|)
: (Intercept) 51.68687    3.05019  16.945  < 2e-16 ***
: Salary       0.25703    0.03181   8.079 6.67e-14 ***
: Experience   1.15137    0.30003   3.838 0.000168 ***
: Rank.L       2.73693    1.52151   1.799 0.073591 .
: Rank.Q      -6.47655    1.34038  -4.832 2.73e-06 ***
: ---
: Signif. codes:  0 '***' 0.001 '**' 0.01 '*' 0.05 '.' 0.1 ' ' 1
:
: Residual standard error: 8.418 on 195 degrees of freedom
: Multiple R-squared:  0.3713, Adjusted R-squared:  0.3584
: F-statistic: 28.79 on 4 and 195 DF,  p-value: < 2.2e-16
```

We choose the model with both linear and quadratic patterns as the candidate model. Figure 6.17 illustrates the overall slight increasing and the dominating quadratic patterns: satisfaction keeps increasing until it peaks at about the senior level. The smoothing curve in the figure is simply used to visualize the quadratic pattern of the predicted values.

```
# Predict values to see the trend
data$Predicted <- predict(model_L_Q)
set.seed(111)
ggplot(data, aes(x = as.numeric(Rank), y = Predicted)) +
  geom_jitter(width = 0.1, height = 0) +
  geom_smooth(method = "lm", formula = y ~ poly(x, 2), color = "black") +
  labs(title = "", x = "Rank (1: Junior)", y = "Job Satisfaction") +
  theme_bw()
```

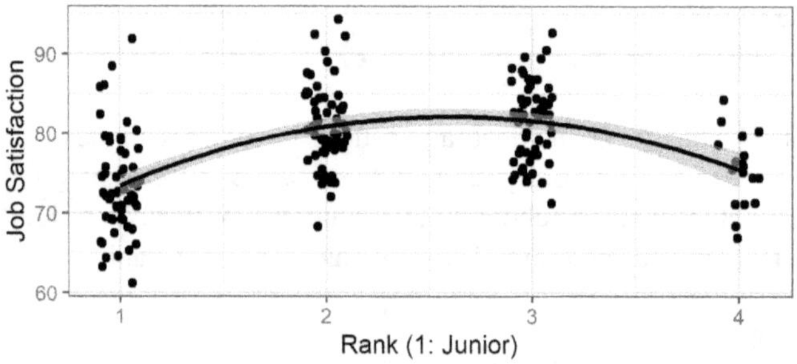

FIGURE 6.17: Predicted Job Satisfaction by Rank.

Conclusions

In this case study, we analyze the relationship between job satisfaction and employee ranks using multiple linear regression with polynomial contrasts for the ordinal predictor "Rank", while controlling for salary and experience. We find that the linear trend is marginally significant (p-value $= 0.07$), suggesting a slight overall increase in satisfaction with rank, and job satisfaction tends to increase with rank, peaking around the senior level, as indicated by the significant quadratic trend (p-value $< 10^{-5}$).

Note 6.19 (Polynomial Contrasts for Ordinal Predictors). In multiple linear regression, we can use polynomial contrasts to encode an ordinal predictor to capture trends such as linear, quadratic, and cubic trends across ordered levels of the ordinal predictor. The process assigns specific contrasts to the levels of the ordinal predictor, and use these contrasts as pseudo data in the regression: $y \sim \beta_0 + \sum_i \beta_i x_i + \sum_k \gamma_k c_k$, where x_i's are other predictors, c_k's are contrasts assigned for the ordinal predictor, and the coefficients γ_k test specific trends, with γ_1 testing linear effect and γ_2 testing quadratic effect, etc.

For example, for an ordinal predictor with four levels: Junior, Intermediate, Senior, and Lead, we can use R function `contr.poly(4)` to generate the polynomial contrasts as follows: the linear contrast is represented as $-0.67, -0.22, 0.22, 0.67$ indicating a steady trend, the quadratic contrast is $0.5, -0.5, -0.5, 0.5$ indicating curvature, and the cubic contrast is $0.22, -0.67, 0.67, -0.22$ indicating an "S" shape (see Figure 6.16). Then the regression model will use $(-0.67, 0.5, 0.22)$ as the pseudo data for the "Junior" level, and so on for the other levels. These contrasts are created orthogonal and normalized and are used as linearly independent pseudo data in regression.

Note that although the polynomial contrasts assume equal spacing between levels of the ordinal predictor, the non-linear trends allowed in the contrasts make it not a concern even if the actual spacing is not equal. Moreover, the use of polynomial contrasts respects the ordinal nature of data, tests specific hypotheses about trends, and requires fewer parameters when there are many levels (otherwise $k-1$ for treating it as nominal).

However, we should avoid using polynomial contrasts for ordinal predictors if we believe that the trend is not polynomial-shaped, such as exponential or logarithmic trends, or if we are more interested in differences between specific levels than trends then treating the ordinal predictor as nominal is more appropriate. We refer to [75] for more details about polynomial contrasts.

6.1.9 Repeated Measures ANOVA

In Section 6.1.4, we have introduced one-way ANOVA to compare means across multiple groups when the observations are independent. However, in some situations, the same subjects are measured under different conditions, leading to correlated observations. In this case study, we will illustrate how to use repeated measures ANOVA to analyze such data.

Case 9: Evaluating the Effectiveness of Study Techniques on Quiz Performance

A university's educational research team is interested in assessing how different study techniques impact student quiz performance. 20 students are randomly selected from a pool of volunteers to ensure a diverse mix of academic abilities. Those participants vary in age, gender, and previous academic performance, although these factors are not explicitly analyzed in this study. Determine whether there are significant differences in quiz scores between the three techniques.

For answers to the initial questions and some initial thoughts about this case study, please refer to Section 5.10. We now prepare the data and assign the student ID and techniques as factors. The R function `suppressPackageStartupMessages` is used to suppress package loading messages to keep the output clean. In practice, do not suppress these messages, as they may contain important information about package versions and dependencies.

```
# Evaluating the Effectiveness of Study Techniques on Quiz Performance
suppressPackageStartupMessages({
library(dplyr)
library(tidyr)
library(ggplot2)
library(ez)          # For Repeated Measures ANOVA
library(lme4)        # For Random Intercept Model
})

data <- read.csv("data/student-technique-quizscores.csv")

# Convert characters to factors
data$Student <- as.factor(data$Student)
data$Technique <- as.factor(data$Technique)
```

We may use a plot to visualize the difference among different techniques. We add some small random values horizontally in the plot using `geom_jitter()` to make the points non-overlapped and easier to see. This plot suggests that there could be some difference between A and C.

```
# Visualize means by Technique
set.seed(111)
ggplot(data, aes(x = Technique, y = Score)) +
  stat_summary(fun = "mean", geom = "point", size = 3) +
  stat_summary(fun = "mean", geom = "line", aes(group = 1)) +
  stat_summary(fun.data = "mean_se", geom = "errorbar", width = 0.2) +
  geom_jitter(width = 0.1, alpha = 0.5) +
  labs(title = "", x = "Technique", y = "Score") + theme_bw()
```

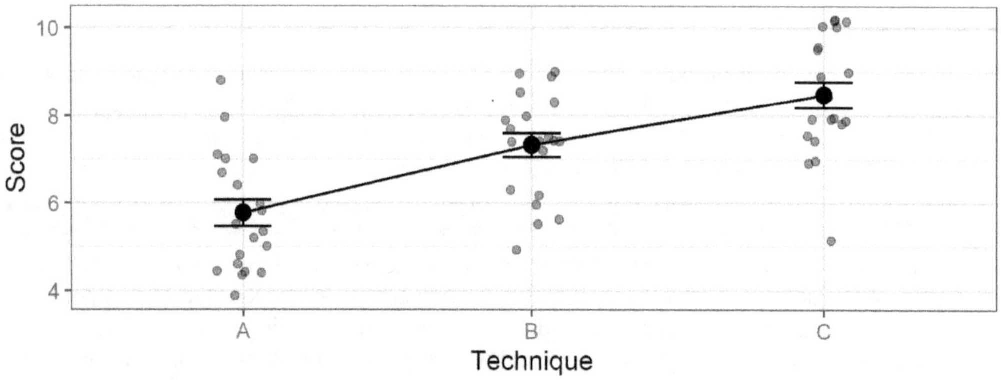

FIGURE 6.18: Mean Quiz Scores by Technique (20 Students).

Unlike the case when observations are independent we can use ANOVA directly, for this case we have to account for the dependency brought into by the repeated measures. Here we conduct a repeated measure ANOVA, and please Note 6.20 for more details. The *p*-values are much smaller than 5%, suggesting that there are significant difference between the scores associated with the three techniques.

```
# Repeated Measures ANOVA
re_anova <- ezANOVA(data = data, dv = Score, wid = Student,
  within = Technique, , detailed = TRUE)
print("Repeated Measures ANOVA:")
: [1] "Repeated Measures ANOVA:"
re_anova$ANOVA
:          Effect DFn DFd        SSn     SSd         F            p p<.05       ges
: 1 (Intercept)   1  19 3103.20417 69.7125 845.77198 3.227053e-17     * 0.9698069
: 2   Technique   2  38   73.43333 26.9000  51.86741 1.373516e-11     * 0.4318444
```

Before we move on, we need to conduct model diagnostics. The key assumption for one-way repeated measures ANOVA is that: All pairwise differences between conditions are approximately normally distributed, and their variances are equal (sphericity); see Note 6.20 for details of the sphericity assumption.

First, check the normality assumption. The p-values are all greater than 5%, suggesting that the normality assumption holds.

```
# Check normality using Shapiro-Wilk test for each pairwise difference
wide_data <- data %>%
  pivot_wider(names_from = Technique, values_from = Score)

diffs <- wide_data %>% mutate( diff_AB = A - B, diff_AC = A - C,
    diff_BC = B - C)

normality_test <- data.frame(comparison = c("A - B", "A - C", "B - C"),
  shapiro_p = c(shapiro.test(diffs$diff_AB)$p.value,
                shapiro.test(diffs$diff_AC)$p.value,
                shapiro.test(diffs$diff_BC)$p.value))
print(normality_test)
:   comparison shapiro_p
: 1      A - B 0.1058712
: 2      A - C 0.1031909
: 3      B - C 0.1082364
```

Now, check the sphericity condition. The p-value (0.66) for the Mauchly's Test of sphericity suggests that sphericity assumption holds.

```
# Check sphericity (Mauchly's test is included in ezANOVA output)
print("Mauchly's Test of Sphericity:")
: [1] "Mauchly's Test of Sphericity:"
print(re_anova$`Mauchly's Test for Sphericity`)
:       Effect         W         p p<.05
: 2 Technique 0.9550264 0.6609041
```

Although the case study does not specifically ask for which technique works the best, we can still conduct multiple comparisons using paired t tests, and here a simple Bonferroni adjustment is applied. All p-values are smaller than 5%, suggesting that B is significantly greater than A, and C is significantly greater than both A and B.

```
# Post hoc pairwise comparisons (if ANOVA is significant)
post_hoc <- pairwise.t.test(data$Score, data$Technique,
                            paired = TRUE, p.adjust.method = "bonferroni",
                            alternative="greater")
print("Post hoc Pairwise Comparisons:")
: [1] "Post hoc Pairwise Comparisons:"
print(post_hoc)
:
:  Pairwise comparisons using paired t tests
:
: data:  data$Score and data$Technique
:
:   A       B
: B 3.8e-05 -
: C 8.8e-10 0.00087
:
: P value adjustment method: bonferroni
```

Conclusions

In this case study, we use repeated measures ANOVA to analyze quiz scores under three different study techniques. The results suggest that there are significant differences among the techniques (p-value $= 1.37 \times 10^{-11}$), with Technique C performing the best, followed by Technique B, and then Technique A; p-values for all pairwise comparisons after Bonferroni adjustment are the following: C vs. B (8.7×10^{-4}), B vs. A (3.8×10^{-5}), and C vs. A (8.8×10^{-10}). This is an observational study, and causality cannot be established. The population to which these results generalize is the pool of volunteers who participated in this study, however, not necessarily to the population of interest of the study. The population of interest needs to be clearly defined and whether the pool of volunteers is representative of that population should be carefully considered before making any generalizations.

An alternative method is to treat Student as a random effect and then test whether the fixed effect Technique is significant. We first fit a random intercept model. In `lmer()`, the default method is restricted maximum likelihood (REML), and we need to let `REML=FALSE` to test fixed effects. In Note 6.23, we provide more details about REML.

```
# Random Intercept Model
ri_model <- lmer(Score ~ Technique + (1 | Student), data = data,
                 REML = FALSE)
```

An F test based on [63] with adjusted degrees of freedom is recommended to test whether the fixed effect Technique is significant. In Section 6.1.11, we will discuss more about linear mixed-effects models. The p-value is well below the significance level of 5%, suggesting Technique is significant.

```
null_model <- lmer(Score ~ (1 | Student), data = data, REML = FALSE)
pbkrtest::KRmodcomp(ri_model, null_model)
: large : Score ~ Technique + (1 | Student)
: small : Score ~ (1 | Student)
:         stat    ndf    ddf F.scaling   p.value
```

```
: Ftest 51.867  2.000 38.000          1 1.374e-11 ***
: ---
: Signif. codes:  0 '***' 0.001 '**' 0.01 '*' 0.05 '.' 0.1 ' ' 1
```

Note 6.20 (Repeated Measures ANOVA). *Repeated measures ANOVA* is a statistical method to analyze differences in means under multiple conditions when the same subjects are measured under each condition. It is an extension of a paired t test for more than two conditions. The model for one-way repeated measures ANOVA can be expressed as:

$$Y_{ij} = \mu + \alpha_i + S_j + \varepsilon_{ij}, \quad i = 1, \ldots, k; j = 1, \ldots, n,$$

where Y_{ij} is the observation for subject j under condition i, μ is the overall mean, α_i is the fixed effect of condition i, S_j is the random effect of subject j, $S_j \sim N(0, \sigma_s^2)$, ε_{ij} is the random error, with $\varepsilon_{ij} \sim N(0, \sigma^2)$, and $S_j \perp \varepsilon_{ij}$. The null hypothesis is

$$H_0 : \alpha_1 = \alpha_2 = \cdots = \alpha_k.$$

The key assumption for one-way repeated measures ANOVA is that all pairwise differences between conditions are normally distributed, and their variances are equal (*Sphericity*); see [58], [94], and [46]. One can use Mauchly's test of sphericity, and a small p-value indicates a violation of sphericity. When such assumptions are violated, mixed-effects models are recommended as a more flexible alternative (see Note 6.25).

The total variability is partitioned into:

$$SS_{\text{Total}} = SS_{\text{Condition}} + SS_{\text{Within}},$$

where $SS_{\text{Condition}}$ is the variability due to conditions (e.g., Study Techniques), SS_{Within} is the variability due to subjects and such variability can be further split into

$$SS_{\text{Within}} = SS_{\text{Subjects}} + SS_{\text{Error}}.$$

With $\bar{Y}_{i.}$ the mean of condition i, $\bar{Y}_{.j}$ the mean of subject j, and $\bar{Y}_{..}$ the grand mean, a one-Way repeated measures ANOVA table is the following:

TABLE 6.2: General One-Way Repeated Measures ANOVA Table.

Source	df	SS	MS	F
Condition	$k-1$	$k\sum_{i=1}^{k}(\bar{Y}_{i.} - \bar{Y}_{..})^2$	$\frac{SS_{\text{Condition}}}{k-1}$	$\frac{MS_{\text{Condition}}}{MS_{\text{Error}}}$
Subjects	$n-1$	$n\sum_{j=1}^{n}(\bar{Y}_{.j} - \bar{Y}_{..})^2$	$\frac{SS_{\text{Subjects}}}{n-1}$	
Error	$(k-1)(n-1)$	$SS_{\text{Total}} - SS_{\text{Condition}} - SS_{\text{Subjects}}$	$\frac{SS_{\text{Error}}}{(k-1)(n-1)}$	
Total	$nk-1$	$\sum_{i=1}^{k}\sum_{j=1}^{n}(Y_{ij} - \bar{Y}_{..})^2$		

Note 6.21 (One-way Repeated Measures ANOVA vs. Two-Way ANOVA). Notice that the one-way repeated measures ANOVA table is algebraically the same as a two-way ANOVA without interaction terms, where the two factors are subjects and conditions, with one observation per cell. For the quiz score data, if we ignore the dependency among the scores from the same student, then we can treat Student as a fixed effect and run a two-way ANOVA. The p-value for Technique is 1.37×10^{-11}, which is identical to the p-value of the one-way repeated measure ANOVA.

```
# two-way ANOVA (should not be used for this case)
anova(lm(Score ~ Student + Technique, data=data))
: Analysis of Variance Table
:
: Response: Score
:           Df Sum Sq Mean Sq F value    Pr(>F)
: Student   19 69.713   3.669  5.1831 8.266e-06 ***
: Technique  2 73.433  36.717 51.8674 1.374e-11 ***
: Residuals 38 26.900   0.708
: ---
: Signif. codes:  0 '***' 0.001 '**' 0.01 '*' 0.05 '.' 0.1 ' ' 1
```

Although the p-values for Technique are identical between the one-way repeated measures ANOVA and the two-way ANOVA, the two models are not equivalent. First, in this case study, the quiz scores are not independent, which violates a required assumption for using two-way ANOVA. Second, the interpretations are different between these two models: The repeated measures ANOVA generalizes to the pool of volunteers, while the two-way ANOVA only generalizes to the specific students in the dataset. In practice, the one-way repeated measures ANOVA is more commonly used for this type of data, and the two-way ANOVA is not recommended.

6.1.10 Random Intercept Model

In Section 6.1.9, we briefly mentioned that a random intercept model can be used to solve a case that repeated measures ANOVA can handle. We now discuss the random intercept model in more detail using a case study about chip thickness measurements from different machines.

Case 10: Chip Thickness by Machines

A semiconductor manufacturing facility wants to understand the inherent variation in chip thickness produced by different wafer coating machines. They have randomly selected 20 identical machines from the facility and want to quantify the natural machine-to-machine variation in their production line. Each machine is maintained under identical conditions, uses the same materials, and follows the same protocols.

For answers to the initial questions and some initial thoughts about this case study, please refer to Section 5.11. To assess machine-to-machine variability, we first use box plots of each machine to gain some first impression; see Figure 6.19. It seems that there are quite a few differences among different machines.

```
# Exploratory data analysis
library(ggplot2)
library(lme4)

data <- read.csv("data/chip-thickness-machines.csv")
data$machine_id <- as.factor(data$machine_id)

# Box plot of thickness by machines
ggplot(data, aes(x = machine_id, y = thickness)) + geom_boxplot() +
  theme_bw() + labs(title = "", x = "Machine ID", y = "Thickness (nm)")
```

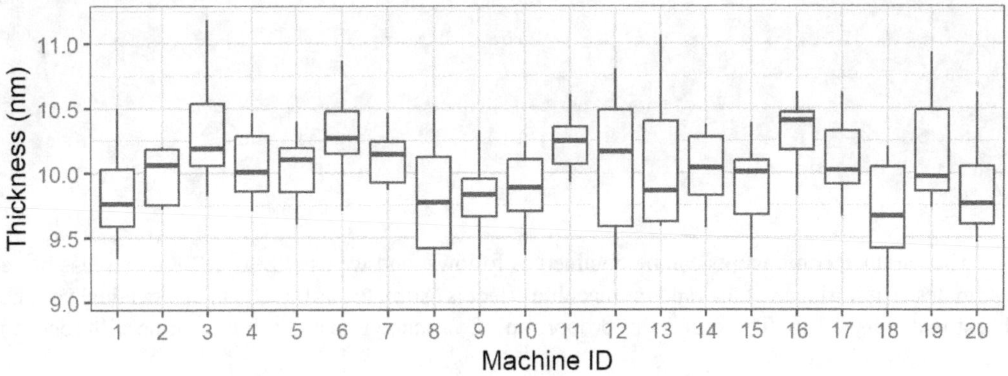

FIGURE 6.19: Distribution of Chip Thickness by Machines.

To account for repeated measurements within each machine, we consider a random intercept model; refer to Note 6.22 for details on random intercept models. Here we use `REML=FALSE` to fit the model with maximum likelihood estimation, because we want to perform a likelihood ratio test later on to test the significance of the random intercept; see Note 6.23 for details on restricted maximum likelihood (REML) estimation.

```
ri_model <- lmer(thickness ~ 1 + (1|machine_id), data = data, REML=FALSE)
```

To test whether the random intercept has significantly improved the modeling, we first explore whether a likelihood ratio test can be performed. The models fitted by `lm()` and `lmer()` are not compatible, so we manually calculate the likelihood ratio test as follows:

```
null_model <- lm(thickness ~ 1, data = data)
null_logLik <- logLik(null_model)
ri_logLik <- logLik(ri_model)

# Likelihood ratio test
LR_stat <- 2 * (as.numeric(ri_logLik) - as.numeric(null_logLik))
df <- attr(ri_logLik, "df") - attr(null_logLik, "df")
p_value <- pchisq(LR_stat, df, lower.tail = FALSE)
cat("LRT stat =", LR_stat, ", df =", df, ", p =", p_value, "\n")
: LRT stat = 10.00783 , df = 1 , p = 0.001558763
```

The above likelihood ratio test uses an approximated χ^2 distribution under the null. For random effects models, the asymptotic test using approximations is not valid, because under the null the variance of the random effect is 0 which is on the boundary of the parameter space (variance is ≥ 0). A parametric bootstrap method is preferred to get a more accurate p-value (see [40] for details). The following R codes show how to perform a parametric bootstrap manually, and it is expected that there would be messages like " `boundary (singular) fit`" as we fit the random-effects model under the null. Although there are "singular" warnings, the likelihood calculated can still be used. The p-value (0.001) from the parametric bootstrap is similar to that from the asymptotic test, indicating that the random effect is significant.

```
LR_stat_vec <- numeric(1000)
set.seed(123)
for(i in 1:1000){y <- unlist(simulate(null_model))
  bnull <- lm(y ~ 1)
  balt <- lmer(y ~ 1 + (1|machine_id), data, REML=FALSE)
  LR_stat_vec[i] <- as.numeric(2*(logLik(balt)-logLik(bnull)))}
mean(LR_stat_vec > LR_stat)
: [1] 0.001
```

The variance components can be obtained as follows, and we use `REML=TRUE` to get less biased estimates. The variance of the random machine effect is estimated as 0.021, and the residual variance is estimated as 0.126. Therefore, the proportion of variance explained by machine differences is about 14.3%.

```
# Extract variance components
ri_model_REML <- lmer(thickness ~ 1 + (1|machine_id), data = data)
vc <- VarCorr(ri_model_REML)
(machine_var <- attr(vc[[1]], "stddev")^2)
: (Intercept)
:  0.02108608
(residual_var <- attr(vc, "sc")^2)
: [1] 0.1259502
# Proportion of variance explained by machine differences:
machine_var / (machine_var + residual_var)
: (Intercept)
:   0.1434073
```

The 95% bootstrap confidence interval for σ_α the standard deviation of the random machine effect can be derived using parametric bootstrap. The confidence interval $(0.054, 0.219)$ does not include 0, showing some evidence that the random effect exists.

```
set.seed(123)
confint(ri_model_REML, method="boot")
:                   2.5 %     97.5 %
: .sig01       0.05434369  0.2189952
: .sigma       0.31787286  0.3919902
: (Intercept) 9.95544440 10.0968593
```

Now we perform some model diagnostic. First, check the normality of the random intercept. Figure 6.20 is the QQ plot and the normality looks reasonable. A Shapiro-Wilk test also suggests that there is no violation of the normality assumption for the random intercept.

```
# Normality of Random Effects
random_effects <- ranef(ri_model)$machine_id[,1]

ggplot(data.frame(re = random_effects), aes(sample = re)) +
  stat_qq() + stat_qq_line() + theme_bw() +
  labs(x = "Theoretical Quantiles",
       y = "Empirical Quantiles", title = "")
```

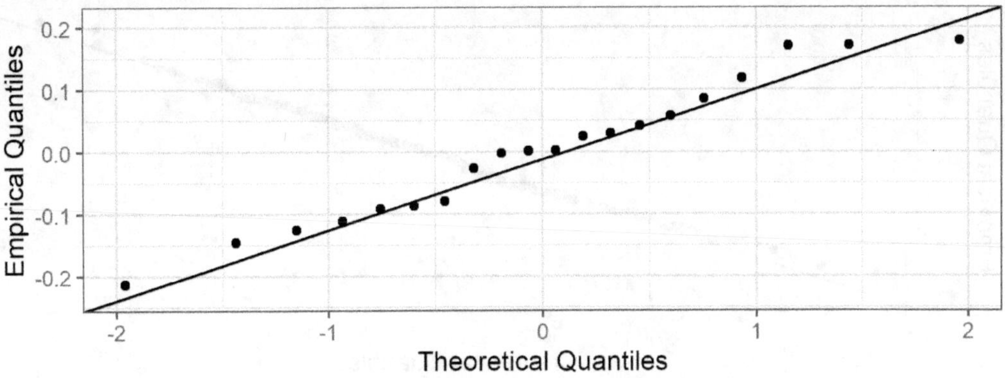

FIGURE 6.20: QQ Plot of Random Effects.

```
# Shapiro-Wilk test for random effects
shapiro.test(random_effects)
:
:  Shapiro-Wilk normality test
:
: data:  random_effects
: W = 0.96676, p-value = 0.6856
```

Similarly, we can test the normality of residuals, and there are no violations from the following plot and test.

```
# Normality of Residuals
residuals <- resid(ri_model)

ggplot(data.frame(res = residuals), aes(sample = res)) +
  stat_qq() +   stat_qq_line() + theme_bw() +
  labs(x = "Theoretical Quantiles",
       y = "Empirical Quantiles", title = "")
```

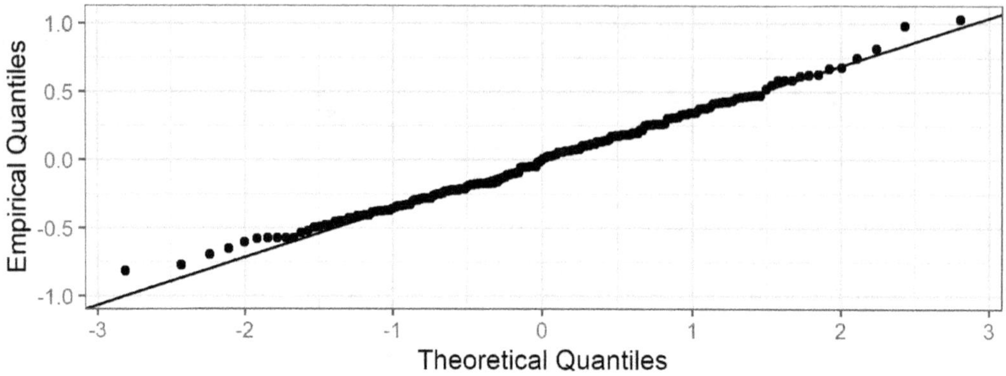

FIGURE 6.21: QQ Plot of Residuals.

```
# Formal test
shapiro.test(residuals)
:
:   Shapiro-Wilk normality test
:
: data:  residuals
: W = 0.99262, p-value = 0.4127
```

The homoscedasticity of residuals can be verified by Figure 6.22 and the following Levene's test.

```
# Plot residuals vs. fitted values
fitted_vals <- fitted(ri_model)

ggplot(data.frame(fitted = fitted_vals, resid = residuals),
       aes(x = fitted, y = resid)) + geom_point() +
  geom_smooth(method = "loess", color = "black",
              formula = y ~ x, linewidth = .5) + theme_bw() +
              labs(title = "", x = "Fitted Values", y = "Residuals")
```

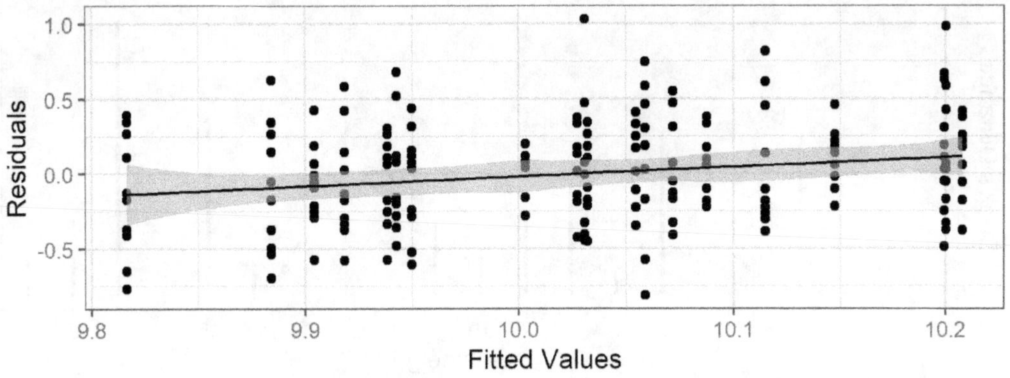

FIGURE 6.22: Residuals vs. Fitted Values.

```
# Levene's test for each machine
car::leveneTest(residuals ~ machine_id, data = data)
: Levene's Test for Homogeneity of Variance (center = median)
:        Df F value Pr(>F)
: group  19  1.2287 0.2388
:       180
```

Highly influential machines can be identified by the following plots showing Cook's distances. Figure 6.23 suggests that Machine 18 could be a highly influential machine.

```
# Calculate Cook's distance
infl <- influence.ME::influence(ri_model, group = "machine_id")
cd <- cooks.distance(infl)
group_labels <- as.numeric(levels(data$machine_id))

# Put Cook's distance into a data frame
cd_df <- data.frame(machine_id = group_labels,
                    cooks_distance = as.numeric(cd))

cd_df$machine_id <- factor(cd_df$machine_id, levels = 1:20)

# Plot with ggplot
ggplot(cd_df, aes(x = machine_id, y = cooks_distance)) +
  geom_linerange(aes(ymin = 0, ymax = cooks_distance), linewidth = 0.8) +
  labs(title = "", x = "machine_id", y = "Cook's Distance") +  theme_bw()
```

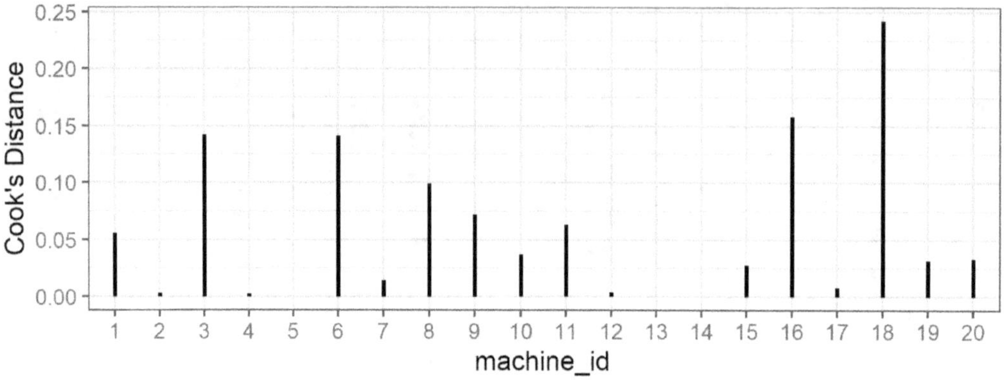

FIGURE 6.23: Cook's Distance per Machine.

Conclusions

In this case study, we use a random intercept model to assess the machine-to-machine variability in chip thickness measurements. The likelihood ratio test (p-value of parametric bootstrap: 0.001) and the bootstrap confidence interval (0.054, 0.219) for the standard deviation of the random machine effect both suggest that there is significant machine-to-machine variability. The variance component analysis shows that the machine-to-machine variability accounts for 14.3% of variability among the chip thickness. This is an observational study and no causality can be claimed. As the 20 machines are randomly selected from the facility, we expect that the result can be generalized to all the machines in the facility.

6.1.11 Linear Mixed-Effects Model

Linear mixed-effects models (LMMs) is an extension of linear models that incorporate both fixed effects and random effects. In this section, we will illustrate the application of LMMs using three case studies, from which we will learn how to fit LMMs, assess the significance of fixed and random effects, perform model diagnostics, and interpret the results. Through the comparison of these case studies, we will also highlight the differences in model specification and interpretation when dealing with different data structures. LMMs is a very powerful tool for analyzing data with complex structures, and we will cover some of the most common scenarios in practice. When appropriate, we will point out some references for further reading on more advanced topics in LMMs.

6.1.11.1 Random Effects for Repeated Measures

We first use a case study to discuss many standard aspects of using a linear mixed-effects model such as model specification, testing fixed and random effects, model diagnostics, and interpretation of results. The case study is about the math performance of students in a metropolitan area, and each student is measured at two time points so that a linear mixed-effects model can be used to analyze the data, where the random effect if for the repeated measures within each student.

Case 11: Examining Student Math Performance

A researcher is studying math test scores of students. The study involves 200 randomly selected students from a large metropolitan area, with each student sampled independently. Each student is tested twice (Time 0 and Time 1) using standardized tests. The following variables are collected: Math score, Time, Student ID, and student socioeconomic status (SES). The researcher wants to study whether math scores improve over time and how SES affects math performance.

For answers to the initial questions and some initial thoughts about this case study, please refer to Section 5.12. We first use a spaghetti plot to visualize the math scores across different individuals; see Figure 6.24. It seems that there is an overall improvement over time, but there is a lot of variability between students, and how ses affects the math scores is not clear from the plot. We will use a linear mixed-effects model to analyze the data and try to answer the questions of interest.

```
# Examining Student Math Performance in a Metropolitan Area
library(lme4)
library(ggplot2)
library(dplyr)
data <- read.csv("data/math-scores.csv")

data$student_id <- as.factor(data$student_id)
data$time <- as.factor(data$time)

# Spaghetti plot by schools
ggplot(data, aes(x = time, y = math_score, group = student_id,
       color = ses)) +
  geom_line(alpha = 0.4) +  # Individual student
  geom_smooth(aes(group = 1), method = "lm", formula = y ~ x,
              se = FALSE, color = "black", linewidth = 0.5,
              linetype="dashed") + # overall trend
  theme_bw() + scale_color_gradient(low = "grey99", high = "grey10") +
  labs(title = "", x = "Time", y = "Math Score")
```

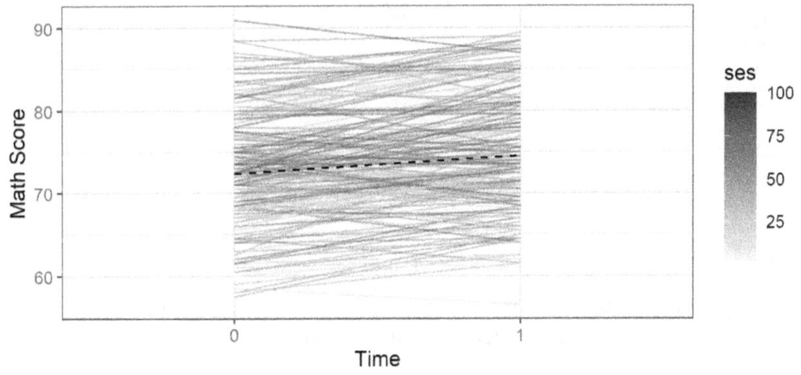

FIGURE 6.24: Spaghetti Plot of Math Scores by Student. Gray lines: Individual students with darker color for higher SES; Dashed line: Overall trend.

A linear mixed-effects model can be fitted as follows. As each student has two measurements, we can include a random intercept for each student to account for the correlation between repeated measures within the same student. The fixed effects include Time and SES. In `lmer()`, the default method is restricted maximum likelihood estimations (REML). To test fixed effects, we need to use the ML method, that is, `REML = FALSE`; refer to Note 6.23 for more details about REML.

```
# Fit linear mixed-effects model
model <- lmer(math_score ~ time*ses + (1|student_id),
              data = data, REML=TRUE)
model_ML <- lmer(math_score ~ time*ses + (1|student_id),
              data = data, REML=FALSE)
```

To assess whether fixed effects are significant, we can conduct a series of likelihood ratio tests. The following is to test whether the interaction between time and ses is significant, and the large p-value (0.754) indicates that the interaction term is not significant. Note that when testing fixed effects in LMM using likelihood ratio tests, it is recommended to fit the models using maximum likelihood estimation (i.e., `REML = FALSE` in `lmer()`). Also, keep in mind that the Chi-square approximation may not be accurate when the sample size is small. Here the sample size is fairly large, so the Chi-square approximation should be reasonable.

```
model_ML_timeses <- lmer(math_score ~ time + ses + (1|student_id),
                         data = data, REML=FALSE)
anova(model_ML, model_ML_timeses)
: Data: data
: Models:
: model_ML_timeses: math_score ~ time + ses + (1 | student_id)
: model_ML: math_score ~ time * ses + (1 | student_id)
:                  npar    AIC    BIC  logLik -2*log(L)  Chisq Df Pr(>Chisq)
: model_ML_timeses    5 2482.9 2502.9 -1236.5    2472.9
: model_ML            6 2484.8 2508.8 -1236.4    2472.8 0.0985  1     0.7537
```

The following is to test whether the main effect of time is significant, and the small p-value indicates that time is significantly associated with math scores.

```
model_ML_ses <- lmer(math_score ~ ses + (1|student_id),
                     data = data, REML=FALSE)
anova(model_ML_timeses, model_ML_ses)
: Data: data
: Models:
: model_ML_ses: math_score ~ ses + (1 | student_id)
: model_ML_timeses: math_score ~ time + ses + (1 | student_id)
:                  npar    AIC    BIC  logLik -2*log(L)  Chisq Df Pr(>Chisq)
: model_ML_ses        4 2528.6 2544.6 -1260.3    2520.6
: model_ML_timeses    5 2482.9 2502.9 -1236.5    2472.9 47.699  1   4.97e-12 ***
: ---
: Signif. codes:  0 '***' 0.001 '**' 0.01 '*' 0.05 '.' 0.1 ' ' 1
```

The following is to test whether the main effect of ses is significant, and the small p-value indicates that ses is significantly associated with math scores.

```
model_ML_time <- lmer(math_score ~ time + (1|student_id),
                      data = data, REML=FALSE)
anova(model_ML_timeses, model_ML_time)
: Data: data
```

```
: Models:
: model_ML_time: math_score ~ time + (1 | student_id)
: model_ML_timeses: math_score ~ time + ses + (1 | student_id)
:                  npar    AIC    BIC  logLik -2*log(L)  Chisq Df Pr(>Chisq)
: model_ML_time       4 2489.9 2505.9 -1241.0    2481.9
: model_ML_timeses    5 2482.9 2502.9 -1236.5    2472.9 8.9983  1   0.002702 **
: ---
: Signif. codes:  0 '***' 0.001 '**' 0.01 '*' 0.05 '.' 0.1 ' ' 1
```

To test whether the random effect is significant, we can also use the likelihood ratio test; see Note 6.25 for more details. However, for this particular case study, there is only one random effect (`student_id`), and under the null hypothesis, the model reduces to a simple linear model without any random effects. The R function `pbkrtest::PBmodcomp()` requires both the alternative model and the null model to be fitted using `lmer()`. Therefore, we can not use `pbkrtest::PBmodcomp()` for this case, and we will manually perform the likelihood ratio test. In the next case study (Case 12), we will use `pbkrtest::PBmodcomp()` to perform the parametric bootstrap LRT for random effects in LMM.

The following likelihood ratio test has a p-value (0), suggesting that the random effect is significant.

```
# Likelihood ratio test
null_model <- lm(math_score ~ time + ses, data = data)
null_logLik <- logLik(null_model)
model_logLik <- logLik(model_ML_timeses)

LR_stat <- 2 * (as.numeric(model_logLik) - as.numeric(null_logLik))
LR_stat_vec <- numeric(1000)
set.seed(123)
for(i in 1:1000){
  y <- unlist(simulate(null_model))
  bnull <- lm(y ~ time + ses, data = data)
  balt <- lmer(y ~  time + ses + (1 |student_id),
               data = data, REML=FALSE)
  LR_stat_vec[i] <- as.numeric(2*(logLik(balt)-logLik(bnull)))
}
mean(LR_stat_vec > LR_stat)
: [1] 0
```

Alternatively, we can use the `RLRsim` package to perform the exact LRT for random effects in LMM; see Note 6.26 for more details. The R package provides two functions: `RLRsim::exactLRT()` for the exact LRT and `RLRsim::exactRLRT()` for the exact restricted LRT. The former can handle testing between a model fitted with `lmer()` and a null model fitted with either `lm()` or `lmer()`, while the latter is specifically designed for testing random effects between two `lmer()` models. The former is for the ML method (`REML=FALSE`), and the latter is for the REML method (`REML=TRUE`).

The following uses `RLRsim::exactLRT()` to perform the exact LRT for the random effect because we have a null model fitted with `lm()`. The small p-value is consistent with the manual calculation above.

```
RLRsim::exactLRT(model_ML_timeses, null_model)
:
:  simulated finite sample distribution of LRT. (p-value based on 10000 simulated values)
```

```
:
: data:
: LRT = 235.76, p-value < 2.2e-16
```

The `RLRsim::exactRLRT()` function can be used to perform the exact restricted LRT for the random effect for two `lmer()` models of which one is nested within the other. When there is only one random effect in LMM, the reduced model is not needed in the function call; this applies to our example here. The small p-value is again similar to the previous calculations.

```
model_timeses <- lmer(math_score ~ time + ses + (1|student_id),
                      data = data, REML=TRUE)
RLRsim::exactRLRT(model_timeses)
:
:  simulated finite sample distribution of RLRT.
:
:  (p-value based on 10000 simulated values)
:
: data:
: RLRT = 235.18, p-value < 2.2e-16
```

The following extract the standard deviations of the random effect and the residuals. The variability of the random effect is much larger than that of the residuals, suggesting that there is a lot of variability between students.

```
# Extract variance of random effects
VarCorr(model_timeses)
:  Groups     Name        Std.Dev.
:  student_id (Intercept) 6.5552
:  Residual               2.9367
```

Model diagnostics can be performed as follows. Normality of the random effect can be checked by the following plot and the normality test. From both the QQ plot and the Shapiro-Wilk test, there is evidence to reject the normality assumption for the random effect.

```
# Normality of Random Effects
random_effects <- ranef(model_timeses)$student_id[,1]

ggplot(data.frame(re = random_effects), aes(sample = re)) + stat_qq() +
  stat_qq_line() + theme_bw() + labs(x = "Theoretical Quantiles",
            y = "Empirical Quantiles", title = "")
```

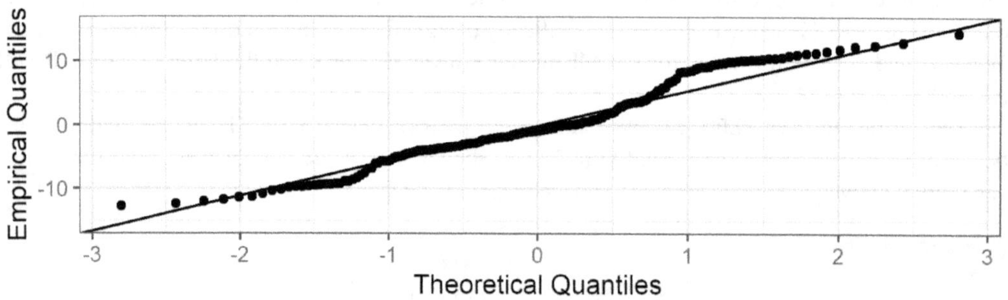

```
# Shapiro-Wilk test for normality assumption of random effects
shapiro.test(random_effects)
:
:  Shapiro-Wilk normality test
:
: data:  random_effects
: W = 0.96627, p-value = 0.0001004
```

The following histogram (Figure 6.25) of the random effects suggests that the distribution of the random effects does not follow a normal distribution, and it might be better modeled by a mixture of normal distributions. This violation suggests that the students may come from different schools and there are variability between schools, which is not captured by the current data and model. A random effect that follows a mixture of normal distributions may be more appropriate for this case, but it is beyond the scope of this book; we refer interested readers to [112] for more details about using mixture of normals for random effects in linear mixed-effects models.

A lesson learned from this case study is that, it is important to collect data at more granular levels with more variables that are relevent to the subjects. Then a more comprehensive model can be fitted to better capture the underlying data structure. Actually, this case study is an abridged version of the next case study (Case 12) where we have data on schools and thus variability between schools can be modeled by including a random effect for schools. In Case 12, the normality assumption for the random effects is not violated, and the model fit is much better than this case study.

```
ggplot(data.frame(re = random_effects), aes(x = re)) +
  geom_histogram(binwidth = 1) + theme_bw() +
  labs(title = "Histogram of Random Effects", x = "", y = "")
```

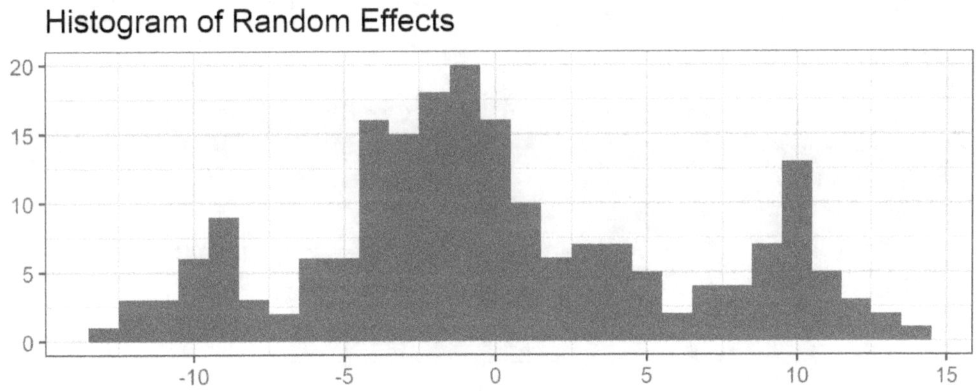

FIGURE 6.25: Histogram of Random Effects.

Normality of the residuals can be verified by the plot and the test as follows, and both the QQ plot and the Shapiro-Wilk test suggest that the normality assumption for the residuals is not violated.

```
# Normality of Residuals
residuals <- resid(model_timeses)
ggplot(data.frame(res = residuals), aes(sample = res)) + stat_qq() +
```

```
  stat_qq_line() + theme_bw() +
  labs(x = "Theoretical Quantiles",
       y = "Empirical Quantiles", title = "")
```

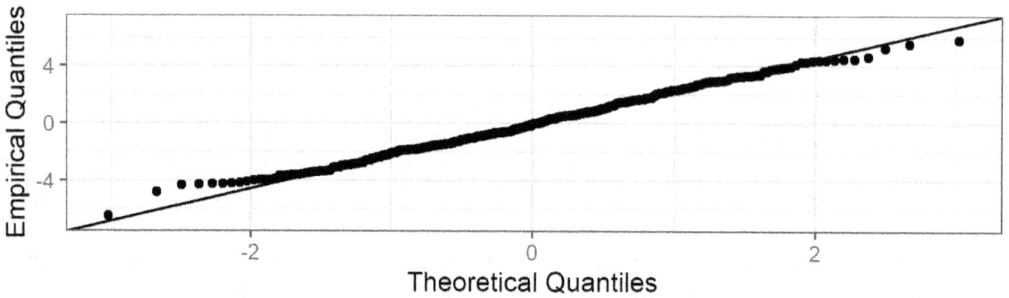


```
# Formal test
shapiro.test(residuals)
:
:   Shapiro-Wilk normality test
:
: data:  residuals
: W = 0.99479, p-value = 0.1953
```

To interpret the result, we can use `ggeffects::ggpredict()` to obtain various predicted values and plots. The following plot shows the overall improvement in math scores from Time 0 to Time 1.

```
# prediction
library(ggeffects)
pred <- ggpredict(model_timeses, terms = c("time"))
names(pred)[names(pred) == "x"] <- "Time"
ggplot(pred, aes(x=Time, y = predicted, color = Time)) +
  geom_point(size = 1) +
  geom_errorbar(aes(ymin = conf.low, ymax = conf.high), width = 0.1) +
  scale_color_manual(values = c("black", "grey70")) + theme_bw()
```

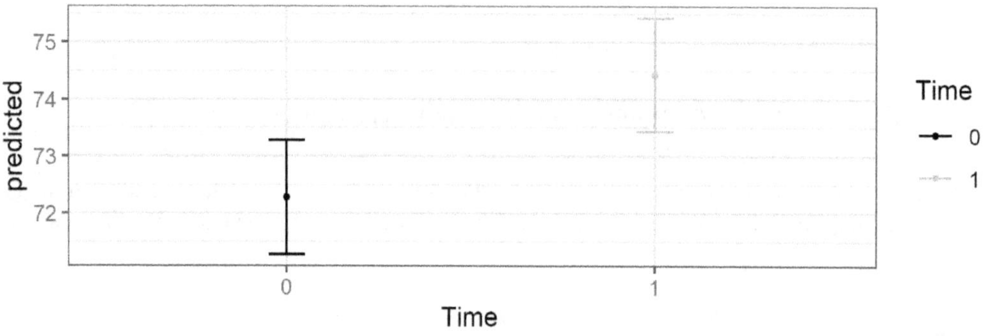

FIGURE 6.26: Improvement at Different Schools.

The following plot shows the improvement in math scores from Time 0 to Time 1 at different SES levels.

```
# prediction
pred <- ggpredict(model_timeses, terms = c("ses", "time"))
names(pred)[names(pred) == "group"] <- "Time"
names(pred)[names(pred) == "x"] <- "ses"

ggplot(pred, aes(x = ses, y = predicted, color = Time,
                 linetype = Time, group = Time)) + geom_line() +
  geom_ribbon(aes(ymin = conf.low, ymax = conf.high, fill = Time),
              alpha = 0.2) +
  scale_color_manual(values = c("black", "grey70")) +
  scale_fill_manual(values = c("black", "grey70")) + theme_bw()
```

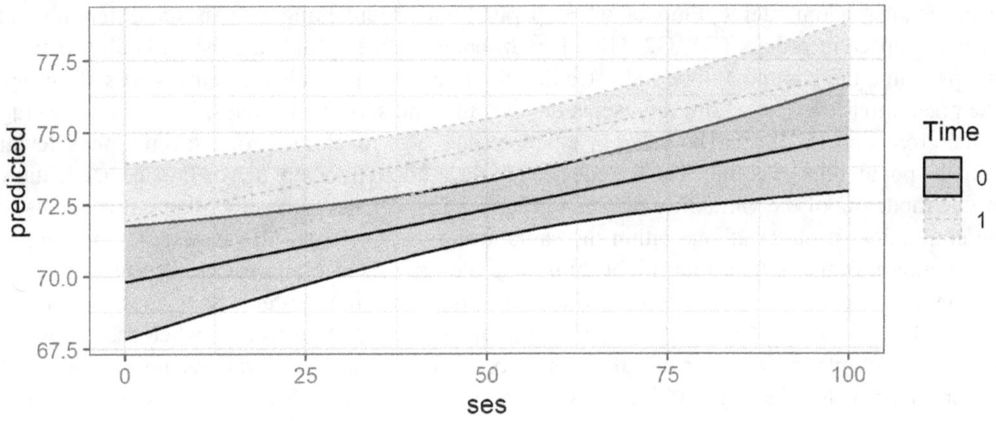

FIGURE 6.27: Improvement at Different SES Scores.

Conclusions

In this case study, we fit a linear mixed-effects model to analyze the math scores of students in a metropolitan area over two time points, accounting for both fixed effects (time, SES) and random effects (student). The analysis reveals that math scores have improved over time (p-value: 4.97×10^{-12}), and that a higher SES is associated with a higher math score (p-value: 0.003). As subjects are randomly selected, the result can be generalized to all students in the metropolitan area. However, model diagnostics suggest that the normality assumption for the random effects is violated, which may be due to the fact that students come from different schools and there is between school variability that is not captured by the current model. Moreover, this is an observational study and causality cannot be established. For example, we cannot conclude that higher SES causes higher math scores, or that the passage of time causes an increase in math scores automatically, as confounding factors may be responsible for these associations.

Note 6.22 (Linear Mixed-Effects Model)**.** *Linear mixed-effects models* are an extension of traditional linear regression by incorporating both fixed and random effects, making them useful for analyzing hierarchical data, clustered observations, repeated measures, and longitudinal data, where observations are not independent.

The standard simple linear regression model is described by the following equation: for subject j,

$$Y_j = \beta_0 + \beta_1 x_j + \varepsilon_j,$$

where $\varepsilon_j \overset{iid}{\sim} \mathcal{N}(0, \sigma^2)$. If we assume that subjects are nested within groups (e.g., students nested within schools), we can extend the simple linear regression model to a linear mixed-effects model by including random effects to account for the variability between groups. Letting Y_{ij} be the response variable for subject j in group i, and x_{ij} be the corresponding predictor variable, we will introduce three common types of linear mixed-effects models: random intercept models, random slope models, and random intercept and slope models.

Random intercept models include a random intercept (b_{0i}) for each group i, and assume a common slope across all groups:

$$Y_{ij} = (\beta_0 + b_{0i}) + \beta_1 x_{ij} + \varepsilon_{ij},$$

where $b_{0i} \overset{iid}{\sim} \mathcal{N}(0, \sigma_0^2)$ is the group-specific intercept and $\varepsilon_{ij} \overset{iid}{\sim} \mathcal{N}(0, \sigma^2)$ is the observation-level error, with $\varepsilon_{ij} \perp b_{0i}$. The total variance of Y_{ij} is given by $\text{Var}(Y_{ij}) = \sigma_0^2 + \sigma^2$. In Figure 6.28, we visualize the random intercept model, where parallel lines intercepting the y-axis at different points represent different groups (G1, G2, G3). The different intercepts indicate the variability between groups, while the common slope indicates that the fixed effect (β_1) is the same across all groups. The population-level line (Pop) represents the overall trend across all groups, which is the average of the group-specific lines. The value of σ_0^2 represents how much the group-specific lines deviate from the population-level line. The introduction of the random intercept b_{0i}, on the one hand, allows for the modeling of the variability between groups, i.e., σ_0^2, and on the other hand, it accounts for the correlation of observations within the same group, as they share the same random intercept. A common mistake is that, one might think that adding a fixed effect for each group will make the random effect unnecessary. However, adding group-level fixed effects does not automatically account for the correlation of observations within the same group, especially when each subject is measured multiple times under different conditions / time points, observations are more likely to be correlated within the same subject. In such cases, adding fixed effects for each condition / time points will not capture the correlation of repeated measures within the same subject, and thus the random effect is still needed. Moreover, if groups are considered as random samples from a larger population, then the random effects are still needed to account for the variability between groups and to make inferences about the population.

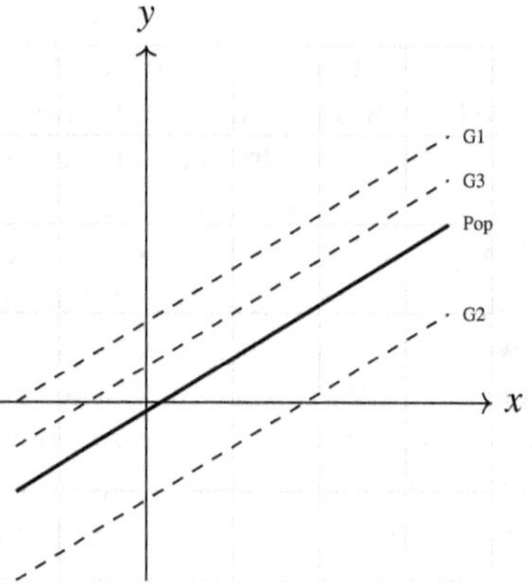

FIGURE 6.28: Visualization for Random Intercept Models.

Random slope (only) models are less common, and have a fixed intercept with random slopes:

$$Y_{ij} = \beta_0 + (\beta_1 + b_{1i}) x_{ij} + \varepsilon_{ij},$$

where $b_{1i} \overset{iid}{\sim} \mathcal{N}(0, \sigma_1^2)$, $\varepsilon_{ij} \overset{iid}{\sim} \mathcal{N}(0, \sigma^2)$, and $\varepsilon_{ij} \perp b_{1i}$. All groups share the same intercept β_0. This model is less common because it assumes that all groups have the same baseline level (intercept), which may not be realistic in many applications. The total variance is given by $\text{Var}(Y_{ij}) = \sigma_1^2 x_{ij}^2 + \sigma^2$. In Figure 6.29, we illustrate the random slope (only) model, where the lines have different slopes but intersect at the same point on the y-axis.

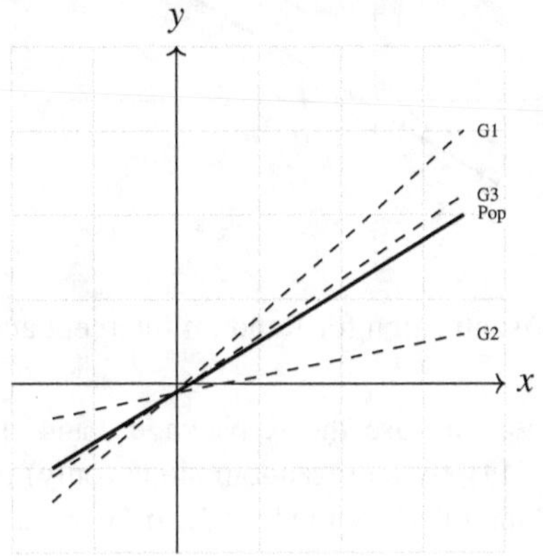

FIGURE 6.29: Visualization for Random Slopes Models.

Random intercept and slope models include a random intercept and a random slope:

$$Y_{ij} = (\beta_0 + b_{0i}) + (\beta_1 + b_{1i})x_{ij} + \varepsilon_{ij},$$

with

$$\begin{pmatrix} b_{0i} \\ b_{1i} \end{pmatrix} \overset{iid}{\sim} \mathcal{N}\left(\begin{pmatrix} 0 \\ 0 \end{pmatrix}, \begin{pmatrix} \sigma_0^2 & \sigma_{01} \\ \sigma_{01} & \sigma_1^2 \end{pmatrix} \right),$$

$$\varepsilon_{ij} \overset{iid}{\sim} \mathcal{N}(0, \sigma^2), \quad \begin{pmatrix} b_{0i} \\ b_{1i} \end{pmatrix} \perp \varepsilon_{ij}.$$

The total variance decomposition is $\text{Var}(Y_{ij}) = \sigma_0^2 + 2\sigma_{01}x_{ij} + \sigma_1^2 x_{ij}^2 + \sigma^2$. In Figure 6.30, we visualize the random intercept and slope model, where the lines have different intercepts and slopes. The different intercepts indicate the variability of baseline levels between groups, while the different slopes indicate that the effect from the predictor x varies across groups. The population-level line (Pop) represents the overall trend across all groups, which is the average of the group-specific lines. The value of σ_0^2 represents how much the group-specific lines deviate from the population-level line in terms of intercepts, while σ_1^2 represents how much the group-specific lines deviate from the population-level line in terms of slopes. The covariance σ_{01} captures the relationship between the random intercepts and random slopes; for example, a positive covariance indicates that groups with higher intercepts tend to have steeper slopes.

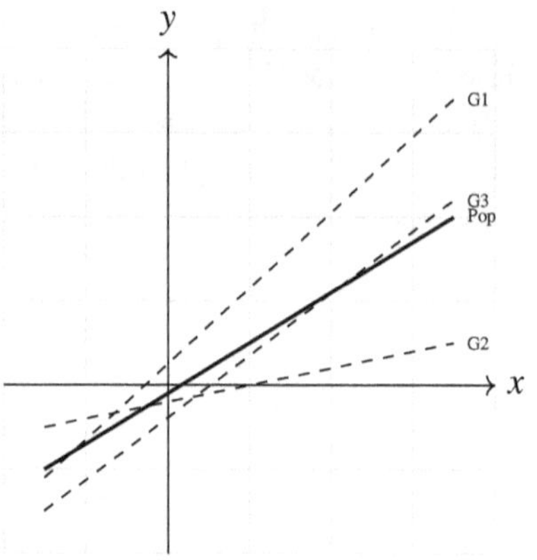

FIGURE 6.30: Visualization for Random Intercept and Slope Models.

For any of these models, we can use the R package `lme4` ([8]) as follows: random intercept models: `lmer(y ~ x + (1|group))`, random slope (only) models: `lmer(y ~ x + (0 + x|group))`, and random slope and intercept models: `lmer(y ~ x + (1 + x|group))`.

Note 6.23 (Restricted Maximum Likelihood (REML)). *Restricted maximum likelihood (REML)* method works by maximizing the likelihood of a transformed version of the data that has been adjusted to remove the fixed effects. In contrast, the maximum likelihood (ML) method maximizes the likelihood of the original data. The `lmer()` function in the `lme4` package estimates linear mixed-effects models using either ML or REML methods. By default, `lmer()` fits models using `REML = TRUE`, which generally produces less biased estimates by cancelling out fixed effects before maximizing the likelihood with only random effects parameters involved. REML is preferred when comparing models with the same fixed-effects structure, as it provides unbiased estimates of variance components. For example, when testing random effects, both the full and reduced models have the same fixed effects, and then REML is more appropriate. However, when comparing models differing in their fixed effects (e.g., via likelihood ratio tests or AIC), one should use `REML = FALSE` (i.e., use ML estimation), since likelihoods obtained from REML are not comparable across models with different fixed-effects structures; refer to [49, 40] and the references therein for more details on REML and ML methods in linear mixed-effects models.

Note 6.24 (Testing Fixed Effects in LMM). When testing fixed effects in linear mixed-effects models (LMM), the following methods can be considered: Likelihood ratio tests (LRT), F tests with Kenward-Roger approximation, and parametric bootstrap methods. LRT is commonly used when the sample size is large, while Kenward-Roger and parametric bootstrap methods are more suitable for smaller sample sizes. When using LRT for testing fixed effects in LMM, the R function `anova()` can be used, and it is recommended to fit the models using maximum likelihood estimation (i.e., `REML = FALSE` in `lmer()`) to ensure valid comparisons between models with different fixed effects. Kenward-Roger method is needed to adjust the degrees of freedom for the F tests, as the standard F tests may not be accurate in LMM due to the presence of random effects. The `pbkrtest` package ([48]) in R provides functions for both Kenward-Roger and parametric bootstrap methods. The R function `pbkrtest::KRmodcomp()` can be used to perform Kenward-Roger tests for comparing nested models fitted with `lmer()`, and REML is always used by the R function `pbkrtest::KRmodcomp()` even if `REML = FALSE` is specified. The parametric bootstrap method involves simulating data under the null hypothesis and then comparing the observed test statistic to the distribution of test statistics obtained from the simulated data. This method can provide more accurate p-values and is the golden standard for testing fixed effects in LMM, especially when the sample size is small or the model is complex. The R function `pbkrtest::PBmodcomp()` can be used to perform parametric bootstrap tests for comparing nested models fitted with `lmer()`, and ML is always used by the R function `PBmodcomp()` even if `REML = TRUE` is specified. See [70] and [48] for more details, and Table 6.3 for a summary of the methods for testing fixed effects in LMM.

TABLE 6.3: Comparison of Methods for Testing Fixed Effects in Linear Mixed-effects Models.

Method	Main Idea	Recommended Sample Size	R Function(s)	Estimation Method
LRT	Compares nested models using differences in log-likelihoods	Large	`lmer()` `anova()`	ML
KR's F Test	Adjusts F tests by correcting degrees of freedom for random effects	Small to moderate	`KRmodcomp()`	REML
Parametric Bootstrap	Simulates data under the null hypothesis and compares observed test statistic to simulated distribution	Small samples or complex models	`PBmodcomp()`	ML

Note 6.25 (Testing Random Effects in LMM)**.** For testing random effects in LMM, the likelihood ratio test (LRT) can be used. However, it will have the following issue: under the null hypothesis that the variance of the random effect is zero, which is on the boundary of the parameter space, and thus the standard Chi-square approximation does not hold; refer to [98] and [104] for more details. A preferred approach is to use the parametric bootstrap method to obtain a more accurate p-value for the LRT; refer to [48] and Note 6.22 for more details. The R function `pbkrtest::PBmodcomp()` can be used to perform the parametric bootstrap LRT for random effects in LMM, and ML is always used by the R function `PBmodcomp()` even if `REML = TRUE` is specified.

Note 6.26 (Exact LRT and Exact RLRT for LMM)**.** For testing either fixed effects or random effects in linear mixed-effects models (LMMs), likelihood ratio test (LRT) is commonly used. However, when the sample size is small or moderate, the standard Chi-square approximation for the null distribution of the LRT statistic may not be accurate. Also, when testing random effects the null distribution has variance 0 that is on the boundary of the parameter space and consequently, the usual Chi-square distribution is invalid. We have discussed such issues in above Note 6.24 and 6.25, and suggested the use of parametric bootstrap methods to obtain more accurate p-values for the LRT. Alternatively, exact LRT and exact restricted LRT (RLRT) can be used to obtain accurate p-values using the exact finite-sample null distribution of the LRT statistic, without relying on large-sample approximations. The R package `RLRsim` ([97]) provides functions for performing exact LRT and exact restricted LRT for LMMs. The `RLRsim::exactLRT()` function can be used to perform the exact LRT for testing either fixed effects or random effects in LMMs, and the `RLRsim::exactRLRT()` function is specifically designed for testing random effects between models fitted using the REML method (`REML=TRUE`).

6.1.11.2 Random Effects for Clustered Data

Now we slightly modify the previous case study (Case 11) by including school variables and treat `school_id` as a random effect in this case study.

Case 12: Examining Student Math Performance with School Effects

A researcher is studying math test scores of students in a large metropolitan area that has 20 schools. The researcher randomly selects 200 students from six different schools randomly selected from the metropolitan area. Each student is tested twice (Time 0 and Time 1) using standardized tests and the following variables are collected: Math score, Time, Student ID, School ID, and student socioeconomic status (SES). The researcher wants to study whether math scores improve over time, whether the rate of improvement varies between schools, and how SES affects math performance.

For answers to the initial questions and some initial thoughts about this case study, please refer to Section 5.13. We first use a spaghetti plot to visualize the math scores across different schools; see Figure 6.31. It seems that there are differences in both the baseline scores and the improvement rates across different schools.

```
# Examining Student Math Performance Across Schools
library(lme4)
library(ggplot2)
library(dplyr)
library(ggeffects)
data <- read.csv("data/math-scores-among-schools.csv")

data$school_id <- as.factor(data$school_id)
data$student_id <- as.factor(data$student_id)
data$time <- as.factor(data$time)

# Spaghetti plot by schools
ggplot(data, aes(x = time, y = math_score, group = student_id)) +
  geom_line(alpha = 0.2, color = "gray") +  # Individual student
  geom_smooth(aes(group = school_id), method = "lm", formula = y ~ x,
              se = FALSE, color = "black", linewidth = .5,
              linetype="dashed") + # School trend
  facet_wrap(~school_id) + theme_bw() +
  labs(title = "", x = "Time", y = "Math Score")
```

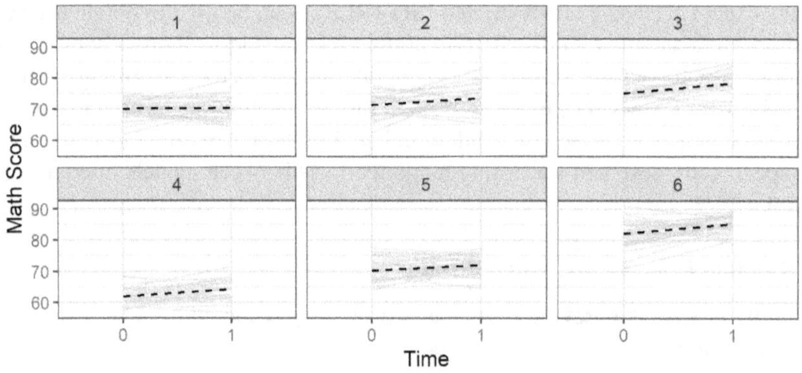

FIGURE 6.31: Spaghetti Plot of Math Scores by School. Gray lines: Individual students; Dashed line: School trend.

Although the Student effect is nested in the School effect, the data use unique student id from 1 to 200 for the students sampled, so there is no difference in using either `school_id:student_id` or `student_id` in the R functions below. If there were students having the same ID from different schools, then we must use `school_id:student_id` in the model to give each student a unique ID; see Section 10.8 of [40] for an example with nested effects.

We want to assess whether math scores have improved over time, so we add `time` as a fixed effect. In addition, we want to know whether improvement rates vary between schools, so we use a random slope of `time` in different schools; that is, allowing different schools have different improvement rates over time, although we do not model each one of them specifically, and instead, we model the variation of the improvement rates across schools. We also include `SES` as a fixed effect to control for its influence on math scores. Finally, we include `student_id` as a random intercept to account for the dependence between repeated measurements on the same student. Therefore, the model includes three random effects: a random slope of `time` and random intercepts of `student_id` and `school_id`, respectively; see Note 6.22 for details about different types of linear mixed-effects models.

A linear mixed-effects model is then constructed as follows. In `lmer()`, the default method is restricted maximum likelihood estimations (REML). To test fixed effects, we need to use the ML method, that is, `REML = FALSE`. To assess whether fixed effects are significant, we can conduct a series of likelihood ratio tests. The following tests whether there is an interaction between Time and SES, and the larger p-value (0.763) indicates that there is no significant interaction between Time and SES in the model, and thus we can remove the interaction term from the model.

```
# Fit linear mixed-effects model
model0_ML <- lmer(math_score ~ time*ses + (1 +time|school_id) +
                  (1|student_id), data = data, REML=FALSE)
model_ML <- lmer(math_score ~ time + ses + (1 +time|school_id) +
                 (1|student_id), data = data, REML=FALSE)
model <- lmer(math_score ~ time + ses + (1 + time|school_id) +
              (1|student_id), data = data)
anova(model0_ML, model_ML)
: Data: data
: Models:
: model_ML: math_score ~ time + ses + (1 + time | school_id) + (1 | student_id)
: model0_ML: math_score ~ time * ses + (1 + time | school_id) + (1 | student_id)
:             npar    AIC    BIC  logLik -2*log(L)  Chisq Df Pr(>Chisq)
: model_ML       8 2042.2 2074.1 -1013.1    2026.2
: model0_ML      9 2044.1 2080.0 -1013.0    2026.1 0.0912  1     0.7626
```

The following is to test whether the fixed effect SES is significant, and the small p-value indicates its significance. Notice that we let `REML=FALSE` to test fixed effects, and the sample size is fairly large, so the regular likelihood ratio test is used here.

```
model_ML_time <- lmer(math_score ~ time + (1 + time|school_id) +
                      (1|student_id), data = data, REML=FALSE)
anova(model_ML, model_ML_time)
: Data: data
: Models:
: model_ML_time: math_score ~ time + (1 + time | school_id) + (1 | student_id)
: model_ML: math_score ~ time + ses + (1 + time | school_id) + (1 | student_id)
:                 npar    AIC    BIC  logLik -2*log(L)  Chisq Df Pr(>Chisq)
: model_ML_time      7 2142.0 2169.9 -1064.0    2128.0
: model_ML           8 2042.2 2074.1 -1013.1    2026.2 101.81  1  < 2.2e-16 ***
: ---
: Signif. codes:  0 '***' 0.001 '**' 0.01 '*' 0.05 '.' 0.1 ' ' 1
```

To test whether there is an overall time effect, we can use the following LRT on the fixed effect of `time`. The LRT has a *p*-value (0.001) less than the common significance levels, indicating that, there is an overall significant time effect.

```
model_ML_ses <- lmer(math_score ~ ses + (1+time|school_id) +
                        (1|student_id), data = data, REML=FALSE)
anova(model_ML, model_ML_ses)
: Data: data
: Models:
: model_ML_ses: math_score ~ ses + (1 + time | school_id) + (1 | student_id)
: model_ML: math_score ~ time + ses + (1 + time | school_id) + (1 | student_id)
:              npar    AIC    BIC  logLik -2*log(L)  Chisq Df Pr(>Chisq)
: model_ML_ses    7 2050.4 2078.4 -1018.2    2036.4
: model_ML        8 2042.2 2074.1 -1013.1    2026.2 10.236  1   0.001377 **
: ---
: Signif. codes:  0 '***' 0.001 '**' 0.01 '*' 0.05 '.' 0.1 ' ' 1
```

To test whether the improvement rates vary between schools, we can use the following LRT on the random slope of `time`. Since the null hypothesis is on the boundary of the parameter space, the regular LRT is not valid, and a parametric bootstrap approach is used. The parametric bootstrap LRT is performed using `pbkrtest::PBmodcomp()`. The *p*-value (0.057) indicates that there is some evidence that the improvement rates are different between schools, although it is not significant at the common 5% level. Note that here we use the model fitted with REML, as `model` and `model_timeschool` have exact same fixed effects, and therefore, REML does not introduce bias in the variance components estimation; see Note 6.23 for details.

```
model_timeschool <- lmer(math_score ~ time + ses + (1|school_id) +
                          (1|student_id), data = data)
set.seed(123)
pbkrtest::PBmodcomp(model, model_timeschool)
: Bootstrap test; time: 31.56 sec; samples: 1000; extremes: 56;
: large : math_score ~ time + ses + (1 + time | school_id) + (1 | student_id)
:          stat df p.value
: LRT    3.8596  2 0.14518
: PBtest 3.8596    0.05694 .
: ---
: Signif. codes:  0 '***' 0.001 '**' 0.01 '*' 0.05 '.' 0.1 ' ' 1
```

The following likelihood ratio test has a *p*-value (0.60) greater than the common significance levels, suggesting that the random effect `student_id` is not significant. Although it is not significant, we can still keep it in the model to respect the design and the dependence between paired data. Again, here we use REML since both models have the same fixed effects.

```
# Likelihood ratio test
model_student <- lmer(math_score ~ time + ses + (1 + time|school_id),
                      data = data)
set.seed(123)
pbkrtest::PBmodcomp(model, model_student)
: Bootstrap test; time: 28.97 sec; samples: 1000; extremes: 420;
: Requested samples: 1000 Used samples: 702 Extremes: 420
: large : math_score ~ time + ses + (1 + time | school_id) + (1 | student_id)
:          stat df p.value
: LRT    0.0574  1  0.8107
: PBtest 0.0574     0.5989
```

Using `model` as the candidate model for further analysis, the following extracts the fixed effects, the random effect, and the residuals. There is a lot of variability between schools. The standard deviation of the random slope of `time` is 0.844, which suggests a mild variability in the improvement rates across schools compared to the fixed effect of `time` (2.13).

```
# fixed effects
fixef(model)
: (Intercept)       time1         ses
: 68.78420744  2.12629769  0.05858362
# Extract variance-covariance of random effects
VarCorr(model)
:  Groups     Name        Std.Dev. Corr
:  student_id (Intercept) 0.41145
:  school_id  (Intercept) 6.69850
:             time1       0.84371  0.620
:  Residual               2.88347
```

Model diagnostics can be performed as follows. The normality of the random effect can be verified by the plot and the test. The QQ plots for the random effects are shown in Figure 6.32, and the Shapiro-Wilk tests for normality are also performed. The normality assumption seems reasonable for both random effects.

```
# Normality of Random Effects
random_effects_student <- ranef(model)$`student_id`[,1]
random_effects_school <- ranef(model)$school_id[,1]

qq_student <- ggplot(data.frame(re = random_effects_student),
                     aes(sample = re)) + stat_qq() + stat_qq_line() +
              theme_bw() + labs(x = "Theoretical Quantiles",
                                y = "Empirical Quantiles",
                                title = "Student Random Effect")

qq_school <- ggplot(data.frame(re = random_effects_school),
                    aes(sample = re)) + stat_qq() + stat_qq_line() +
              theme_bw() + labs(x = "Theoretical Quantiles",
                                y = "Empirical Quantiles",
                                title = "School Random Effect")
gridExtra::grid.arrange(qq_student, qq_school, ncol = 2)
```

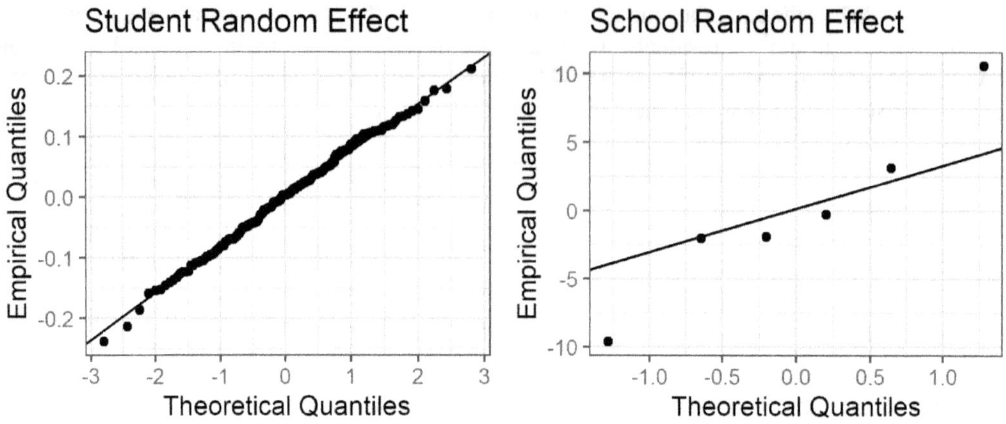

FIGURE 6.32: QQ Plots for Random Effects.

```
# Shapiro-Wilk test for random effects
shapiro.test(random_effects_student)
:
:   Shapiro-Wilk normality test
:
: data:  random_effects_student
: W = 0.99659, p-value = 0.9399
shapiro.test(random_effects_school)
:
:   Shapiro-Wilk normality test
:
: data:  random_effects_school
: W = 0.95719, p-value = 0.7978
```

The normality of the residuals can be verified by the plot and the test as follows, with the QQ plot shown in Figure 6.33. The normality assumption seems reasonable for the residuals.

```
# Normality of Residuals
residuals <- resid(model)

# Visual check
ggplot(data.frame(res = residuals), aes(sample = res)) +
       stat_qq() + stat_qq_line() + theme_bw() +
        labs(x = "Theoretical Quantiles",
             y = "Empirical Quantiles", title = "")
```

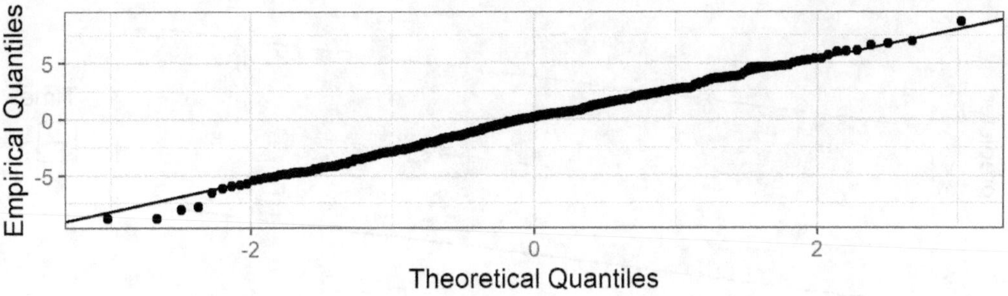

FIGURE 6.33: QQ Plot for Residuals.

```
# Formal test
shapiro.test(residuals)
:
:  Shapiro-Wilk normality test
:
: data:  residuals
: W = 0.99631, p-value = 0.4842
```

The six schools are randomly sampled from the large metropolitan area, and one usually is not interested in comparisons between these specific schools. Instead, we focus on the `time` and `SES` effects, and the variability of improvement rates across schools.

To visualize the improvement over `SES`, the following can be used. The plot suggests that there is an improvement in math scores over time, and higher SES is associated with higher math scores.

```
# prediction
pred <- ggpredict(model, terms = c("ses", "time"))
names(pred)[names(pred) == "group"] <- "Time"
names(pred)[names(pred) == "x"] <- "SES"
ggplot(pred, aes(x = SES, y = predicted, color = Time, linetype = Time,
       group = Time)) + geom_line() + geom_ribbon(aes(ymin = conf.low,
       ymax = conf.high, fill = Time), alpha = 0.2) + theme_bw() +
  scale_color_manual(values = c("black", "grey70")) +
  scale_fill_manual(values = c("black", "grey70"))
```

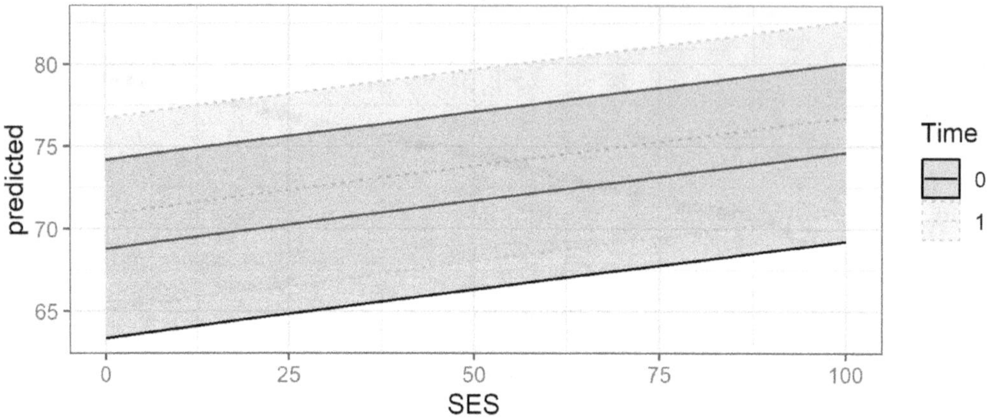

FIGURE 6.34: Improvement at Different SES Scores.

Conclusions

In this case study, we analyze the math scores of students from six randomly selected schools in a large metropolitan area at two time points, while controlling for socioeconomic status (SES). A linear mixed-effects model is employed, incorporating random effects to account for students clustered within schools and repeated measures on the same student. The analysis reveals that for a randomly selected school in the large metropolitan area, math scores have improved significantly over time (p-value is 0.001), and there is some evidence that the rate of improvement varies between schools (p-value is 0.057 for the test of random slopes). In addition, a higher SES is significantly associated with a higher score (p-value $< 2.2 \times 10^{-16}$). Subject are randomly selected and schools are randomly sampled from the large metropolitan area, so the results can be generalized to all students in the area. However, both Time and SES are observational variables, and their effects can only be interpreted as associations; causality cannot be established from the study. For example, we cannot conclude that increasing a student's SES will lead to a higher math score as there may be other confounding variables that are associated with both SES and math scores. Similarly, we cannot conclude that the passage of time causes an increase in math scores automatically, as other factors (e.g., additional tutoring) that occur over time may be responsible for the observed improvement.

6.1.11.3 Partially Paired Data

For longitudinal data analysis where participants are measured at multiple time points, we can consider the participants as a random sample from a larger population, and use random effects to account for the variability across participants. Due to the fact that longitudinal studies often last for a long time, it is common for participants to have missing data at some time points, and some participants may only have data at one time point. In this case, we can still use all available data to estimate the overall trend or before-after difference, while accounting for the variability across participants using a random effect. This approach allows us not to exclude data that have missing values, and generally provides more accurate estimates of the overall trend or before-after difference. We refer to [43] for more details on the analysis of unbalanced longitudinal data.

In what follows, we will use a case study with partially paired data to illustrate the use of a linear mixed-effects model for studying a typical question on before-after difference in a stress-reduction program.

Case 13: Effectiveness of a Stress-Reduction Program

A health researcher evaluates a stress-reduction program designed to help individuals manage stress more effectively. The program includes mindfulness exercises, physical activity, and cognitive-behavioral techniques. The researcher collects data from 142 volunteers on stress levels before and after the program using a standardized stress assessment tool with stress scores range from 0-100, with higher scores indicating more stress. Due to incomplete participation, there are 55 participants completed both before and after measurements, 39 participants completed only before measurements, and 48 participants completed only after measurements. The objective is to assess whether the stress-reduction program is effective in reducing stress levels.

For answers to the initial questions and some initial thoughts about this case study, please refer to Section 5.14. First, we load the data, specify appropriate data types, and check the structure of the data. The following R codes verify that there are 142 distinct participants, among which 39 participants have only before measurements, 48 have only after measurements, and 55 have both before and after measurement.

```
library(lme4)
library(ggplot2)
data <- read.csv("data/stress-reduction-program.csv")
data$id <- as.factor(data$id)
data$time <- factor(data$time, levels = c("Before","After"))
cat("unique participants:", length(unique(data$id)), "\n")
: unique participants: 142
tab <- table(data$id, data$time)
before_only <- sum(tab[, "Before"] == 1 & tab[, "After"] == 0)
after_only  <- sum(tab[, "After"] == 1 & tab[, "Before"] == 0)
paired      <- sum(tab[, "Before"] ==1 & tab[, "After"] == 1)
c(before_only, after_only, paired)
: [1] 39 48 55
```

We use a spaghetti plot to visualize the stress levels across different participants; see Figure 6.35. We use jittering to avoid overlapping of points, and notice that some participants only have data at one time point, and some participants have data at both time points. It seems that there is a slight overall decrease in the stress levels after the program, but there is also a lot of variability across participants.

```
# Spaghetti plot by participants
set.seed(123)
ggplot(data, aes(x = time, y = stress, group = id)) +
  geom_line(alpha = 0.5, color = "gray") +
  geom_point(position = position_jitter(width = 0.04, height = 0),
             alpha = 0.5, size = 1) + theme_bw() +
  labs(title = "", x = "Time", y = "Stress")
```

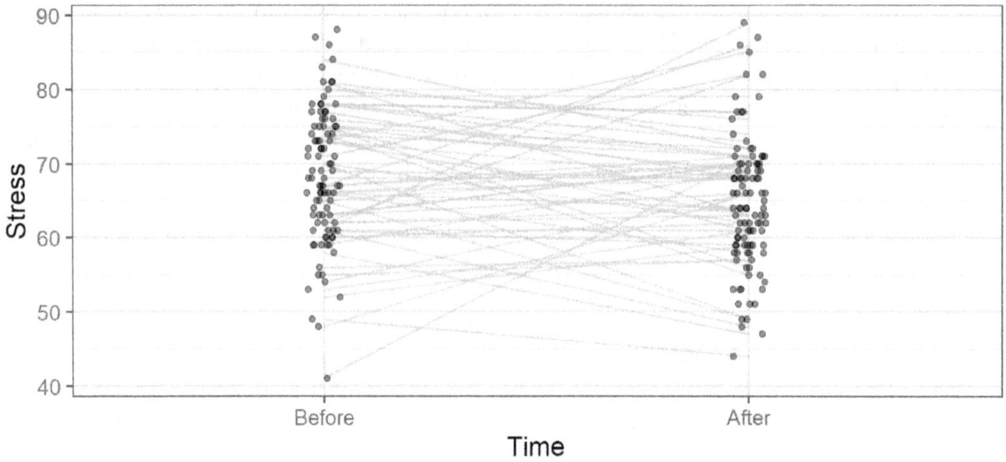

FIGURE 6.35: Spaghetti Plot of Stress Levels by Participant. Gray lines: Individual participants who have both before and after measurements; Points: Individual measurements with some jittering.

To assess whether the stress-reduction program is effective, we can fit a linear mixed-effects model with `time` as a fixed effect and participant `id` as a random effect, and the fixed effect of time will indicate whether there is a significant improvement in stress levels after the program. The following R codes fit the data with and without the fixed effect of time, respectively, and then a likelihood ratio test is conducted to assess the significance of the fixed effect of time. Here we use `REML = FALSE` to fit the models using maximum likelihood estimation, which is recommended when conducting likelihood ratio tests for fixed effects in LMMs. The sample size is moderate to large, so the likelihood ratio test is reasonably accurate; of course other methods such as the Kenward-Roger approximation can also be used to test fixed effects in LMMs. See Note 6.24 for details on testing fixed effects in LMMs. The small p-value 0.01 indicates that there is a significant difference in stress levels after the program, and the estimated fixed effect of time is -3.02, which indicates that on average the stress level is reduced by 3.02 units after the program.

```
model_ML <- lmer(stress ~ time + (1|id), data=data, REML = FALSE)
model_ML_notime <- lmer(stress ~ (1|id), data=data, REML = FALSE)
anova(model_ML, model_ML_notime)
: Data: data
: Models:
: model_ML_notime: stress ~ (1 | id)
: model_ML: stress ~ time + (1 | id)
:                 npar    AIC    BIC  logLik -2*log(L)  Chisq Df Pr(>Chisq)
: model_ML_notime    3 1431.4 1441.2 -712.69    1425.4
: model_ML           4 1426.8 1439.9 -709.39    1418.8 6.6008  1    0.01019 *
: ---
: Signif. codes:  0 '***' 0.001 '**' 0.01 '*' 0.05 '.' 0.1 ' ' 1
fixef(model_ML)
: (Intercept)   timeAfter
:   67.813403   -3.018892
```

Model diagnostics can be conducted, and the following R codes check the normality of the random effects and residuals, respectively; the normality assumptions are reasonably satisfied based on the QQ plots and the Shapiro-Wilk tests.

```
random_effects <- ranef(model_ML)$id[,1]

ggplot(data.frame(re = random_effects), aes(sample = re)) + stat_qq() +
  stat_qq_line() + theme_bw() + labs(x = "Theoretical Quantiles (Normal)",
             y = "Empirical Quantiles", title = "")
```

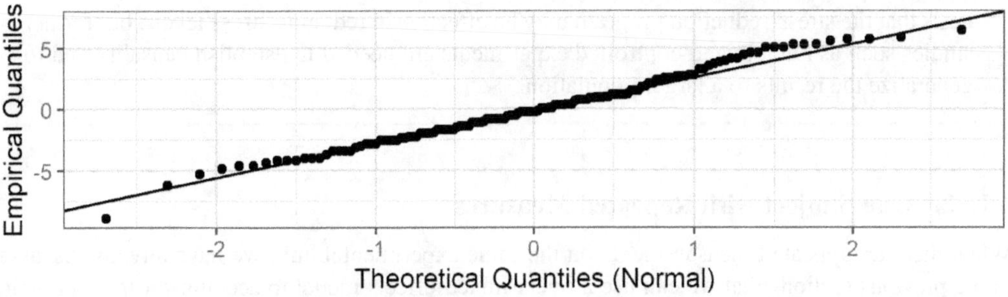


```
shapiro.test(random_effects)
:
:  Shapiro-Wilk normality test
:
: data:  random_effects
: W = 0.98727, p-value = 0.2173
```

```
# Normality of Residuals
residuals <- resid(model_ML)
ggplot(data.frame(res = residuals), aes(sample = res)) + stat_qq() +
  stat_qq_line() + theme_bw() +
  labs(x = "Theoretical Quantiles (Normal)",
       y = "Empirical Quantiles", title = "")
```

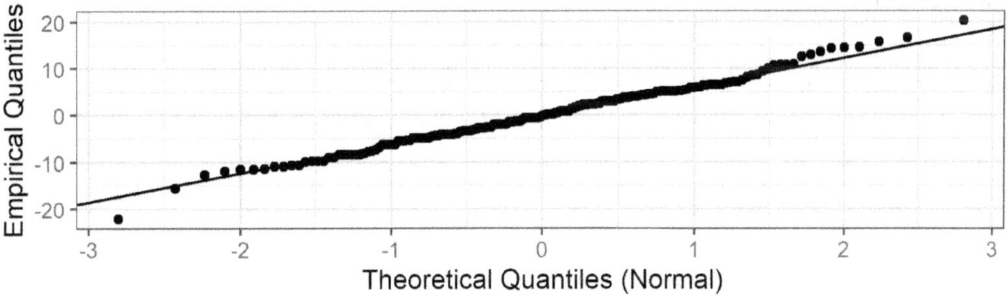


```
shapiro.test(residuals)
:
:  Shapiro-Wilk normality test
:
: data:  residuals
: W = 0.99288, p-value = 0.4573
```

Conclusions

In this case study, we use a linear mixed-effects model to analyze partially paired data from a stress-reduction program. The results indicate that there is a significant reduction in stress levels after the program (p-value is 0.01 for the fixed effect of time). However, the subjects are not randomly selected, so there is a limitation in generalizing the results to a larger population. Also, this is an observational study, so there is no causality between the stress-reduction program and the reduction in stress levels can be established. This study suggests that the stress-reduction program may be effective in reducing stress levels, but further studies such as randomized controlled experiments are needed to establish causality and to generalize the results to a larger population.

6.1.12 One-Subject with Repeated Measures

When there are repeated measurements on the same experimental unit, we have already discussed in the previous sections that we can use a linear mixed-effects model to account for the correlation among the repeated measurements. However, in practice, we may face situations where repeated measurements are taken, but there is only one group of experimental units where individuals cannot be identified. In this section, we discuss how to analyze such data.

Case 14: Environmental Factors on Antler Length

In ecological research, it is often necessary to study a group of wild animals in a nonintrusive way, such as through camera traps or direct observation, where individual animals cannot be distinguished due to their similarity and elusive behavior. Researchers are investigating the antler length (in centimeters) of a small, isolated group of animals that belong to the Bovidae family over several months. The objective is to analyze the antler length data to assess whether environmental factors, such as habitat quality (e.g., forest density), influence antler length. The data includes repeated measures from multiple sightings, collected over time with an associated predictor: forest density (low or high).

For answers to the initial questions and some initial thoughts about this case study, please refer to Section 5.15. First, some exploratory data analysis can be performed. The box plot (Figure 6.36) indicates that there could be no significant differences between antler lengths due to the high and low forest densities.

```
library(ggplot2)
data <- read.csv("data/forest-density-antler-length.csv")

# Descriptive Statistics
summary_stats <- tapply(data$antler_length, data$forest_density, summary)
print(summary_stats)
: $high
:    Min. 1st Qu.  Median    Mean 3rd Qu.    Max.
:   22.00   26.15   26.85   27.27   28.52   33.40
:
: $low
:    Min. 1st Qu.  Median    Mean 3rd Qu.    Max.
:   20.70   23.73   27.00   26.45   29.07   31.80
ggplot(data, aes(y = forest_density, x = antler_length)) + geom_boxplot() +
```

```
stat_summary(fun = "mean", geom = "point", shape = 23, size = 2) +
labs(title = "", y = "Forest Density", x = "Antler Length (cm)") +
theme_bw() + theme(legend.position = "none")
```

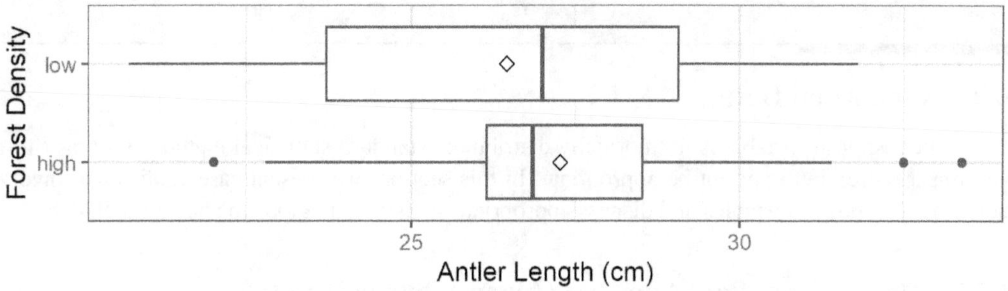

FIGURE 6.36: Antler Length by Forest Density.

As we have discussed in the initial thoughts (Section 5.15) of the case study, we perform a descriptive analysis but not an inference analysis for this finite population, using a simple linear regression. The following results suggest that forest density does not significantly affect antler length for this group of animals at the significance level of 5%. Some model diagnostics for linear regression need to be performed and are skipped here.

```
# Linear Fixed-Effects Model
fixed_model <- lm(antler_length ~ forest_density, data = data)
summary(fixed_model)
:
: Call:
: lm(formula = antler_length ~ forest_density, data = data)
:
: Residuals:
:     Min      1Q  Median      3Q     Max
: -5.7550 -1.5663  0.2875  1.5662  6.1300
:
: Coefficients:
:                    Estimate Std. Error t value Pr(>|t|)
: (Intercept)         27.2700     0.7148  38.153   <2e-16 ***
: forest_densitylow   -0.8150     1.0108  -0.806    0.425
: ---
: Signif. codes:  0 '***' 0.001 '**' 0.01 '*' 0.05 '.' 0.1 ' ' 1
:
: Residual standard error: 3.197 on 38 degrees of freedom
: Multiple R-squared:  0.01682, Adjusted R-squared:  -0.009054
: F-statistic: 0.6501 on 1 and 38 DF,  p-value: 0.4251
```

Conclusions

In this case study, we analyze the effect of forest density on antler length using a simple linear regression model. There is no evidence that forest density affects antler length for this group of animals (p-value is 0.425), and this study is a descriptive study and does not generalize to other groups of animals. Also, no causality can be established from this observational study.

6.2 Nonnormal Data

When the response variable is not normally distributed, standard statistical methods such as *t* tests and linear regression may not be appropriate. In this section, we present case studies that involve nonnormal response variables and discuss appropriate statistical methods to analyze such data.

6.2.1 The Wilcoxon Rank Sum Test / Mann-Whitney U Test

For comparing two independent groups with nonnormal continuous response variables, we can consider using data transformation to make the response variable approximately normal, or use nonparametric methods that do not assume normality. We illustrate these approaches using the following case study.

Case 15: Salary Comparisons Between Technology and Finance Sectors

To analyze the salary differences between the technology and finance sectors, a random sample comprising 40 employees is chosen, as illustrated in Table 5.5. Is there a significant disparity in the salaries of these two industries?

TABLE 5.5: Salary by Industry.

Industry	Salary	Industry	Salary
Tech	105522	Finance	83877
Tech	196808	Finance	154135
Tech	127812	Finance	339442
Tech	151245	Finance	57313
Tech	131364	Finance	16052
Tech	130426	Finance	163410
Tech	264158	Finance	49931
Tech	119028	Finance	165694
Tech	284428	Finance	66215
Tech	61428	Finance	91758
Tech	178493	Finance	81148
Tech	141765	Finance	105841
Tech	148445	Finance	126888
Tech	161106	Finance	69263
Tech	103657	Finance	174170
Tech	112803	Finance	109604
Tech	80074	Finance	64334
Tech	77972	Finance	180623
Tech	155091	Finance	162016
Tech	166726	Finance	118643

For answers to the initial questions and some initial thoughts about this case study, please refer to Section 5.16. We first perform some exploratory data analysis. Figure 6.37 shows the box plots of salary by industry, and it suggests that the salary data may not follow a normal distribution.

```
library(ggplot2)
data <- read.csv("data/salary-between-industry.csv")
data$Industry <- as.factor(data$Industry)

set.seed(123)
ggplot(data, aes(y = Industry, x = Salary)) +
  geom_boxplot() +
  stat_summary(fun = "mean", geom = "point", shape = 23, size = 2) +
  scale_x_continuous(labels = scales::dollar_format()) + theme_bw() +
  labs(x = "Annual Salary")
```

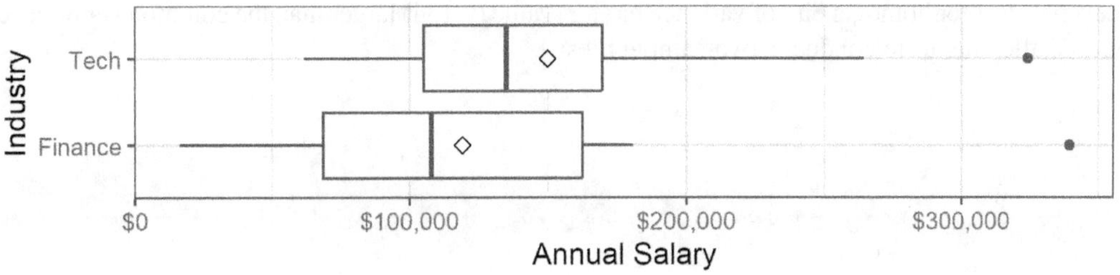

FIGURE 6.37: Salary Distribution by Industry.

We can use QQ plots and a normality test to verify the nonnormality of the salary data. Figure 6.37 suggests that the data are heavy-tailed, and Shapiro-Wilk normality tests lead to p-values of 0.015 and 0.049 for the finance and technology industries, respectively, indicating that there is evidence to reject the null hypothesis of normality at the significance level of 5%.

```
# Visual checks for heavy tails
# QQ plots
ggplot(data, aes(sample = Salary)) + geom_qq() + geom_qq_line() +
  facet_wrap(~Industry) + theme_bw() +
  labs(y="Sample quantiles of salary", x="Theoretical quantiles")
```

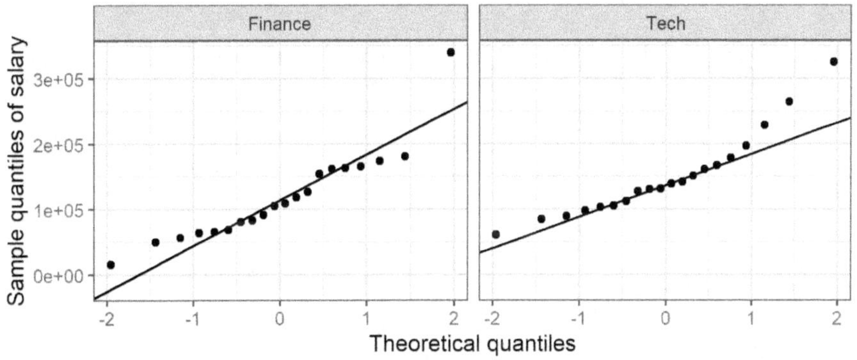

FIGURE 6.38: QQ Plots of Salary by Industry.

```
shapiro.test(data[data$Industry=="Finance", "Salary"])
:
:  Shapiro-Wilk normality test
:
: data:  data[data$Industry == "Finance", "Salary"]
: W = 0.87522, p-value = 0.01453
shapiro.test(data[data$Industry=="Tech", "Salary"])
:
:  Shapiro-Wilk normality test
:
: data:  data[data$Industry == "Tech", "Salary"]
: W = 0.90391, p-value = 0.04885
```

To address the nonnormality issue, we can first try a logarithm transformation of the salary. Both QQ plots and normality tests suggest that the log-salary follows a normal distribution. Furthermore, Levene's test for homogeneity of variance has a p-value (0.134) larger than the common significance levels, allowing us to conduct a two-sample t test.

```
data$Salary_log <- log(data$Salary)
ggplot(data, aes(sample = Salary_log)) + geom_qq() + geom_qq_line() +
  facet_wrap(~Industry) + theme_bw() +
  labs(y="Sample quantiles of log-salary", x="Theoretical quantiles")
```

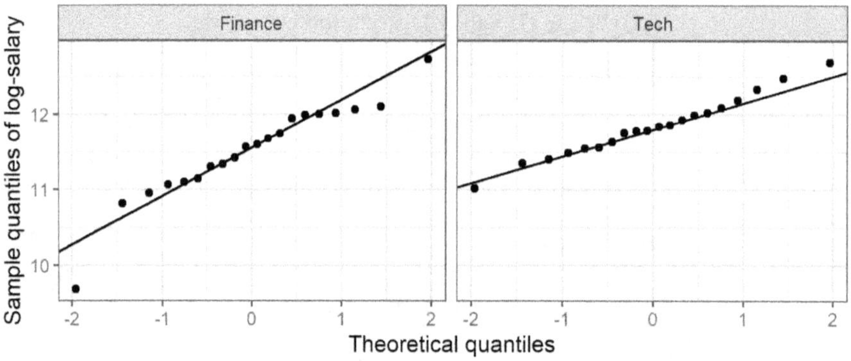

FIGURE 6.39: QQ Plots of Log-Salary by Industry.

```
shapiro.test(log(data[data$Industry=="Finance", "Salary"]))
:
:  Shapiro-Wilk normality test
:
: data:  log(data[data$Industry == "Finance", "Salary"])
: W = 0.93229, p-value = 0.1709
shapiro.test(log(data[data$Industry=="Tech", "Salary"]))
:
:  Shapiro-Wilk normality test
:
: data:  log(data[data$Industry == "Tech", "Salary"])
: W = 0.98981, p-value = 0.9979
```

```
car::leveneTest(Salary_log ~ Industry, data = data)
: Levene's Test for Homogeneity of Variance (center = median)
:       Df F value Pr(>F)
: group  1  2.3443  0.134
:       38
```

As the research question is to find out whether there is a difference between the salaries of the two industries, not directional difference, we can perform a two-sided two-sample t test on the log-salary.

```
t_test_log <- t.test(Salary_log ~ Industry, data = data, var.equal=TRUE)
cat("Two-sample t test:", t_test_log$alternative,
    "p-value=", t_test_log$p.value)
: Two-sample t test: two.sided p-value= 0.06437826
```

The p-value is 0.064 for a two-sided two-sample t test, so there is no strong evidence to reject the null hypothesis. However, as we have already discussed in Section 4.4, we shall not rely solely on one p-value to make conclusions. We also need to consider the effect size to assess the practical significance of the difference.

Here, we can calculate Cliff's Delta as a nonparametric effect size measure. Refer to Note 6.28 for more details about Cliff's Delta. The calculated Cliff's Delta is -0.3, suggesting a small effect size.

The following is the manual calculation of Cliff's Delta.

```
# Calculate effect size (Cliff's Delta)
cliff_delta <- function(x, y) {
  n1 <- length(x); n2 <- length(y); dominance <- 0
  for(i in x) {for(j in y) { if(i > j) dominance <- dominance + 1
                            else if(i < j) dominance <- dominance - 1}}
  return(dominance / (n1 * n2))
}
cliff_delta(data$Salary[data$Industry == "Finance"],
                     data$Salary[data$Industry == "Tech"])
: [1] -0.3
```

We can also use the `effsize` package to calculate it.

```
effsize::cliff.delta(data$Salary[data$Industry == "Finance"],
                     data$Salary[data$Industry == "Tech"])
:
: Cliff's Delta
:
: delta estimate: -0.3 (small)
: 95 percent confidence interval:
:       lower       upper
: -0.60121515  0.07586277
```

If logarithm transformation can not make the response variable normally distributed, then we can consider using a nonparametric alternative with Wilcoxon rank sum test (see Note 6.27). The p-value is 0.108, indicating that there is no statistically significant difference between the salaries of the two industries in terms of their distributions. Refer to Note 6.27 for more details about the Wilcoxon rank sum test / Mann-Whitney U test.

```
# Wilcoxon rank sum test (nonparametric)
wilcox.test(Salary ~ Industry, data = data)
:
:  Wilcoxon rank sum exact test
:
: data:  Salary by Industry
: W = 140, p-value = 0.1081
: alternative hypothesis: true location shift is not equal to 0
```

Conclusions

In this case study, we compare the salaries between the finance and technology industries. After transforming the salary data using logarithm transformation, we found no strong evidence to suggest a difference between the salaries of the finance and technology industries based on a two-sided two-sample t test (p-value = 0.064). Using Wilcoxon rank sum test on the original salary data also leads to the same conclusion (p-value = 0.108). Furthermore, the calculated Cliff's Delta is -0.3, suggesting a small effect size. Subjects are randomly sampled from both industries, and the sample sizes are relatively small (20 for each industry). Therefore, we should be cautious when generalizing the results to the entire population. Causality cannot be established due to the observational nature of the data.

Note 6.27 (The Wilcoxon Rank Sum Test / Mann-Whitney U Test)**.** These two tests are essentially the same test with different names, and both are used to compare distributions of two independent samples. These tests are alternatives to the two-sample t test when assumptions of the t test are violated. The hypotheses are the following:

$$H_0 : F(x) = G(x), \forall x,$$

$$H_1 : F(x) = G(x - \Delta), \forall x, \Delta \geq 0.$$

That is, the null hypothesis states that the two samples come from the same distribution, while the alternative hypothesis suggests that one distribution is shifted to the right of the other by a nonnegative amount Δ. The nonparametric nature of these tests allows them to be used in situations where the data do not meet the assumptions of parametric tests, such as normality, and they work for ordinal or continuous data and compare the ranks of observations rather than their actual values.

For two independent random samples: $\{X_1, \ldots X_{n_1}\}$ and $\{X_{n_1+1}, \ldots X_{n_1+n_2}\}$, with $n = n_1 + n_2$, let R_j be the rank of X_j in the combined sample, where the smallest observation has rank 1 and the largest observation has rank n. Then the test statistic of Wilcoxon rank sum test is the rank sum of the second random sample:

$$W = \sum_{j=1}^{n_2} R_{n_1+j}.$$

For large samples, the null distribution of W can be approximated by a normal distribution with mean and variance given by $\mathrm{E}(W|H_0) = n_2(n+1)/2$ and $\mathrm{Var}(W|H_0) = n_1 n_2(n+1)/12$. For small samples, the null distribution of W can be calculated exactly.

The Mann-Whitney U test is based on the number of times an observation from one sample exceeds an observation from the other sample, and the test statistic is

$$U = \sum_{i=1}^{n_1} \sum_{j=1}^{n_2} I(X_i \leq X_{n_1+j}),$$

where n_1 and n_2 are sample sizes, and I is the indicator function. The test statistic U can be calculated from the rank sum W as follows:

$$U = W - \frac{n_2(n_2+1)}{2}.$$

We refer to [34] for more details on the Wilcoxon rank sum test and the Mann-Whitney U test.

Note 6.28 (Cliff's δ). *Cliff's* δ is a robust, nonparametric effect size measure that is particularly useful for nonnormal data. For two groups X and Y, Cliff's δ is defined as

$$\delta = \mathrm{P}(X > Y) - \mathrm{P}(X < Y),$$

which ranges from -1 to $+1$. For samples of size m and n,

$$\delta = \frac{1}{mn}\sum_{i=1}^{m}\sum_{j=1}^{n} \text{dominance}(x_i, y_j),$$

where the dominance function is defined as

$$\text{dominance}(x_i, y_j) = \begin{cases} +1 & \text{if } x_i > y_j \\ -1 & \text{if } x_i < y_j \\ 0 & \text{if } x_i = y_j. \end{cases}$$

$\delta = +1$ indicates that all values in X are greater than all values in Y, $\delta = -1$ means that all values in X are smaller than all values in Y, and $\delta = 0$ suggests that there is no difference between the two groups. The following thresholds are suggested by [90]: $|\delta| = 0.147$: small effect, $|\delta| = 0.33$: medium effect, and $|\delta| = 0.474$: large effect.

For example, consider two groups: $X : [8,7,6,2,5]$ and $Y : [4,3,1,2,6]$, the dominance matrix is

X/Y	4	3	1	2	6
8	+1	+1	+1	+1	+1
7	+1	+1	+1	+1	+1
6	+1	+1	+1	+1	0
2	-1	-1	+1	0	-1
5	+1	+1	+1	+1	-1

Then $\delta = 15/25 = 0.6$ (suggesting a relatively large effect size). For more details about Cliff's δ, please refer to [23] and [90].

6.2.2 The Wilcoxon Signed-Rank Test

When we have paired data, but the differences do not follow a normal distribution, we can consider using nonparametric methods such as the Wilcoxon signed-rank test to compare the two related groups. We illustrate this approach using the following case study.

Case 16: Sleep Hygiene Practices on Sleep Quality

A sleep research institute develops a nonpharmacological intervention program focusing on sleep hygiene practices, and the study evaluates the effectiveness of this 4-week intervention program by measuring participants' sleep quality scores before and after the intervention. Twenty participants with mild to moderate sleep difficulties are recruited for the study. The Pittsburgh Sleep Quality Index (PSQI) is administered before and after the 4-week intervention. The PSQI provides a score ranging from 0 to 21, where higher scores indicate poorer sleep quality. Does the sleep hygiene intervention program significantly improve participants' sleep quality as measured by PSQI?

For answers to the initial questions and some initial thoughts about this case study, please refer to Section 5.17. The data provided are in a wide format, and we first convert them to a long format so that we can use `ggplot()` to plot box plots prior to and after the intervention, respectively. Figure 6.40 suggests that PSQI scores are significantly lower after the intervention.

```
# Case Study: Impact of Mindfulness Training on Employee Stress Levels
library(ggplot2)
library(dplyr)
library(tidyr)
data <- read.csv("data/sleep-quality.csv")
data$difference <- data$post - data$pre

# Descriptive statistics
data %>% summarise(mean_pre = mean(pre), mean_post = mean(post),
            median_pre = median(pre), median_post = median(post),
            sd_pre = sd(pre), sd_post = sd(post))
:   mean_pre mean_post median_pre median_post   sd_pre  sd_post
: 1     12.6      9.65       12.5         9.5 1.930367 1.496487
# make wide format to long format
sleep_data <- data %>% pivot_longer(cols = c("pre", "post"),
              names_to = "time", values_to = "score")
sleep_data$difference <- NULL # delete difference column

# Create box plot
ggplot(sleep_data, aes(y = time, x = score)) + geom_boxplot() +
  stat_summary(fun = "mean", geom = "point", shape = 23, size = 2) +
  theme_bw() + labs(x = "PSQI Score", y = "Time Point")
```

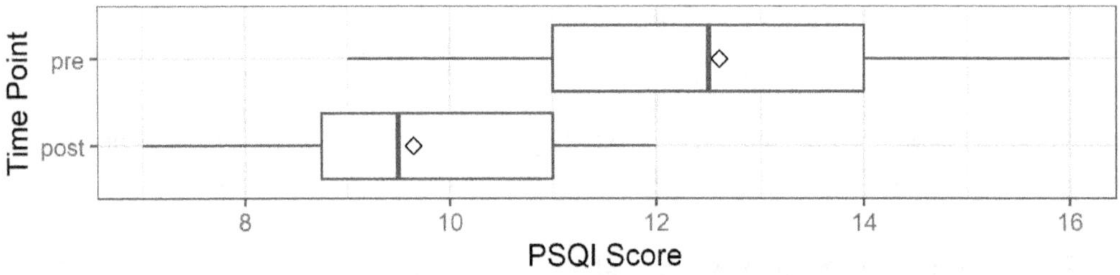

FIGURE 6.40: Sleep Quality Scores Before and After Intervention.

A subject-level comparison in Figure 6.41 also suggests that the scores are lower after the intervention.

```
# Create spaghetti plot
sleep_data$time <- factor(sleep_data$time, levels = c("pre", "post"))
ggplot(sleep_data, aes(x = time, y = score, group = participant)) +
  geom_line(alpha = 0.5, color = "gray") +
  geom_point(position = position_jitter(width = 0.03, height = 0),
             alpha = 0.4, size = 1) + theme_bw() +
  labs(y = "PSQI Score", x = "Time") +
  scale_x_discrete(labels = c("pre" = "Pre", "post" = "Post"))
```

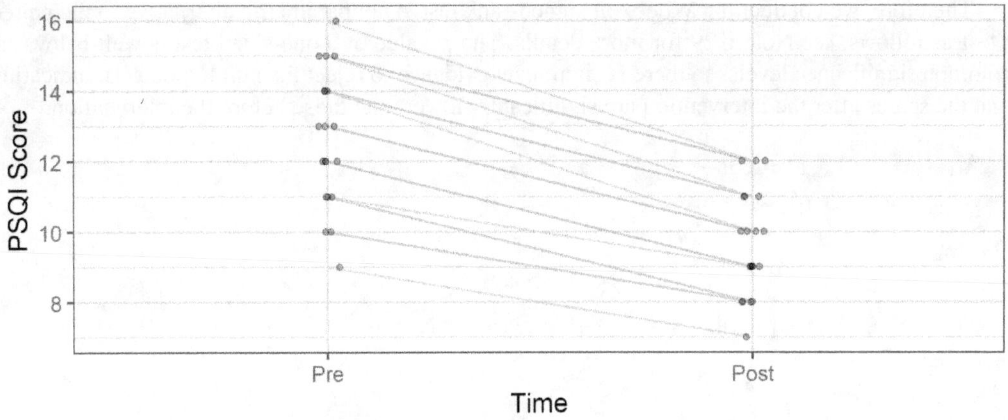

FIGURE 6.41: Individual Changes in Sleep Quality Scores.

For paired data, we first need to check the distribution of the difference. Figure 6.42 clearly shows that the difference does not follow the normal distribution, and the Shapiro-Wilk normality test leads to a p-value (0.0003) well below the common significance levels, so there is significant evidence to reject the normality assumption.

```
ggplot(data, aes(sample = difference)) + geom_qq() + geom_qq_line() +
        theme_bw() + labs(y="Sample quantiles of difference",
                          x="Theoretical quantiles")
```

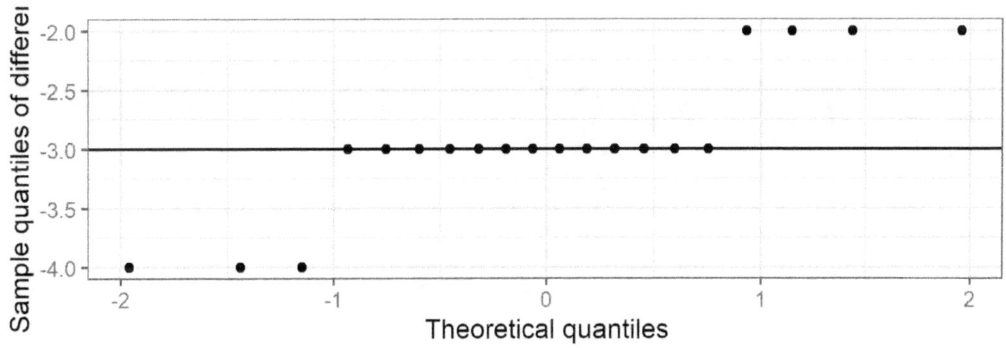

FIGURE 6.42: QQ Plots of difference.

```
# normal test
shapiro.test(data$difference)
:
:  Shapiro-Wilk normality test
:
: data:  data$difference
: W = 0.76768, p-value = 0.000296
```

Therefore, we conduct the Wilcoxon signed-rank test, a nonparametric alternative to the paired t test as follows; see Note 6.29 for more details. The p-value of a one-sided test is well below the common significance levels, so there is significant evidence to reject the null hypothesis, indicating that the scores after the intervention are significantly lower than those before the intervention.

```
# Perform the Wilcoxon signed-rank test
wilcox.test(data$pre, data$post, paired = TRUE,
  exact = FALSE, # Use the asymptotic method due to ties
  alternative = "greater")
:
:  Wilcoxon signed rank test with continuity correction
:
: data:  data$pre and data$post
: V = 210, p-value = 2.713e-05
: alternative hypothesis: true location shift is greater than 0
```

Without having to rely on any specific tests, a nonparametric bootstrap can help us compare the scores before and after the intervention. For example, if we are interested in assessing whether the mean difference of the scores between before and after the intervention is significantly different, we can do the following nonparametric bootstrap. The function `boot_mean_diff()` returns the mean of differences for each re-sampled data, where sampling by replacement is applied. For constructing the bootstrap confidence interval, we use a more robust method: bias-corrected and accelerated interval (`type = "bca"`); see [38] for more details on bias-corrected and accelerated intervals for bootstrap confidence intervals. The 95% confidence interval $(2.65, 3.15)$ is above 0, suggesting that the scores before and after are significantly different.

```
# Calculate confidence intervals using bootstrap
library(boot)

# Bootstrap function for median difference
boot_mean_diff <- function(data, indices) { d <- data[indices,]
  return(mean(d$pre - d$post))}

# Perform bootstrap
set.seed(123)
boot_results <- boot(data, boot_mean_diff, R=5000)

# two-sided bounds with bca
boot.ci(boot_results, type = "bca", conf = 0.95)
: BOOTSTRAP CONFIDENCE INTERVAL CALCULATIONS
: Based on 5000 bootstrap replicates
:
: CALL :
: boot.ci(boot.out = boot_results, conf = 0.95, type = "bca")
:
: Intervals :
: Level       BCa
: 95%    ( 2.65,  3.15 )
: Calculations and Intervals on Original Scale
```

For a one-sided test, say H_0 : the scores before the intervention are higher than those after the intervention, we can simply find the lower bound using the `quantile()` function. The following calculation shows that the 95% confidence lower bound is 2.75, well above 0, and therefore there is significant evidence to reject the null hypothesis, indicating that the scores are significantly higher after the intervention at a significance level of 5%.

```
# one-sided bound
quantile(boot_results$t, 0.05)
:    5%
: 2.75
```

Conclusions

In this case study, we compare sleep quality scores before and after an intervention using the Wilcoxon signed-rank test and nonparametric bootstrap methods. There is strong evidence that the sleep quality index after the intervention are lower than those before the intervention, with p-value 2.713×10^{-5} from the Wilcoxon signed-rank test, indicating an improvement in sleep quality after the intervention. A nonparametric bootstrap 95% confidence interval $(2.75, \infty)$ for the mean difference (before - after) also supports this conclusion. It is unclear whether the subjects are randomly selected; therefore, we shall be cautious when making inferences to a larger population. Moreover, as this is a pre-post study without a control group, causality can not be established, and there may be other confounding factors that lead to the improvement in sleep quality other than the intervention.

Note 6.29 (The Wilcoxon Signed-Rank Test)**.** The *Wilcoxon signed-rank test* is a powerful nonparametric alternative to the paired t test when the normality assumption is violated. The null hypothesis for the Wilcoxon signed-rank test is

$$H_0 : \text{the differences are symmetrically distributed about } 0.$$

For a random sample of n paired observations (x_{1i}, x_{2i}), $i = 1, \ldots, n$, the following steps are performed to conduct the Wilcoxon signed-rank test: (1) Calculate differences $d_i = x_{1i} - x_{2i}$ for $i = 1, \ldots, n$. (2) Rank absolute values of the differences, and when there are ties, assign average ranks. (3) Assign original signs of the differences back to the ranks. (4) Compute the test statistic: $W = \sum_{d_i > 0} \text{rank}(|d_i|)$. (5) Compare the calculated W with the critical values of the Wilcoxon signed-rank table to obtain the p-value. For large samples, an asymptotic normal approximation can be used to calculate the p-value. We refer readers to [115] for the original paper on the Wilcoxon signed-rank test and to [34] for more details on nonparametric statistical methods, including the Wilcoxon signed-rank test.

Table 6.4 provides an example of how to perform the Wilcoxon signed-rank test to compare scores before and after a new study method is implemented for a group of students.

TABLE 6.4: Example of the Wilcoxon Signed-Rank Test.

Student	Before	After	Difference	Abs. Difference (Rank)
1	75	82	+7	7 (5)
2	68	65	–3	3 (2)
3	90	85	–5	5 (3)
4	72	78	+6	6 (4)
5	83	81	–2	2 (1)

The test statistic $W = 5+4 = 9$. For a one-sided test with $n = 5$ and $W = 9$, the p-value can be calculated using the `psignrank()` function in R, which gives a p-value of 0.6875, indicating that there is no evidence to reject the null hypothesis at common significance levels, suggesting that the scores after the new study method are not significantly higher than those before the new study method.

```
psignrank(9, 5)
: [1] 0.6875
```

In R, this can be done directly using `wilcox.test()` by letting `paired = TRUE` and specifying the alternative hypothesis. The following code performs the Wilcoxon signed-rank test for the data in Table 6.4, and the p-value is consistent with the manual calculation.

```
wilcox.test(y=c(75,68,90,72,83), x=c(82, 65, 85, 78, 81), paired = TRUE,
            alternative = "less")
:
:  Wilcoxon signed rank exact test
:
: data:  c(82, 65, 85, 78, 81) and c(75, 68, 90, 72, 83)
: V = 9, p-value = 0.6875
: alternative hypothesis: true location shift is less than 0
```

6.2.3 The Kruskal-Wallis Test

When we have more than two groups to compare, but the response variable does not follow a normal distribution, we can consider using nonparametric methods such as the Kruskal-Wallis test to compare the groups. We illustrate this approach using the following case study.

Case 17: Impact of Different Training Methods on Employee Performance

A multinational company wants to evaluate the effectiveness of three different training methods (Traditional Classroom, Online Self-Paced, and Blended Learning) on employee performance improvement percentages. The HR department collects performance improvement data (percentage increase in performance after training) from randomly selected employees who are randomly assigned to carry out one of these different training approaches. Is there a significant difference in the percentages of improvement in employee performance between the three training methods, and which method is most effective?

For answers to the initial questions and some initial thoughts about this case study, please refer to Section 5.18. To compare the means of groups, we can first consider ANOVA.

```
library(dplyr)
library(ggplot2)
data <- read.csv("data/training-employee-performance.csv")
data$Method <- as.factor(data$Method)
```

The following are QQ plots and normality tests to verify the key assumption used by ANOVA, and they suggest that the normality assumption does not hold.

```
# Check for normality using QQ plots
ggplot(data, aes(sample = Improvement)) + geom_qq() + geom_qq_line() +
  facet_wrap(~Method) + theme_bw() +
  labs(y = "Sample quantiles of improvements", x = "Theoretical quantiles")
```

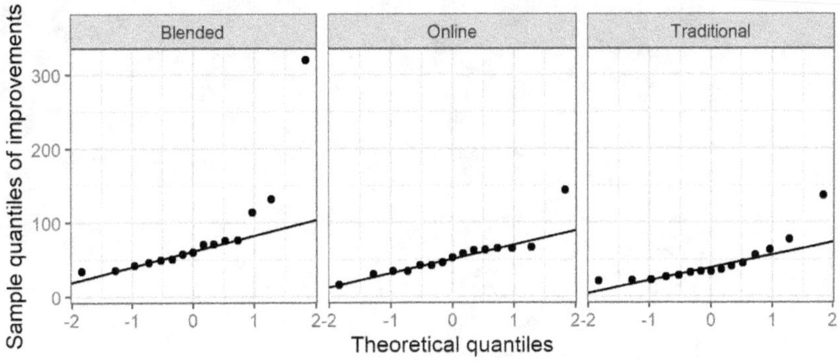

FIGURE 6.43: QQ Plots of Performance Improvements.

```
# Shapiro-Wilk test for normality
data %>% group_by(Method) %>% summarise(
    shapiro_stat = shapiro.test(Improvement)$statistic,
    shapiro_p = shapiro.test(Improvement)$p.value)
: # A tibble: 3 x 3
:   Method      shapiro_stat shapiro_p
:   <fct>              <dbl>     <dbl>
: 1 Blended            0.612 0.0000326
: 2 Online             0.789 0.00267
: 3 Traditional        0.738 0.000648
```

We can then use the logarithm to transfer the data and again check the normality assumption. The p-value for the transformed data of "Blended" is 0.052, suggesting that it may not be the ideal case for using parametric ANOVA that relied on the normality assumption. Therefore, we consider nonparametric methods to compare the groups.

```
# Shapiro-Wilk test for normality after log transformation
data$logImprovement <- log(data$Improvement)
data %>% group_by(Method) %>% summarise(
    shapiro_stat = shapiro.test(logImprovement)$statistic,
    shapiro_p = shapiro.test(logImprovement)$p.value)
: # A tibble: 3 x 3
:   Method      shapiro_stat shapiro_p
:   <fct>              <dbl>     <dbl>
: 1 Blended            0.883    0.0518
: 2 Online             0.934    0.314
: 3 Traditional        0.918    0.182
```

We consider the Kruskal-Wallis test, a nonparametric alternative to parametric ANOVA. Please find Note 6.30 for details. The Kruskal-Wallis test generally tests whether the distributions of multiple groups are identical. If the shapes of the distributions across groups are similar, then it tests whether the group medians are equal. So, we first check whether the groups have similar distributional shapes and scales. The following are some basic statistics that tell the scales and shapes of the three groups, and these values are not remarkably different.

```
# Descriptive statistics
data %>% group_by(Method) %>%
  summarise(CV = sd(Improvement) / mean(Improvement) * 100,
    IQR = IQR(Improvement),
    skewness = moments::skewness(Improvement),
    kurtosis = moments::kurtosis(Improvement))
: # A tibble: 3 x 5
:   Method         CV   IQR skewness kurtosis
:   <fct>       <dbl> <dbl>    <dbl>    <dbl>
: 1 Blended      86.6  28.4     2.70     9.62
: 2 Online       53.2  25.9     1.88     7.17
: 3 Traditional  67.1  23.3     2.07     6.86
```

Violin plots and box plots can be used to further check whether shapes and scales are similar. The Blended group seems to have an outlier, and the shapes and scales are not remarkably different. An outlier like this will be less of an issue as the ranks will be used for the test.

```
# Check distributions with box plots and violin plots
ggplot(data, aes(y = Method, x = Improvement)) +
  geom_violin() + geom_boxplot(width = 0.2) +
  stat_summary(fun = "mean", geom = "point", shape = 23, size = 2) +
  theme_bw() + theme(legend.position = "none")
```

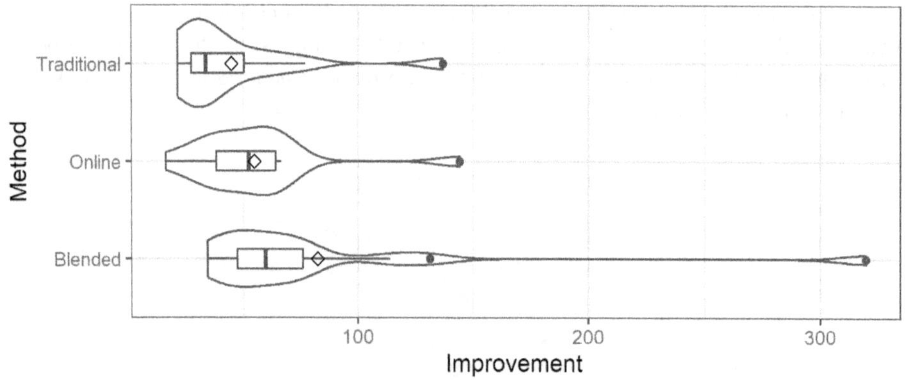

FIGURE 6.44: Distribution of Performance Improvements.

Furthermore, Levene's test for homogeneity of variances can be employed to check difference in scales / spread across the groups. The p-value is 0.418, so we do not reject equal variance between the groups.

```
levene_test <- car::leveneTest(Improvement ~ Method, data = data)
print(levene_test)
: Levene's Test for Homogeneity of Variance (center = median)
:       Df F value Pr(>F)
: group  2  0.8904 0.4181
:       42
```

Now, we proceed to conduct the Kruskal-Wallis test as follows. The p-value (0.015) is less than the common significance levels, indicating that there is a significant difference in the group medians. But we do not know where the difference is, and we need some appropriate post hoc tests to find it out.

```
# Perform Kruskal-Wallis test
kw_test <- kruskal.test(Improvement ~ Method, data = data)
print(kw_test)
:
:  Kruskal-Wallis rank sum test
:
: data:  Improvement by Method
: Kruskal-Wallis chi-squared = 8.4081, df = 2, p-value = 0.01494
```

We can use Mann-Whitney U tests for pairwise comparisons with some multiple comparisons techniques, say the Bonferroni adjustment. Alternatively, the following Dunn's test is often used for post hoc analysis after a significant Kruskal-Wallis test. From the tests, we conclude that the improvements are significantly different in group medians between the Traditional and Blended methods (the p-value is 0.011), and we fail to reject the null hypothesis of equal medians for the other two pairs at the family-wise error rate of 5%.

```
# Post-hoc analysis
library(rstatix)

result <- dunn_test(data, Improvement ~ Method,
                    p.adjust.method = "bonferroni")
data.frame(result)[,1:8]
:           .y.  group1      group2 n1 n2 statistic           p      p.adj
: 1 Improvement Blended      Online 15 15 -1.376240 0.168747198 0.50624159
: 2 Improvement Blended Traditional 15 15 -2.898446 0.003750171 0.01125051
: 3 Improvement  Online Traditional 15 15 -1.522205 0.127957632 0.38387290
```

We can also calculate Cliff's Delta as a nonparametric effect size measure for the pairwise comparisons. The calculated Cliff's Deltas are shown below.

```
rcompanion::cliffDelta(x=data$Improvement[data$Method == "Blended"],
                       y=data$Improvement[data$Method == "Online"])
: Cliff.delta
:       0.329
rcompanion::cliffDelta(x=data$Improvement[data$Method == "Blended"],
                       y=data$Improvement[data$Method == "Traditional"])
: Cliff.delta
:       0.582
rcompanion::cliffDelta(x=data$Improvement[data$Method == "Online"],
                       y=data$Improvement[data$Method == "Traditional"])
: Cliff.delta
:        0.36
```

Conclusions

In this case study, we use the Kruskal-Wallis test to compare the median performance improvements across three training methods when the normality assumption required by ANOVA is violated. There is a strong evidence that the median improvements are different in different methods (p-value is 0.015 of the Kruskal-Wallis test). At the FWER of 5% of post hoc tests, the blended method has a significantly higher median improvement than the traditional method (p-value is 0.0113, Cliff's delta is 0.582), but does not have a significantly higher median improvement than the online method (p-value is 0.506, Cliff's delta is 0.33). Online and traditional methods do not have significantly different median improvements (p-value is 0.384, Cliff's delta is 0.36). Subjects are randomly selected and randomly assigned to one of the three methods, and therefore, the results can be generalized to all employees of the company, and causality can be inferred. That is, the different methods lead to the difference in performance improvement.

Note 6.30 (The Kruskal-Wallis Test). The *Kruskal-Wallis test* ([65]) is an extension of Wilcoxon's rank sum test / Mann-Whitney U test when there are more than two groups, and it is nonparametric alternative to one-way ANOVA when assumptions of parametric ANOVA are violated. It works with ordinal or continuous data, and tests whether multiple groups have statistically different distributions. The hypotheses are the following:

$$H_0 : \text{The populations have identical distributions;}$$
$$H_1 : \text{At least one population distribution differs from the others.}$$

If the shapes of the distributions across groups are similar, then the test can be used to compare group medians, and the hypotheses can be restated as

$$H_0 : \text{The population medians of all groups are equal;}$$
$$H_1 : \text{At least one group median is different from the others.}$$

The test statistic is

$$H = \frac{12}{N(N+1)} \sum_{i=1}^{k} \frac{R_i^2}{n_i} - 3(N+1),$$

where k = number of groups, n_i = sample size of group i, R_i = sum of ranks in group i, and N = total number of observations. It is equivalent to a more intuitive form of the test statistic as follows:

$$H = \frac{12}{N(N+1)} \sum_{i=1}^{k} n_i(\bar{R}_i - \bar{R})^2,$$

where $\bar{R}_i$ is the mean rank of the ith group, and $\bar{R} = (N+1)/2$ is the overall mean rank. Under the null hypothesis, for moderate to large sample sizes, the distribution of the test statistic H approximately follows a Chi-square distribution with $k-1$ degrees of freedom when there are no ties. When there are ties, assign average ranks to the tied values, and the test statistic is calculated using the average ranks. However, when there are ties, the presence of tied ranks reduces the variability of the ranks, and therefore, the distribution of the test statistic is affected. In this case, a correction factor is applied to adjust the test statistic; refer to [34] for details.

To compare groups medians, the shape of distribution across groups should be similar. Otherwise, the test may be more sensitive to differences in distributional shapes rather than differences in medians. Therefore, it is important to check the distributional shapes across groups before using the Kruskal-Wallis test to compare group medians. If Kruskal-Wallis test is significant, then use Dunn's test (see Note 6.31) or Mann-Whitney U tests with correction to help identify which specific groups differ.

Note 6.31 (Dunn's Test for Multiple Comparisons)**.** *Dunn's test* ([36]) is a nonparametric post hoc test following a significant Kruskal-Wallis test to identify specific group differences in multiple comparisons. To test group medians, we need the similar-shaped distributions across groups to hold under the null. The test statistic for comparing groups i and j is (no ties):

$$Q_{ij} = \frac{\bar{R}_i - \bar{R}_j}{\sqrt{\frac{N(N+1)}{12}\left(\frac{1}{n_i} + \frac{1}{n_j}\right)}},$$

where $\bar{R}_i$ and $\bar{R}_j$ are the average ranks for groups i and j, respectively, N is the total sample size, and n_i and n_j are the sample sizes of groups i and j, respectively. For large sample sizes, the distribution of Q_{ij} approximately follows a standard normal distribution under the null hypothesis when there are no ties. When there are ties, assign average ranks to the tied values, and the test statistic is calculated using the average ranks; a correction factor similar to the one used in the Kruskal-Wallis test is applied to adjust the test statistic; refer to [34] for details.

To control the family-wise error rate in multiple comparisons using Dunn's test, various methods can be used, such as the Bonferroni correction and Holm's method. The choice of method depends on the desired balance between Type I error control and statistical power. Note that, Dunn's test is different from pairwise Wilcoxon rank sum / Mann-Whitney U tests with corrections, as the former uses the ranks from all groups to calculate the test statistics, while the latter uses only the ranks from the two groups being compared. Therefore, Dunn's test is more appropriate for post hoc analysis following a significant Kruskal-Wallis test.

6.2.4 Permutation Test

When the assumptions for ANOVA or Kruskal-Wallis test are violated, we can use permutation tests to compare groups. The following case study illustrates how to use permutation tests to compare multiple groups when the normality assumption is violated and the group distributions have different shapes.

Case 18: Teaching Methods and Students Engagement Scores

A high school implemented three different teaching methods in their biology classes to improve student engagement. The study is carried out among 45 randomly selected students, who are then randomly divided into three groups of 15 each, each group receiving one of the following teaching methods: Traditional Lecture (Control), Project-Based Learning, and Flipped Classroom. The school measures student engagement through a standardized observation protocol that scores engagement on a scale from 0 to 100, as shown in Table 5.6. Engagement scores are collected over a six-week period, and trained observers rate student engagement using standardized criteria.

TABLE 5.6: Teaching Methods and Engagement Scores.

Method	Score	Method	Score	Method	Score
Traditional	40	Project_Based	60	Flipped	50
Traditional	41	Project_Based	61	Flipped	51
Traditional	42	Project_Based	62	Flipped	52
Traditional	42	Project_Based	63	Flipped	53
Traditional	43	Project_Based	64	Flipped	54
Traditional	44	Project_Based	90	Flipped	55
Traditional	45	Project_Based	91	Flipped	56
Traditional	46	Project_Based	92	Flipped	57
Traditional	47	Project_Based	93	Flipped	58
Traditional	48	Project_Based	94	Flipped	60
Traditional	50	Project_Based	95	Flipped	65
Traditional	55	Project_Based	96	Flipped	75
Traditional	70	Project_Based	97	Flipped	85
Traditional	85	Project_Based	98	Flipped	90
Traditional	95	Project_Based	99	Flipped	95

The following are the research questions:

- Is there a significant difference in student engagement levels across the three teaching methods?
- Which teaching methods, if any, show significantly different levels of engagement?

For answers to the initial questions and some initial thoughts about this case study, please refer to Section 5.19. For comparing group means, we might be able to use ANOVA. We first check the normality assumption. Both the QQ plots and the Shapiro-Wilk test suggest that the normality assumption does not hold. A logarithm transformation can also not fix the issue.

```
library(ggplot2)
data <- read.csv("data/teaching-methods-scores.csv")
data$Method <- as.factor(data$Method)

# QQ plot
ggplot(data, aes(sample = Score)) + geom_qq() + geom_qq_line() +
  facet_wrap(~Method) + theme_bw() + labs()
```

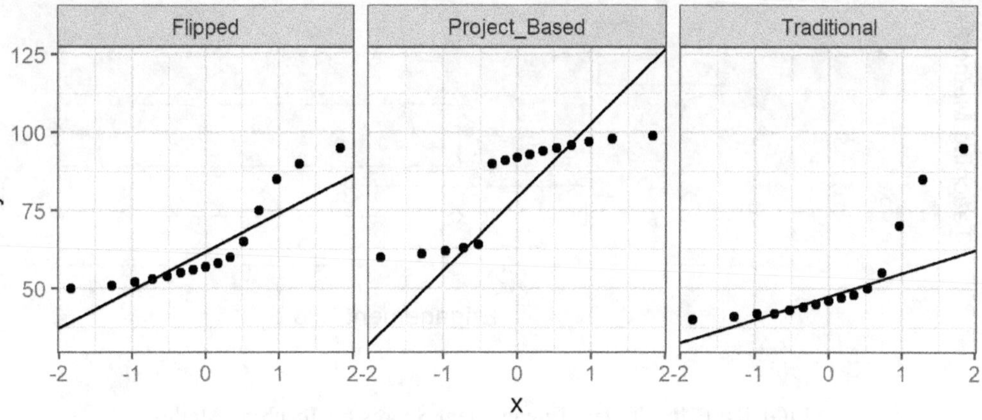

FIGURE 6.45: QQ Plots by Teaching Method.

```
# Shapiro-Wilk test for normality
data %>% group_by(Method) %>% summarise(
  shapiro_stat = shapiro.test(Score)$statistic,
  shapiro_p = shapiro.test(Score)$p.value)
: # A tibble: 3 x 3
:   Method        shapiro_stat shapiro_p
:   <fct>                <dbl>     <dbl>
: 1 Flipped              0.802  0.00385
: 2 Project_Based        0.745  0.000785
: 3 Traditional          0.721  0.000416
```

Therefore, we need to use some nonparametric alternatives to ANOVA. The Kruskal-Wallis test can be used to test whether there is a group effect when the normality assumption does not hold. However, for this case, we are interested in whether the scores are different (higher / lower) between groups, for which the distributions for the three groups should have similar shapes. Figure 6.46 suggests that the distributions of the three groups are very different, and thus a significant Kruskal-Wallis test result can only tell us that there is a significant difference between the groups, but it may not necessarily be due to location shifts.

```
# Create box plot
ggplot(data, aes(y = Method, x = Score)) + geom_boxplot() +
  stat_summary(fun = "mean", geom = "point", shape = 23, size = 2) +
  theme_bw() + theme(legend.position = "none") +
  labs(y = "Teaching Method", x = "Engagement Score")
```

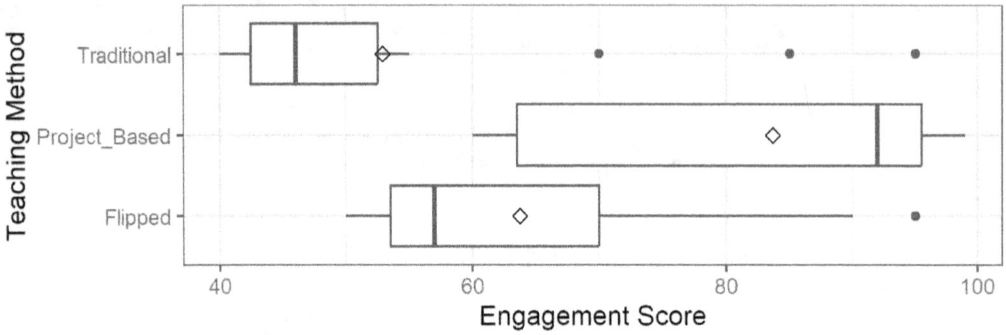

FIGURE 6.46: Student Engagement Scores by Teaching Method.

This is a difficult case with some commonly used tests with specific assumptions. Instead, we use a permutation test to replace the Kruskal-Wallis test as well as its post hoc Wilcoxon rank sum / Mann-Whitney U tests. Because the data have extreme values and very skewed distributions, it is more robust to compare the medians of the groups. To detect whether there is an overall group difference in medians, we use median gap, the difference of maximum median and minimum median as the statistic for an omnibus test.

Under the null assumption of no group effects, teaching methods can be permuted and assigned to any of the subjects. We permute the methods for many times, and for each permutation we calculate the median gap so that we have a pretty good estimation of the distribution of the median gap under the null. Then we compare the observed median gap to its estimated distribution under the null to get the p-value. The following codes are used to perform the permutation test, and the p-value is 0, well below common significance levels, suggesting that the observed median gap (46) is not by chance and is due to the group effect.

```
median_observed <- tapply(data$Score, data$Method, median)
median_gap_observed <- max(median_observed) - min(median_observed)
P <- 1000
median_gap_perm <- numeric(P)
set.seed(123)
data_perm <- data
for(i in 1:P){
  data_perm$Method <- sample(data$Method)
  median_perm <- tapply(data_perm$Score, data_perm$Method, median)
  median_gap_perm[i] <- max(median_perm) - min(median_perm)
}

p_val <- mean(median_gap_perm >= median_gap_observed)
df_plot <- data.frame(median_gap_perm = median_gap_perm)

ggplot(df_plot, aes(x = median_gap_perm)) +
  geom_histogram(binwidth = 1, fill = "grey80", color = "black") +
  geom_vline(xintercept = median_gap_observed, color = "black",
             linetype = "dashed", linewidth = 0.5) +
  labs(x = "Median Gaps for Permuted Samples", y = "Count") + theme_bw()
```

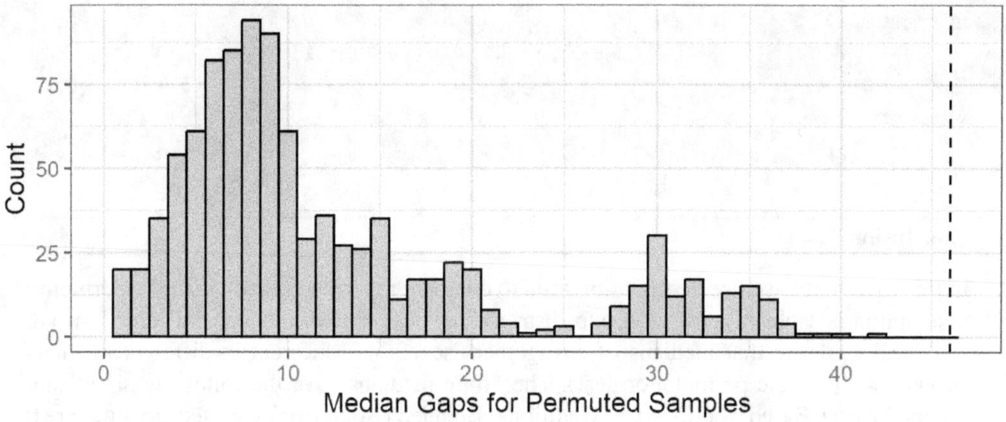

FIGURE 6.47: Histogram of Median Gaps for Permuted Samples (the Vertical Dashed Line is the Observed Median Gap).

For post hoc tests, we can still use permutation tests for each pair of groups and then adjust the p-values for multiple comparisons. We perform three one-sided tests as follows, and all adjusted p-values are well below common significance levels, suggesting the following significant orders in the group medians: Project_Based > Flipped > Traditional.

```
comps <- list(c("Project_Based", "Flipped"),
              c("Project_Based", "Traditional"),
              c("Flipped", "Traditional"))
results <- data.frame(group1 = character(), group2 = character(),
    median_diff = numeric(), p_value = numeric(), stringsAsFactors = FALSE)

set.seed(123)
for(pair in comps) {g1 <- pair[1]; g2 <- pair[2]
    data_pair <- subset(data, data$Method %in% pair)
    data_pair$Method <- factor(data_pair$Method, levels=pair)
    median_observed <- tapply(data_pair$Score, data_pair$Method, median)
    median_gap_observed <- median_observed[1] - median_observed[2]
    P <- 5000
    median_gap_perm <- numeric(P)
    set.seed(123)
    data_perm <- data_pair
    for(i in 1:P)
    {
      data_perm$Method <- sample(data_pair$Method)
      median_perm <- tapply(data_perm$Score, data_perm$Method, median)
      median_gap_perm[i] <- median_perm[1] - median_perm[2]
    }
    p_val <- mean(median_gap_perm >= median_gap_observed)
    results <- rbind(results, data.frame(group1 = g1, group2 = g2,
                             median_diff = median_gap_observed,
                             p_value = p_val, row.names = NULL))
  }
results$adj_p <- p.adjust(results$p_value, method = "bonferroni")
```

```
print(results)
:          group1       group2 median_diff p_value  adj_p
: 1 Project_Based      Flipped          35  0.0000 0.0000
: 2 Project_Based Traditional           46  0.0000 0.0000
: 3       Flipped Traditional           11  0.0034 0.0102
```

Conclusions

In this case study, we use permutation tests to compare multiple groups when the normality assumption is violated and the group distributions have different shapes. There is strong statistical evidence that medians of engagement scores are different in different methods (p-value is 0 for the permutation test). The group medians have the following significant orders: Project_Based > Flipped > Traditional Blended (Bonferroni adjusted p-values are 0 and 0.01, respectively). The subjects are randomly selected and randomly assigned to one of the three methods. Therefore, the result can be generalized to all the students in the biology classes of the high school. Causality can also be claimed, and the different teaching methods lead to the engagement difference.

Note 6.32 (Permutation Test). A *permutation test* is a nonparametric statistical procedure that evaluates the significance of an observed test statistic by comparing it to the distribution obtained from all (or many) possible rearrangements of the data labels. It does not rely on specific distributional assumptions, making it a flexible tool for hypothesis testing, especially when traditional parametric assumptions are violated. The key idea is to assess how likely the observed test statistic is under the null hypothesis by simulating the distribution of the test statistic through random permutations of the data. It is very flexible for choices of test statistics, and it can be used for various types of data and hypotheses, even with small sample sizes. However, it can be computationally intensive, especially for large datasets or complex test statistics, as the number of possible permutations grows factorially with the sample size. To use permutation tests, observations must be exchangeable under H_0 and independent within and between groups. We refer to [47] for more details about permutation tests.

6.2.5 Reliability and Validity of Survey

For surveys using Likert scales, one often needs to validate the intended structure of the survey and check the reliability of the feedback. The following case study illustrates how to use confirmatory factor analysis (CFA) to validate the intended structure of a survey and use Cronbach's Alpha to check the reliability of the feedback.

Case 19: Student Psychological Well-Being Survey

A researcher is developing a survey to measure psychological well-being among university students. The survey consists of two subscales:

- Emotional Stability (ES): Measured by 5 items on a 5-point Likert scale (1 = Strongly Disagree, 5 = Strongly Agree).
- Life Satisfaction (LS): Measured by 5 items on a 10-point scale (1 = Not At All Satisfied, 10 = Completely Satisfied).

The researcher wants to evaluate whether the results are reliable and whether the survey measures what it is intended to measure.

For answers to the initial questions and some initial thoughts about this case study, please refer to Section 5.20. We first validate the designed structure of the survey using confirmatory factor analysis (CFA) (see Note 6.34). The survey is designed to measure two latent constructs: Emotional Stability and Life Satisfaction, each assessed through five items on Likert scales.

We use the R package `lavaan` to perform a CFA, and use `semPlot` to obtain relevant graphs for the analysis.

```
library(lavaan)
library(semPlot)

data <- read.csv("data/student-psychological-well-being-survey.csv")
```

For CFA, the task is to confirm a certain construct, and thus we specify the structure as follows:

```
model <- '
  # Intrinsic Motivation factor
  Emotion =~ ES1 + ES2 + ES3 + ES4 + ES5

  # Extrinsic Motivation factor
  Life =~ LS1 + LS2 + LS3 + LS4 + LS5
'
```

If it is reasonable to treat Likert scales of 1 to 5 and 1 to 10 as continuous variables, then we can fit the following model directly.

```
fit_mle <- cfa(model, data = data, estimator = "ML", std.lv = TRUE)
```

Figure 6.48 shows the factor loadings of CFA, and higher loadings generally suggest that the item is a valid indicator of the factor, with loadings greater than 0.7 indicating an excellent indicator of the factor. The correlation between the two hidden variables is –0.14.

```
semPaths(fit_mle, what = "std",
         intercepts = FALSE, residuals = FALSE, edge.width = 0.3,
         edge.color = "black", node.width = 1.5, edge.label.cex = 0.9,
         nodeLabels = c("ES1","ES2","ES3","ES4","ES5",
                        "LS1","LS2","LS3","LS4","LS5","Emotion","Life"))
```

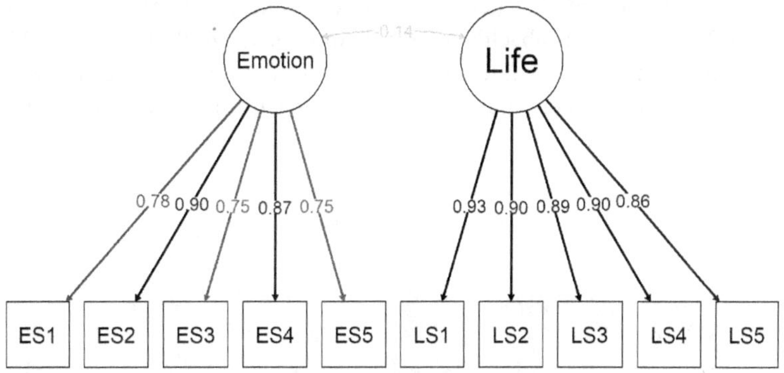

FIGURE 6.48: Factor Loading from Confirmatory Factor Analysis (ML Method).

Sometimes, it is not appropriate to treat Likert scales as continuous variables. In such cases, it is more appropriate to treat it as ordinal data. We can then use a weighted least squares mean and variance adjusted estimator (WLSMV) for CFA of ordinal data (see Note 6.35). In what follows, there is a warning from the `cfa()` function "The smallest eigenvalue (= -1.870091e-16) is smaller than zero", and the negative value is minimal and may be due to round errors, so we can proceed after checking for some obvious errors such as nonconvergence.

```
fit <- suppressWarnings(cfa(model, data = data, estimator = "WLSMV",
                            ordered = names(data), std.lv = TRUE))
```

The factor loadings are generally higher than those obtained from the model that treats the Likert scales as continuous variables. The vertical lines in each rectangle are the estimated thresholds that separate the categories for each observed ordinal variable. The correlation between the two hidden variables is -0.14.

```
semPaths(fit, what = "std",
         intercepts = FALSE, residuals = FALSE, edge.width =0.3,
         edge.color = "black", node.width = 1.5, edge.label.cex = 0.9,
         nodeLabels = c("ES1","ES2","ES3","ES4","ES5",
                        "LS1","LS2","LS3","LS4","LS5","Emotion","Life"))
```

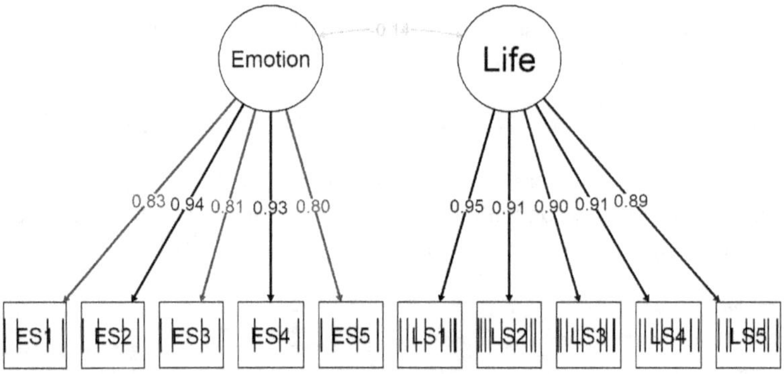

FIGURE 6.49: Factor Loading from Confirmatory Factor Analysis (WLSMV).

After we validate the intended structure, we need to check reliability of the feedback using Cronbach's α (see Note 6.36). Cronbach's α for the emotional and life factors are 0.904 and 0.954, respectively, indicating an excellent internal consistency between those feedbacks. However, the high value of Cronbach's α (> 0.95) may indicate that some questions for Life Satisfaction may be redundant.

```
# Cronbach's Alpha for each subscale
alpha_es <- psych::alpha(data[, c("ES1", "ES2", "ES3", "ES4", "ES5")])
cat("Cronbach's Alpha for Emotional Stability:", alpha_es$total$raw_alpha)
: Cronbach's Alpha for Emotional Stability: 0.9038782
alpha_ls <- psych::alpha(data[, c("LS1", "LS2", "LS3", "LS4", "LS5")])
cat("Cronbach's Alpha for Life Satisfaction:", alpha_ls$total$raw_alpha)
: Cronbach's Alpha for Life Satisfaction: 0.9537175
```

Conclusions

In this case study, we use confirmatory factor analysis to validate the intended structure of a survey using Likert scales and use Cronbach's Alpha to check the reliability of the feedback. A confirmatory factor analysis has validated the underlying structure designed to measure the intended concepts, with all factor loadings above 0.7. The two latent constructs, Emotional Stability and Life Satisfaction, show a slight negative correlation of −0.14. Cronbach's α suggests that the feedback is consistent and reliable to use for further analysis. However, the very high Cronbach's α for Life Satisfaction (0.954) indicates that some questions may be measuring the same aspects and thus could be redundant. Future survey designs may consider reducing the number of items for Life Satisfaction to enhance efficiency without compromising reliability.

Note 6.33 (Reliability and Validity of Survey Data). *Reliability* is the extent to which a measurement procedure yields the same results on repeated trials under the same conditions. In surveys, it refers to the consistency or stability of responses over time, across items, or between different observers. The most common measure of reliability for surveys is Cronbach's α (see Note 6.36), which assesses internal consistency. For more details on reliability for surveys, please refer to [84]. *Validity* is the extent to which a measurement procedure actually measures what it is intended to measure. Confirmatory Factor Analysis (see Note 6.34) is a common method to assess the construct validity of a survey, which evaluates whether the data fit the expected factor structure based on theoretical constructs. For more details on validity for surveys, please refer to [84].

Note 6.34 (Factor Analysis). *Factor analysis* is a statistical method used to describe variability between observed correlated variables in terms of potentially lower-dimensional unobserved variables called factors. It is primarily used to identify underlying relationships between measured variables so that a more interpretable structure can be derived. In the context of surveys, factor analysis can help identify latent constructs that explain the patterns of correlations among survey items. Depending on whether the underlying factor structure is known or hypothesized, there are mainly two types of factor analysis: exploratory factor analysis (EFA) and confirmatory factor analysis (CFA). EFA is used when the underlying factor structure is unknown, while CFA is used to test a hypothesized factor structure based on prior knowledge or theory; the latter is more common in survey validation.

The main idea of factor analysis is to model the observed variables as linear combinations of a smaller number of latent factors plus error terms. The mathematical representation of a factor analysis model can be expressed as:

$$\mathbf{Y} = \mathbf{L}\mathbf{F} + \varepsilon,$$

where $\mathbf{Y}$ is a random vector representing observed variables, $\mathbf{L}$ is a matrix of factor loadings, $\mathbf{F}$ is a vector of common factors that are usually assumed to be uncorrelated and have a mean of zero, and ε is a random vector for error terms. The factor loadings in $\mathbf{L}$ indicate the strength and direction of the relationship between each observed variable and the underlying factors. Higher absolute values of factor loadings suggest that the observed variable is a stronger indicator of the corresponding factor. In CFA, we specify a model based on theoretical expectations and test how well the data fit this model, while in EFA, we let the data determine the factor structure without imposing a predefined model.

Because factor analysis is based on the correlation structure of the data, it is important to ensure that the observed variables are measured on a similar scale. If the variables are measured on different scales, it can lead to distorted factor loadings and misinterpretation of the underlying factor structure. Therefore, in practice, we need to standardize the data if the variables are measured on different scales, although some software packages handle this automatically. For more details on factor analysis, please refer to [109] and [18].

Note 6.35 (WLSMV Estimator in Confirmatory Factor Analysis). In confirmatory factor analysis (CFA), when observed variables are ordinal or categorical, standard maximum likelihood estimation assuming continuous and normal variables may be inappropriate. The *weighted least squares mean and variance adjusted (WLSMV)* estimator implemented in the R package `lavaan` ([92]) is designed for CFA with ordinal data. The main idea behind WLSMV is to model the observed ordinal variables as arising from underlying continuous latent variables. Specifically, each observed ordinal variable Y_i is assumed to arise from an unobserved continuous latent response Y_i^*, and its categorical value is determined by thresholds $\tau_{i,k}$ as follows:

$$Y_i = k \quad \text{if} \quad \tau_{i,k-1} < Y_i^* \leq \tau_{i,k}.$$

Then a diagonally weighted least squares (DWLS) method with robust standard errors is used to estimate the parameters of the CFA model, and thresholds for each ordinal variable are estimated as part of the model. The WLSMV estimator is particularly useful when the observed variables are measured on Likert scales, which are common in survey research. For more details on the WLSMV estimator, please refer to [81] and [92].

Note 6.36 (Cronbach's α). *Cronbach's* α ([32]) is a measure of internal consistency or reliability of a set of items in a survey. It is commonly used to assess how well a set of items measures a single unidimensional latent construct. The formula for Cronbach's α is given by

$$\alpha = \frac{k}{k-1}\left(1 - \left(\sum_{i=1}^{k} \sigma_i^2\right) / \sigma^2\right).$$

Here, k is the number of elements in the measure, σ_i^2 is the variance of element i, and σ^2 is the variance of the total score from all elements combined. For example, if we have a survey with 5 items measuring emotional stability, we can calculate Cronbach's α to assess the internal consistency of those items. A high value of α indicates that the items are highly correlated and thus likely measure the same underlying construct.

The values of Cronbach's α range from 0 to 1, where $\alpha \geq 0.9$ is considered excellent, $0.8 \leq \alpha < 0.9$ good, $0.7 \leq \alpha < 0.8$ acceptable, and $\alpha < 0.7$ questionable or poor. To use Cronbach's α appropriately, the following assumptions must hold: all items measure the same underlying latent factor, true score variances are equal, errors between items are uncorrelated, and items should be measured on a continuous scale (although it is often applied to Likert scales). When these assumptions are violated, Cronbach's α may not provide an accurate estimate of reliability. These assumptions essentially suggest that the items have equal factor loadings (tau-equivalence) from the common factor, and an alternative measure of reliability that does not require tau-equivalence is McDonald's ω, which is based on the common factor model and uses the factor loadings to estimate reliability; for more details, see [79].

6.2.6 Nonparametric Repeated Measures ANOVA

When the same subjects are measured under different conditions and the normality assumption in repeated measures ANOVA (Section 6.1.9) is violated, we can use nonparametric alternatives to analyze the data. The following case study illustrates how to use the Friedman rank sum test to compare groups of repeated measurements when the response variable is measured on a Likert scale.

Case 20: Evaluating User Satisfaction with Three Software Interfaces

A tech company wants to evaluate user satisfaction with three different software interfaces (Interface A, Interface B, and Interface C) designed for a new application. They recruit 20 participants and ask each participant to use all three interfaces in randomized order. After using each interface, participants rate their satisfaction on a 5-point Likert scale (1 = Very Dissatisfied, 5 = Very Satisfied). The company wants to determine whether there are significant differences in satisfaction between the three interfaces and which interface has the highest level of satisfaction.

For answers to the initial questions and some initial thoughts about this case study, please refer to Section 5.21. In this section, we treat the Likert scale response variable as continuous data. For more discussion about Likert scale data, please refer to Note 3.1.

The R function `likert::likert()` can provide intuitive graphs for Likert scale data.

```
# Evaluating User Satisfaction with Three Software Interfaces
library(tidyr)
library(ggplot2)
data <- read.csv("data/software-userinterface-survey.csv")

# Convert to likert format
likert_data <- likert::likert(
  as.data.frame(lapply(data[,names(data)!="Participant"],factor,levels=1:5,
                     labels=c("Very Dissatisfied", "Dissatisfied",
                              "Neutral", "Satisfied", "Very Satisfied"))))
# Create visualization
suppressMessages(plot(likert_data)+scale_fill_grey(start = 0.3, end = 0.8))
```

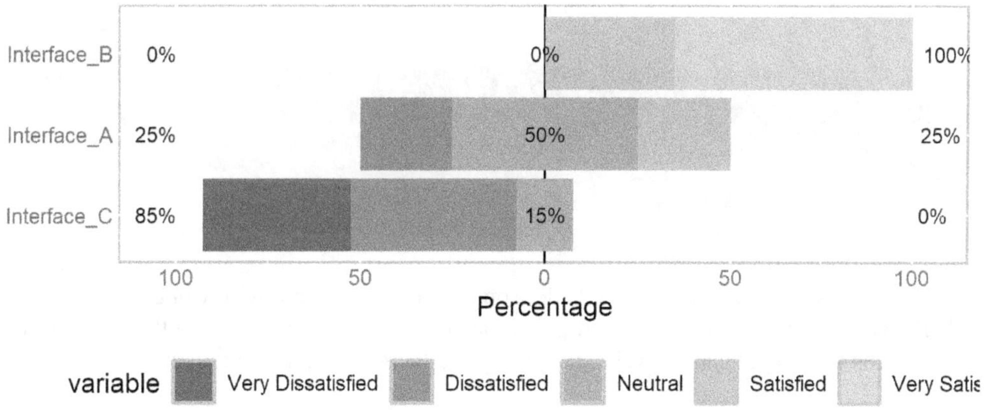

FIGURE 6.50: Satisfaction Scores Across Interfaces.

The original data are in wide format, and we convert it to long format and then visualize it with box plots.

```
# Reshape data into long format
data_long <- data %>%
  pivot_longer(cols = starts_with("Interface"), names_to = "Interface",
```

```
                values_to = "Satisfaction")

# Visualize the data with a box plot
ggplot(data_long, aes(y = Interface, x = Satisfaction)) + theme_bw() +
  geom_boxplot() +
  stat_summary(fun = "mean", geom = "point", shape = 23, size = 2) +
  labs(y = "Interface", x = "Satisfaction (1-5)")
```

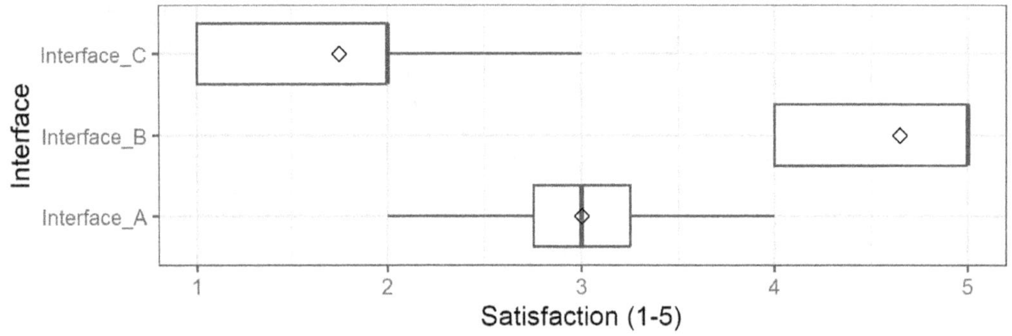

FIGURE 6.51: Satisfaction Scores Across Interfaces.

Clearly, the normality assumption of parametric repeated measures ANOVA does not hold. A nonparametric alternative to repeated measure ANOVA can be considered. We use the Friedman rank sum test, and refer to Note 6.37 for more details about the test.

The following Friedman rank sum test leads to a p-value that is much less than the common significance levels, and it is strong evidence that there is a difference between the interfaces.

```
# Perform Friedman Test
friedman.test(Satisfaction ~ Interface | Participant, data = data_long)
:
:  Friedman rank sum test
:
: data:  Satisfaction and Interface and Participant
: Friedman chi-squared = 37.333, df = 2, p-value = 7.819e-09
```

For post hoc tests, we can use the Wilcoxon signed-rank test pairwise while controlling for the FWER. We first reorder the factor so that we can perform one-sided tests. All the p-values are less than the common significance levels, suggesting that there is a significant order: B > A > C.

```
# perform post hoc Wilcoxon signed-rank tests with Bonferroni correction
# exact = FALSE as there are ties
data_long$Interface <- factor(data_long$Interface,
            levels = c("Interface_B", "Interface_A", "Interface_C"))

test_posthoc <- pairwise.wilcox.test(data_long$Satisfaction,
  data_long$Interface, paired = TRUE, alternative = "less",
  exact = FALSE, p.adjust.method = "bonferroni")
print(test_posthoc$p.value)
:               Interface_B  Interface_A
```

```
: Interface_A 1.008644e-04            NA
: Interface_C 5.998811e-05 0.0005633523
```

Conclusions

In this case study, we use the Friedman rank sum test to compare groups of repeated measurements when the response variable is measured on a Likert scale and the normality assumption of parametric repeated measures ANOVA is violated. The Friedman rank sum test indicates that there are significant differences in user satisfaction across the three software interfaces (p-value is 7.82×10^{-9}). Post hoc pairwise Wilcoxon signed-rank tests reveal that Interface B has significantly higher satisfaction scores than Interface A (Bonferroni adjusted p-value is 1.01×10^{-4}), which in turn has significantly higher scores than Interface C (Bonferroni adjusted p-value < 0.001). Therefore, the order of user satisfaction is Interface B $>$ Interface A $>$ Interface C. Whether the subjects are randomly selected needs to be further clarifies to determine the generalizability of the results to a broader population of users. Also, this is an observational study, so no causality can be inferred from the results.

Note 6.37 (The Friedman Rank Sum Test). The *Friedman rank sum test* ([45]) is a nonparametric statistical test to detect differences in treatments when the same subjects are measured for each treatment. It can be viewed as an extension of the Wilcoxon signed-rank test (see Note 6.29) to more than two groups, or as a nonparametric alternative to repeated measures ANOVA (see Note 6.20). The test is based on ranking the data within each subject and then analyzing the ranks across treatments.

Suppose that we have n subjects and k treatments. For each subject j ($j = 1, \ldots, n$), the k observations x_{ij} ($i = 1, \ldots, k$) are ranked from 1 to k, giving ranks R_{ij}'s for subject j. The sum of ranks for treatment i is $R_{i.} = \sum_{j=1}^{n} R_{ij}$, and the average rank for treatment i is $\bar{R}_{i.} = R_{i.}/n$. The Friedman test statistic is

$$Q = \frac{12}{nk(k+1)} \sum_{i=1}^{k} R_{i.}^2 - 3n(k+1),$$

and alternatively, a more intuitive form of the test statistic is

$$Q = \frac{12n}{k(k+1)} \sum_{i=1}^{k} (\bar{R}_{i.} - \frac{k+1}{2})^2.$$

Under the null hypothesis H_0: all treatments have the same distribution, and for n and k large enough, Q approximately follows a Chi-square distribution with $k-1$ degrees of freedom.

For example, consider $n = 4$ subjects ($j = 1, \ldots, 4$) and $k = 3$ treatments (A, B, C). The observed data are:

Subject	Treatment A	Treatment B	Treatment C
1	8	7	9
2	6	9	5
3	7	6	8
4	5	8	7

Then rank within each row to get

Subject	Treatment A	Treatment B	Treatment C
1	2	1	3
2	2	3	1
3	2	1	3
4	1	3	2

The sum of ranks for each treatment is $R_{A.} = 7$, $R_{B.} = 8$, $R_{C.} = 9$. Then $Q = \frac{12}{4\cdot3\cdot4}(7^2 + 8^2 + 9^2) - 3\cdot4\cdot4 = 0.5$ with $k-1 = 2$ degrees of freedom, and the p-value is 0.779 as calculated by the following R code, so we fail to reject it at common significance levels, suggesting that there is no significant difference among the three treatments.

```
pchisq(.5, df=2, lower.tail = FALSE)
: [1] 0.7788008
```

For post hoc tests after a significant Friedman rank sum test, we can use the Wilcoxon signed-rank test pairwise while controlling for the family-wise error rate (FWER). For example, to compare Treatment A and Treatment B, we can perform a Wilcoxon signed-rank test on the paired differences between the two treatments. We can adjust the p-values using methods such as Bonferroni correction to account for multiple comparisons. For more details about the Friedman rank sum test, please refer to [53].

6.3 Meta-Analysis

When multiple studies have investigated the same research question, we can use meta-analysis to combine the evidence across studies to obtain a more precise estimate of the effect size. We use the following case study to illustrate how to perform a meta-analysis for a continuous response variable.

Case 21: Cognitive Training on Test Scores

Table 5.7 summarizes three existing studies that investigated the effects of short cognitive training programs on standardized test scores. Each study reports the sample sizes, means, and standard deviations for both the treatment (cognitive training) and control (no training) groups, along with the mean age of participants and a quality rating (high/low) for each study.

TABLE 5.7: Existing Studies of Cognitive Training on Test Scores.

Study	Year	n_t	$\overline{X}_t$	SD_t	n_c	$\overline{X}_c$	SD_c	Mean Age
Wong_2017	2017	33	84.5	7.5	32	80.3	8.2	27.6
Patel_2018	2018	38	79.0	9.8	36	74.2	10.1	23.8
Lopez_2019	2019	60	81.4	8.9	58	78.0	9.2	33.9

Research question: Do short cognitive training programs improve standardized test scores compared to no training (control)?

For answers to initial questions and initial thoughts on this case study, see Section 5.22. We first load the R package `metafor` that will be used for meta-analysis, and create a data frame containing the summary statistics from the three studies according to Table 5.7.

```
library(metafor)

# Create the summary data frame
dat <- read.csv(text = "study,year,nt,mt,sdt,nc,mc,sdc,mean_age
Wong_2017,2017,33,84.5,7.5,32,80.3,8.2,27.6
Patel_2018,2018,38,79.0,9.8,36,74.2,10.1,23.8
Lopez_2019,2019,60,81.4,8.9,58,78.0,9.2,33.9", stringsAsFactors = FALSE)
dat
:        study year nt   mt sdt nc   mc  sdc mean_age
: 1  Wong_2017 2017 33 84.5 7.5 32 80.3  8.2     27.6
: 2 Patel_2018 2018 38 79.0 9.8 36 74.2 10.1     23.8
: 3 Lopez_2019 2019 60 81.4 8.9 58 78.0  9.2     33.9
```

We then compute the effect sizes (Hedges' g) and their sampling variances from the group means, standard deviations, and sample sizes. Hedges' g is a bias-corrected standardized mean difference commonly used in meta-analysis of continuous outcomes. For details on Hedges' g, see Note 6.39.

```
# Compute effect sizes (Hedges' g) from group means, SDs, Ns
# measure = "SMD" returns Hedges' g
es <- escalc(measure = "SMD",
             m1i = mt, sd1i = sdt, n1i = nt,    # treatment
             m2i = mc, sd2i = sdc, n2i = nc,    # control
             data = dat, var.names = c("yi", "vi"))
# yi = effect size estimate (Hedges' g), vi = sampling variance
print(es)
:
:        study year nt   mt sdt nc   mc  sdc mean_age     yi     vi
: 1  Wong_2017 2017 33 84.5 7.5 32 80.3  8.2     27.6 0.5285 0.0637
: 2 Patel_2018 2018 38 79.0 9.8 36 74.2 10.1     23.8 0.4775 0.0556
: 3 Lopez_2019 2019 60 81.4 8.9 58 78.0  9.2     33.9 0.3733 0.0345
```

Since there are only three studies, we prefer the fixed-effect model unless there is strong evidence of heterogeneity. We fit a fixed-effect meta-regression model using the `rma()` function from the `metafor` package, with mean age as a covariates / moderator. For details on meta-analysis methods, see Note 6.38. The following results suggest that the mean age is not significantly associated with the effect size (p-value: 0.666).

```
# fix-effects meta-regression
res_re <- rma(yi, vi, mods = ~ mean_age, data = es, method = "FE")
print(res_re)
:
: Fixed-Effects with Moderators Model (k = 3)
:
: I^2 (residual heterogeneity / unaccounted variability): 0.00%
: H^2 (unaccounted variability / sampling variability):   0.09
: R^2 (amount of heterogeneity accounted for):            34.90%
:
: Test for Residual Heterogeneity:
: QE(df = 1) = 0.0901, p-val = 0.7640
:
```

```
: Test of Moderators (coefficient 2):
: QM(df = 1) = 0.1867, p-val = 0.6657
:
: Model Results:
:
:             estimate      se     zval    pval    ci.lb   ci.ub
: intrcpt       0.8089  0.8582   0.9425  0.3459  -0.8732  2.4910
: mean_age     -0.0125  0.0288  -0.4321  0.6657  -0.0690  0.0441
:
: ---
: Signif. codes:  0 '***' 0.001 '**' 0.01 '*' 0.05 '.' 0.1 ' ' 1
```

Therefore, we fit a fixed-effect meta-analysis model without covariates. The `rma()` function with `method="FE"` fits the fixed-effect model, and the same function with different `method` options can be used to fit various random-effects models as well.

Note that although the heterogeneity test is not significant (p-value: 0.871, see below), this is not the main reason for using the fixed-effect model over the random-effect model. The choice of using whether fixed-effect model or random-effects model for meta-analysis is driven by many factors such as the number of studies, study designs, and research objectives. Using the heterogeneity test alone to determine which model to use is not recommended; we refer to [13] for further discussion.

```
# fix-effects meta-regression
res_re <- rma(yi, vi, data = es, method = "FE")
print(res_re)
:
: Fixed-Effects Model (k = 3)
:
: I^2 (total heterogeneity / total variability):   0.00%
: H^2 (total variability / sampling variability):  0.14
:
: Test for Heterogeneity:
: Q(df = 2) = 0.2768, p-val = 0.8707
:
: Model Results:
:
: estimate      se    zval    pval   ci.lb   ci.ub
:   0.4421  0.1263  3.4994  0.0005  0.1945  0.6897  ***
:
: ---
: Signif. codes:  0 '***' 0.001 '**' 0.01 '*' 0.05 '.' 0.1 ' ' 1
```

The above results are consistent with manually calculating the pooled effect size and its standard error using inverse-variance weighting as follows:

```
# manual calculation
w <- 1/es$vi
fixed_mu <- sum(w * es$yi) / sum(w)
fixed_se <- sqrt(1 / sum(w))
c(mu = fixed_mu, se = fixed_se)
:        mu        se
```

```
: 0.4420774 0.1263300
```

We can visualize the meta-analysis results using a forest plot. The following code produces a basic forest plot showing the effect size estimates and their confidence intervals for each study, as well as the overall pooled estimate from the fixed-effect model.

```
# Forest plot (basic)
forest(res_re, slab = es$study, xlab = "Hedges' g", alim=c(-1,2))
```

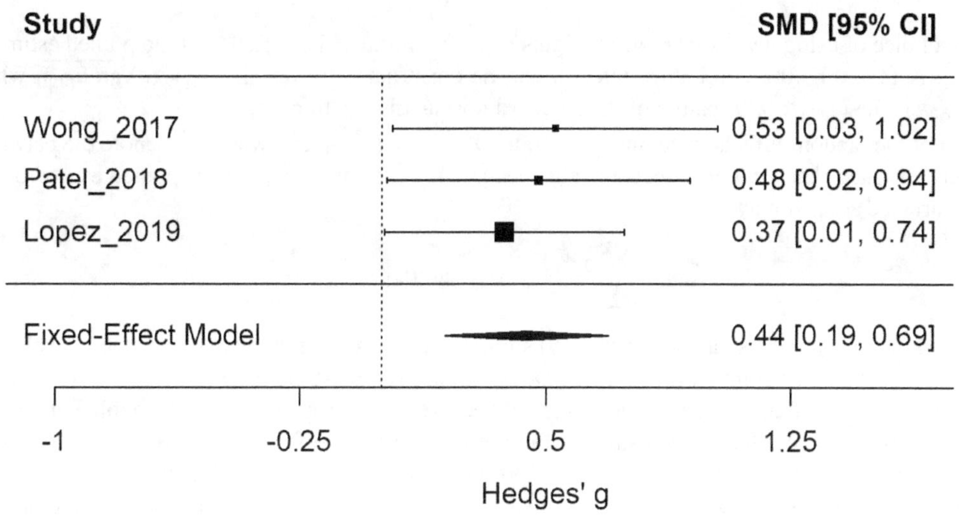

Conclusions

In this case study, we perform a meta-analysis of three studies evaluating the effects of short cognitive training programs on standardized test scores. We compute Hedges' g as the effect size measure and fit a fixed-effect meta-analysis model due to the small number of studies and lack of significant heterogeneity. The results indicate a significant positive effect of cognitive training on test scores, with a pooled Hedges' g of approximately 0.44 (95% CI: 0.19 to 0.69). The mean age of participants is not significantly associated with the effect size. A forest plot is generated to visualize the individual study effects and the overall pooled estimate. Whether the subjects are randomly sampled from each study needs to be further clarified to determine the generalizability of the results to a broader population. Moreover, whether the subjects in each study are randomly assigned to treatment and control groups also needs to be clarified to determine the causal interpretation of the results.

Note 6.38 (Meta-Analysis). *Meta-analysis* is a statistical technique for combining evidence across multiple studies, and it essentially involves calculating a weighted average of effect estimates from individual studies. Depending on the assumptions about between-study heterogeneity, two main models are commonly used: the *fixed-effect* model and the *random-effects* model. If all studies are assumed to estimate the same underlying effect, the fixed-effect model is appropriate. If there is variability in true effects across studies due to factors such as different populations, the random-effects model is more suitable and the summarized effect represents the average effect across studies. In what follows, we outline the key formulations for both models. For details on meta-analysis methods, see [13].

For the fixed-effect meta-analysis model, suppose study i reports an estimate of effect Y_i with known sampling variance $v_i = \text{Var}(Y_i)$ for $i = 1, \ldots, k$. Under the fixed-effect model, all studies are assumed to be independent and estimate a common underlying effect μ, and the pooled estimator is obtained by inverse-variance weights $w_i = 1/v_i$:

$$\hat{\mu}_{\text{FE}} = \frac{\sum_{i=1}^{k} w_i y_i}{\sum_{i=1}^{k} w_i},$$

with estimated variance

$$\text{Var}(\hat{\mu}_{\text{FE}}) = \frac{1}{\sum_{i=1}^{k} w_i}.$$

The choice of using inverse-variance weights w_i is to minimize the variance of the pooled estimator [25]. A $(1-\alpha)100\%$ confidence interval can be constructed by $\hat{\mu}_{\text{FE}} \pm z_{1-\alpha/2}\sqrt{\text{Var}(\hat{\mu}_{\text{FE}})}$, where $z_{1-\alpha/2}$ is the $(1-\alpha/2)$ quantile of the standard normal distribution.

For the random-effects meta-analysis model, $Y_i \sim \mathcal{N}(\mu,\, v_i + \tau^2)$, where τ^2 denote the between-study variance. The random-effects weights become $w_i^* = 1/(v_i + \tau^2)$, and the pooled estimator and its variance are given by

$$\hat{\mu}_{\text{RE}} = \frac{\sum_{i=1}^{k} w_i^* y_i}{\sum_{i=1}^{k} w_i^*}, \qquad \text{Var}(\hat{\mu}_{\text{RE}}) = \frac{1}{\sum_{i=1}^{k} w_i^*},$$

and confidence intervals can be constructed similarly. Since τ^2 is unknown, it needs to be estimated from the data, and a simple method is the DerSimonian-Laird estimator [33].

The random-effects model accounts for between-study heterogeneity, and the pooled effect represents the average effect across studies. In contrast, the fixed-effect model assumes no between-study heterogeneity and the pooled effect represents the common underlying effect. The choice of using fixed-effect or random-effects model should be based on various factors such as the number of studies, study designs, and research objectives, rather than solely relying on the heterogeneity test. We refer to [13] for further discussion on this topic.

Note 6.39 (Hedges' g). *Hedges' g* is an effect size measure used to quantify the standardized mean difference between two groups, while correcting for small-sample bias present in Cohen's d. For two independent groups with sample sizes n_1 and n_2, sample means $\bar{X}_1$ and $\bar{X}_2$, and sample standard deviations s_1 and s_2, the pooled standard deviation is defined as

$$s_p = \sqrt{\frac{(n_1-1)s_1^2 + (n_2-1)s_2^2}{n_1+n_2-2}}.$$

Cohen's d is then defined as

$$d = \frac{\bar{X}_1 - \bar{X}_2}{s_p}.$$

Hedges' g applies a bias-correction factor J to d: $g = Jd$ [50]. In practices, J is approximately calculated as

$$J = 1 - \frac{3}{4(n_1+n_2-2)-1}.$$

The correction factor J approaches 1 as sample sizes increase, so g and d become nearly identical in large samples. Hedges' g is commonly used in meta-analysis because it provides a less biased estimate of the population standardized mean difference, particularly when the total sample size is small.

7

Formal Analysis – Discrete Data

This chapter covers detailed statistical analyses of case studies that have discrete response variables. For each case study, we discuss the background and the data, then present the statistical analysis, and finally conclude with a summary of the findings. When appropriate, discussions on relevant methodologies are included, and references are provided for further reading.

7.1 Binary Data

When the response variable is binary, we can use methods such as logistic regression and methods designed for contingency tables to analyze the data. In this section, we present various case studies that have a binary response variable.

7.1.1 Chi-Square Test of Independence

When both the response variable and the predictor variable are categorical, we can use the Chi-square test of independence to assess whether there is an association between the two variables. The following case study illustrates how to use the Chi-square test of independence. Although the case study involves a 2×2 contingency table, the Chi-square test of independence can be applied to contingency tables with more than two rows and/or columns as well.

Case 22: Sport Brand Preference by Gender

Two sports brands, Brand A and Brand B, compete in the market. To understand the brand preference by gender, a survey is conducted. Each consumer can choose only one of the two brands, and Table 5.8 shows the numbers of male and female consumers who prefer each brand. The total number of consumers surveyed is 750.

TABLE 5.8: Sport Brand Preference by Gender.

Gender	Brand A	Brand B
Male	200	150
Female	180	220

Determine if there is a significant association between gender and brand preference.

DOI: 10.1201/9781003683827-7

For answers to the initial questions and some initial thoughts about this case study, please refer to Section 5.23. From the data summarized in the table, it seems that males prefer Brand A while females prefer Brand B. We need to conduct a statistical test to assess whether this is due to randomness or there is a real association between gender and the two sports brands.

In total, there are 750 consumers and $7/15$ are males. If males do not have preference between brand A and brand B, then we expect the number of males who choose brand A to be $(200+180) \times (7/15) \approx 177$. Similarly, we can calculate the expected numbers for each cell of the table as follows:

TABLE 7.1: Expected Number of Consumers Among Two Sports Brands if There is No Preference by Gender.

Gender	Brand A	Brand B
Male	177	173
Female	203	197

To assess whether there is brand preference by gender, we need to compare how much the observed numbers in Table 5.8 differ from those of Table 7.1. A statistical test can be developed based on the idea, and the Chi-square test of independence (see Note 7.1) is commonly used for this purpose.

The p-value is less than 0.001 and the common significance levels. Therefore, there is strong evidence to reject the null hypothesis that no association between gender and preference among the two sports brands.

```
data <- data.frame("Brand_A" = c(200, 180), "Brand_B" = c(150, 220),
                   row.names = c("Male", "Female"))
data <- as.matrix(data)
data
:          Brand_A Brand_B
: Male         200     150
: Female       180     220
chisq.test(data, correct = FALSE)
:
:   Pearson's Chi-squared test
:
: data:  data
: X-squared = 11.011, df = 1, p-value = 0.0009055
```

Effect size can be measured by Cramér's V (see Note 7.4). The calculation of Cramér's V in R can be done using the `lsr` packages:

```
# Or using lsr package
library(lsr)
cramersV(data)
: [1] 0.1184964
```

Following the common interpretation guidelines in EXHIBIT 10.2 of [88], $V = 0.118$ indicates a weak association between gender and the sports brands.

Conclusions

In this case study, we use the Chi-square test of independence to assess whether there is an association between gender and preference for two sports brands. The results indicate that there is a significant association between gender and the two sports brands (p-value is 0.001). The Cramér's V value quantifies the strength of the association, and $V = 0.118$ indicates a weak association, following the common interpretation guidelines in EXHIBIT 10.2 of [88]. Whether the subjects are randomly selected is not clear, and thus the generalizability of the result is uncertain. Also, no causality can be established from this observational study.

Note 7.1 (Chi-Square Test of Independence). The *Chi-square test of independence* is a nonparametric statistical test used to determine whether two categorical variables are independent of each other. Suppose that we have two categorical variables with r rows and c columns forming a contingency table. Let O_{ij} denote the observed frequency in the ith row and jth column, and let E_{ij} denote the expected frequency under the null hypothesis of independence. The expected frequency is given by

$$E_{ij} = \frac{(\text{total of row } i)(\text{total of column } j)}{\text{grand total}}.$$

The Chi-square test statistic is then computed as

$$\chi^2 = \sum_{i=1}^{r}\sum_{j=1}^{c}\frac{(O_{ij}-E_{ij})^2}{E_{ij}}.$$

Under the null hypothesis of independence, this statistic approximately follows a Chi-square distribution with $(r-1)(c-1)$ degrees of freedom. A large value of χ^2 indicates stronger evidence against the null hypothesis, suggesting a potential association between the two categorical variables. In order to use the Chi-square test of independence, all observations should be independent, and the expected frequencies should be large enough, say, ≥ 5. If the expected frequencies are too small, the Chi-square approximation may not be accurate, and alternative methods such as Fisher's exact test may be more appropriate; see Note 7.5 for more details on Fisher's exact test. For more details on Chi-square tests, please refer to [6].

Note 7.2 (Chi-Square Goodness-of-Fit Test). The *Chi-square goodness-of-fit test* is used to assess whether the distribution of a categorical variable in a sample is consistent with a hypothesized theoretical distribution. Suppose that we have a categorical variable with k possible categories. Let O_i denote the observed frequency for category i, and let E_i denote the expected frequency under the null hypothesis that the true distribution of the categorical variable matches the hypothesized distribution. Then the expected frequency is given by

$$E_i = np_i,$$

where n is the sample size and p_i is the hypothesized percentage of category i. The Chi-square test statistic is then computed as

$$\chi^2 = \sum_{i=1}^{k}\frac{(O_i-E_i)^2}{E_i}.$$

Under the null hypothesis, this statistic approximately follows a Chi-square distribution with $k-1$ degrees of freedom. A large value of χ^2 provides stronger evidence that the observed distribution deviates from the hypothesized distribution. For more details, please refer to [5] and [27].

The following is an example for testing whether a die is fair, where $E_i = 100 \times \frac{1}{6} = 16.67$ for each face.

Face	1	2	3	4	5	6	Total
Observed	14	17	15	18	21	15	100
Expected	16.67	16.67	16.67	16.67	16.67	16.67	100

So $\chi^2 = 2$, with degree of freedom being $6-1=5$ and thus the p-value is 0.849, well above common significance levels. Therefore, we fail to reject the null hypothesis that the die is a fair one. In R, this is done as follows:

```
chisq.test(x=c(14,17,15,18,21,15), p=rep(1/6, 6))
:
: Chi-squared test for given probabilities
:
: data:  c(14, 17, 15, 18, 21, 15)
: X-squared = 2, df = 5, p-value = 0.8491
```

Note 7.3 (Chi-Square Test of Independence vs. Chi-Square Goodness-of-Fit Test)**.** Both the Chi-square test of independence and the Chi-square goodness-of-fit test are nonparametric methods based on the same general test statistic,

$$\chi^2 = \sum \frac{(O-E)^2}{E},$$

but they are applied in different contexts. The *goodness-of-fit test* is used when there is a single categorical variable and the goal is to assess whether the observed distribution of frequencies across the k categories matches a hypothesized probability distribution. In this case, the expected frequencies are calculated as $E_i = np_i$, and the degrees of freedom are $k-1$. In contrast, *test of independence* is applied when two categorical variables are cross-classified into a contingency table with r rows and c columns. The goal is to assess whether the variables are independent, with expected counts given by $E_{ij} = (\text{total of row } i) \times (\text{total of column } j)/\text{grand total}$. The degrees of freedom in this case are $(r-1)(c-1)$. Thus, while both tests rely on comparing observed and expected frequencies, the goodness-of-fit test checks agreement with a theoretical distribution, whereas the test of independence checks association between two categorical variables.

In Table 7.1, the expected numbers of observations are calculated on the basis of the sample. If the expected numbers of observations were given, one can perform a Chi-square goodness-of-fit test for a categorical random variable, using the same formula except that the E_{ij}'s are those given numbers, and the two categorical random variables can be looked at as a univariate categorical random variable that has the following four categories: “Male-Brand A”, “Male-Brand B”, “Female-Brand A”, and “Female-Brand B”.

Note 7.4 (Cramér's V)**.** *Cramér's V* (p.282 of [31]) is a measure of the association between two categorical variables. It is defined as

$$V = \sqrt{\frac{\chi^2}{n \cdot \min(r-1, c-1)}},$$

where χ^2 is the Pearson Chi-square statistic, n is the total sample size, r is the number of rows and c is the number of columns in the contingency table. The value of V lies between 0 (no association) and 1 (perfect association), with higher values indicating a stronger association between the two variables. Unlike the Chi-square statistic, V is not affected by sample size, making it a normalized measure of association. Refer to [31] for the original definition of Cramér's V, and to [88] for more details and especially EXHIBIT 10.2 for common interpretation guidelines of Cramér's V.

7.1.2 Fisher's Exact Test

When the sample size is small, some cells in the contingency table may have very small expected frequencies (e.g., less than 5), and Chi-square approximation used in the Chi-square test of independence may not be good. In this case, Fisher's exact test can be used to test the association between two categorical variables in a 2×2 contingency table.

Case 23: Treatment Protocol and Adverse Reactions

A research team at a medical center investigates the possible relationship between a new treatment protocol and adverse reactions in patients with chronic kidney disease. 40 patients are randomly selected and randomly assigned to the treatment and control groups. The research team organizes their findings in the following contingency table (Table 5.9):

TABLE 5.9: Contingency Table of Treatment Groups and Adverse Reactions.

	Adverse Reaction	No Adverse Reaction
Treatment Group	2	18
Control Group	8	12

Does the new treatment protocol significantly reduce the occurrence of adverse reactions in patients with chronic kidney disease?

For answers to the initial questions and some initial thoughts about this case study, please refer to Section 5.24. The minimum expected number for the cells of the contingency table is 5 (treatment group with adverse reaction), and Chi-square approximation may not be accurate. Fisher's exact test is usually used when there is at least one cell in the contingency table of the expected frequencies less than 5.

First, we create the contingency table in R as follows:

```
# Create the contingency table
adverse_data <- matrix(c(2, 8, 18, 12), nrow = 2, byrow = TRUE,
                      dimnames = list(Group = c("Treatment", "Control"),
                        Reaction = c("Adverse", "No_Adverse")))
print(adverse_data)
:            Reaction
: Group       Adverse No_Adverse
:   Treatment       2          8
:   Control        18         12
```

The R function `fisher.test()` can be used to perform Fisher's exact test. The p-value is 0.032 for a one-sided test, suggesting that there is strong evidence to reject the null hypothesis of equal odds of adverse reaction between the treatment group and the control group in favor of the alternative hypothesis that the treatment group has lower odds of adverse reaction than the control group.

```
# Perform Fisher's exact test
fisher.test(adverse_data, alternative = "less")
:
:   Fisher's Exact Test for Count Data
:
: data:  adverse_data
: p-value = 0.03242
```

```
: alternative hypothesis: true odds ratio is less than 1
: 95 percent confidence interval:
:  0.0000000 0.8658186
: sample estimates:
: odds ratio
:  0.1743672
```

The odds ratios between the treatment group and the control group can be calculated as $(2/18)/(8/12) \approx 0.167$, which is a little different from the output of the R function `fisher.test()`. The reason is that the sample estimates of the odds ratio are based on conditional MLE, that is, the estimated odds ratio $\hat{\theta}$ such that $E(O_{11}|\hat{\theta}) = 2$; see Note 7.5 for details on Fisher's exact test.

Conclusions

In this case study, we use Fisher's exact test to examine the association between treatment protocol and adverse reaction. There is strong evidence that the treatment group has lower odds of adverse reaction than the control group (The p-value is 0.032 for a one-sided Fisher's exact test). Fisher's exact test is more appropriate than Chi-square test of independence in this case because one cell in the contingency table has a very small expected frequency (5). Subjects are randomly selected, so the conclusion can be generalized to the population of interest. Moreover, subjects are randomly assigned to the treatment and control groups, so causal conclusion can be drawn that the treatment protocol reduces the odds of adverse reaction.

Note 7.5 (Fisher's Exact Test)**.** *Fisher's exact test* ([42]) is a statistical test used to determine whether there is an association between two categorical variables in a 2×2 contingency table. Unlike the Chi-square test of independence, Fisher's exact test is valid even for small sample sizes because it is based on exact probabilities rather than asymptotic approximations. Suppose the data can be represented as

	Category 1	Category 2	Row Totals
Group 1	$a(O_{11})$	b	$a+b$
Group 2	c	d	$c+d$
Column Totals	$a+c$	$b+d$	$n=a+b+c+d$

Under the null hypothesis of independence, the conditional distribution of O_{11} given $n, a+b$ and $a+c$ follows a hypergeometric distribution, that is, the probability of observing $O_{11} = a$ in this particular table is given by

$$\mathrm{P}[O_{11} = a] = \frac{\binom{a+b}{a}\binom{c+d}{c}}{\binom{n}{a+c}}.$$

The p-value is obtained by summing the probabilities of O_{11} taking equal or more extreme than the observed a under the null hypothesis. The definition of "more extreme" depends on the alternative hypothesis, and for different alternative hypotheses, the p-value is calculated as follows:

- H_a: Group 1 has *lower* probability of Category 1 than Group 2 does (i.e., odds ratio of Category 1 between Group 1 and 2 is lower than 1):

$$p\text{-value} = \sum_{a' \in \{0,\ldots,a\}} \mathrm{P}[O_{11} = a'].$$

- H_a: Group 1 has *higher* probability of Category 1 than Group 2 does (i.e., odds ratio of Category 1 between Group 1 and 2 is greater than 1):

$$p\text{-value} = \sum_{a' \in \{a,\ldots,a+c\}} \mathrm{P}[O_{11} = a'].$$

- H_a: Group 1 has a probability of Category 1 *different* than Group 2 does (that is, the odds ratio of Category 1 between Group 1 and 2 is not 1):

$$p\text{-value} = \sum_{a' : \mathrm{P}[O_{11}=a'] \le \mathrm{P}[O_{11}=a]} \mathrm{P}[O_{11} = a'],$$

and the condition $\mathrm{P}[O_{11} = a'] \le \mathrm{P}[O_{11} = a]$ literally translates the condition of "equal or more extreme than observed a", and treats "more extreme" as "less possible". Note that Chi-square test of independence does not tell us the direction of effect, but Fisher's exact test allows us to specify the direction of effects for the alternative hypothesis. For more details on Fisher's exact test, please refer to [3].

7.1.3 McNemar's Test

When an intervention is applied to a group of subjects, it is often of interest to assess whether the intervention has a significant effect on some binary outcome for the subjects. McNemar's tests are commonly used for analyzing such paired binary data. In this section, we illustrate the application of McNemar's test through the following case study.

Case 24: Effectiveness of an Educational Intervention on Smoking Cessation

A health organization implements an educational intervention program to promote smoking cessation among a group of 100 participants. The program consists of educational sessions, counseling, and follow-up support over a three-month period. The effectiveness of the program is assessed by comparing the smoking status of the participants before and after the intervention. Before the intervention, the 100 participants are surveyed to determine their smoking status: Smoker or Nonsmoker. After the intervention, the same participants are surveyed again to determine their current smoking status. Table 5.10 summarizes the survey. Does the educational intervention affect smoking cessation?

TABLE 5.10: Smoking Status.

Before \ After	Smoker	Nonsmoker	Total
Smoker	58	12	70
Nonsmoker	2	28	30
Total	60	40	100

For answers to the initial questions and some initial thoughts about this case study, please refer to Section 5.25.

Intuitively, to answer whether an educational intervention affects smoking habit, one only needs to know whether the numbers of smokers are significantly different before and after the intervention. In other words, one needs to assess whether 70 and 60 in Table 5.10 are significantly different, or alternatively whether 12 and 2 are significantly different. McNemar's test is a test designed for this purpose; see Note 7.6. The null hypothesis (H_0) is that there is no significant difference in smoking status before and after the intervention. The alternative hypothesis (H_1) is that there is a significant difference in smoking status before and after the intervention.

First, we construct the contingency table for McNemar's test as follows. The cell counts in the table are $a = 58$, $b = 2$, $c = 12$, and $d = 28$; see Note 7.6 for the definition of a, b, c, and d.

```
## McNemar's test for smoking status
data <- matrix(c(58, 2, 12, 28), nrow = 2, dimnames = list(
                  Before = c("Smoker", "Nonsmoker"),
                  After = c("Smoker", "Nonsmoker")))
print(data)
:             After
: Before      Smoker Nonsmoker
:   Smoker        58        12
:   Nonsmoker      2        28
```

The following are the R codes and outputs of McNemar's test. The p-value associated with this test is 0.008 less than the common significance levels. Therefore, there is sufficient evidence to reject the null hypothesis and this suggests that the educational intervention resulted in a significant change in smoking status among participants. Note that if the test was not significant, it does not necessarily mean that the educational intervention did not affect each individual. It simply means that the proportions of smokers do not change significantly before and after the educational intervention, although some people may have changed from smoker to nonsmoker and some from nonsmoker to smoker.

```
mcnemar.test(data, correct=FALSE) # do not use continuity correction
:
:  McNemar's Chi-squared test
:
: data:  data
: McNemar's chi-squared = 7.1429, df = 1, p-value = 0.007526
```

Conclusions

In this case study, we use McNemar's test to assess whether an educational intervention affects smoking cessation. There is a significant change in smoking status among the participants after the educational intervention (p-value of McNemar's test is 0.008). Whether subjects are randomly selected from the population of interest needs to be carefully considered before generalizing the result to the population. This is an observational study, and no causality can be established; that is, the educational intervention may not necessarily cause the change in smoking status, and the change could be due to confounders that are not provided in the data.

Note 7.6 (McNemar's Test). *McNemar's test* ([80]) is a statistical test used to analyze paired binary data. It is typically employed to assess the changes in responses before and after a treatment or intervention for the same group of subjects. The test is based on the following 2×2 contingency table.

TABLE 7.2: A 2×2 Table for Illustrating McNemar's Test.

Before \ After	Yes	No	Total
Yes	a	c	$a+c$
No	b	d	$b+d$
Total	$a+b$	$c+d$	$a+b+c+d$

The null and alternative hypotheses of McNemar's test are the following:

$$H_0 : p_b = p_c;$$

$$H_a : p_b \neq p_c.$$

Here p_b and p_c denote the theoretical probability of occurrences in cells b and c, respectively; that is, p_b is the probability of having b occurrences of Before:No and After:Yes, and p_c is the probability of having c occurrences of Before:Yes and After:No. McNemar's test statistic is $\chi^2 = (b-c)^2/(b+c)$. Under the null hypothesis, with a sufficiently large number of $b+c$, χ^2 approximately follows a Chi-square distribution with 1 degree of freedom. If the statistics is large enough, then the null hypothesis will be rejected, suggesting that there is a significant difference in the proportions of changes.

The Chi-square test is a two-sided test. To conduct a one-sided test, one can use χ instead of χ^2 as the test statistic. For sufficiently large $n = b+c$, χ approximately follows a normal distribution. In this way, a normal test (z test) can be easily conducted for either one-sided or two-sided McNemar's tests. For more details, please refer to [6].

McNemar's Chi-square test is an approximation test, and we usually require that $b+c > 10$ to have a reasonable approximation. For a small sample size with $b+c <= 10$, we can use a binomial test instead. Let us consider an example similar to Case 24, except that the data have been changed to smaller numbers as shown in Table 7.3.

TABLE 7.3: Does the Educational Intervention Affect Smoking Cessation?

Before \ After	Smoker	Nonsmoker	Total
Smoker	7	3	10
Nonsmoker	1	9	10
Total	8	12	20

The education intervention results in 4 participants changed their smoking status. To answer whether the educational intervention affects smoking cessation, we can assess whether the current outcomes of 1 from nonsmoker to smoker and 3 from smoker to nonsmoker are due to randomness or not.An exact binomial test can be performed. The null and alternative hypotheses are

H_0 : The true probability of smoker to nonsmoker is the same as that of nonsmoker to smoker

H_a : not the same.

The following R codes perform the exact binomial test, and the test suggests that there is insufficient evidence to reject the null hypothesis (p-value is 0.625), indicating that the educational intervention does not affect smoking status.

```
# for small sample size, use exact binomial test
binom.test(x=1, n=4, p=0.5, alternative = "two.sided")
:
:  Exact binomial test
:
: data:  1 and 4
: number of successes = 1, number of trials = 4, p-value = 0.625
: alternative hypothesis: true probability of success is not equal to 0.5
: 95 percent confidence interval:
:  0.006309463 0.805879550
: sample estimates:
: probability of success
:                   0.25
```

7.1.4 McNemar's Test – Revisited

Sometimes, the data presented may look similar to those for McNemar's test, but actually do not have sufficient information to directly apply McNemar's test. The following case study illustrates such a situation.

Case 25: Advertisements and Purchase Intention

In order to study whether advertising can increase consumers' purchase intention, a random sample of 100 people is asked to watch the ads. Before they see anything, they are asked if they intend to buy the product. Then after watching the ads, they are asked again to see if anyone has changed their mind. The following is the contingency table (Table 5.11):

TABLE 5.11: Does Watching Ads Increase Purchase Intention?

	Before	After	Total
Yes	30	10	40
No	70	90	160
Total	100	100	200

Does the ad have significant effects in changing the consumers' purchase intention?

For answers to the initial questions and some initial thoughts about this case study, please refer to Section 5.26. At first glance, this case looks very similar to Case 24. However, the data do not have sufficient information for us to use McNemar's test: the values of a, b, c, and d in Table 7.2 are not unique. For example, both $a = 5, b = 5, c = 25, d = 65$ (Table 7.4) and $a = 2, b = 8, c = 28, d = 62$ (Table 7.5) satisfy Table 5.11. Therefore, McNemar's test can not be applied for this case directly. The main reason is that although we know the total numbers for "before" and "after", respectively, Table 5.11 does not provide the cross-classified information of the paired data, which is essential for McNemar's test.

TABLE 7.4: A Reorganized Table for Table 5.11.

Before \ After	Yes	No	Total
Yes	5	25	30
No	5	65	70
Total	10	90	100

TABLE 7.5: Another Reorganized Table for Table 5.11.

Before \ After	Yes	No	Total
Yes	2	28	30
No	8	62	70
Total	10	90	100

Both cases lead to *p*-values well below the common significance levels.

```
data1 <- data.frame("Yes_After" = c(5, 5), "No_After" = c(25, 65),
                    row.names = c("Yes_Before", "No_Before"))
data1 <- as.matrix(data1)
mcnemar.test(data1, correct=FALSE)
:
:  McNemar's Chi-squared test
:
: data:  data1
: McNemar's chi-squared = 13.333, df = 1, p-value = 0.0002607
```

```
data2 <- data.frame("Yes_After" = c(2, 8), "No_After" = c(28, 62),
                    row.names = c("Yes_Before", "No_Before"))
data2 <- as.matrix(data2)
mcnemar.test(data2, correct=FALSE)
:
:  McNemar's Chi-squared test
:
: data:  data2
: McNemar's chi-squared = 11.111, df = 1, p-value = 0.0008581
```

To assess the effect of watching the ads, we can consider the possible range of the test statistics of McNemar's test by varying a from 0 to 10 while keeping the numbers unchanged in Table 5.11.

First, convert Table 7.2 as follows to the format of the contingency table used in this case study.

TABLE 7.6: A 2×2 Table for Illustrating McNemar's Test.

Before \ After	Yes	No	Total
Yes	a	c	$a+c$
No	b	d	$b+d$
Total	$a+b$	$c+d$	$a+b+c+d$

The converted table looks like the following:

TABLE 7.7: Converted From the Table for McNemar's Test.

	Before	After	Total
Yes	$a+c$	$a+b$	$2a+b+c$
No	$b+d$	$c+d$	$2d+b+c$
Total	$a+b+c+d$	$a+b+c+d$	$2(a+b+c+d)$

Then matching the Tables 7.7 and 5.11, the McNemar's test statistic can be derived as

$$\chi^2 = \frac{(b-c)^2}{b+c} = \frac{400}{40-2a}, \quad 0 \le a \le 10,$$

where $b-c = (2a+b+c) - 2(a+c) = 40 - 2\times 30 = -20$ and $b+c = (2a+b+c) - 2a = 40 - 2a$. Changing a from 0 to 10, we have the following two extreme cases:

TABLE 7.8: A Reorganized Table for Table 5.11 with Minimum $a = 0$.

Before \ After	Yes	No	Total
Yes	0	30	30
No	10	60	70
Total	10	90	100

TABLE 7.9: A Reorganized Table for Table 5.11 with Maximum $a = 10$.

Before \ After	Yes	No	Total
Yes	10	20	30
No	0	70	70
Total	10	90	100

The test statistics of McNemar's test are increasing in a. The largest p-value for the "worst" case scenario of McNemar's test is 0.0016, and therefore the p-values of all the other possibilities will be well below common significance levels, suggesting that there is a significant effect of watching the ads in changing consumers' purchase intention.

```
# worst scenario
data_worst <- data.frame("Yes_After" = c(0, 10), "No_After" = c(30, 60),
                         row.names = c("Yes_Before", "No_Before"))
data_worst <- as.matrix(data_worst)
print(data_worst)
:             Yes_After No_After
: Yes_Before          0       30
: No_Before          10       60
mcnemar.test(data_worst, correct=FALSE)
:
:   McNemar's Chi-squared test
:
: data:  data_worst
: McNemar's chi-squared = 10, df = 1, p-value = 0.001565
```

Just be careful that the intention has actually decreased. McNemar's test is a two-sided test and cannot handle one-sided tests directly. The following binomial test can be used for a one-sided test. Let the alternative hypothesis be

H_a : More customers saying "No" after watching the ads,

then the following binomial test suggests that there is significantly more customers saying "No" after watching the ads with the *p*-value 0.0011.

```
# One-sided binomial test: H1 = b > c
b <- data_worst[1,2]
c <- data_worst[2,1]
binom.test(b, b + c, p = 0.5, alternative = "greater")
:
:   Exact binomial test
:
: data:  b and b + c
: number of successes = 30, number of trials = 40, p-value = 0.001111
: alternative hypothesis: true probability of success is greater than 0.5
: 95 percent confidence interval:
:  0.6129398 1.0000000
: sample estimates:
: probability of success
:                   0.75
```

The above "worst" case leads to the largest possible *p*-value, which has already suggested that watching ads significantly reduces the purchase intention. The following gives the smallest possible *p*-value: 9.53×10^{-7}.

```
# One-sided test: H1 = b > c (smallest p-value case)
data_best <- data.frame("Yes_After" = c(10, 0), "No_After" = c(20, 70),
                         row.names = c("Yes_Before", "No_Before"))
data_best <- as.matrix(data_best)
b <- data_best[1,2]
c <- data_best[2,1]
binom.test(b, b + c, p = 0.5, alternative = "greater")
:
:   Exact binomial test
:
: data:  b and b + c
: number of successes = 20, number of trials = 20, p-value = 9.537e-07
: alternative hypothesis: true probability of success is greater than 0.5
: 95 percent confidence interval:
:  0.8608917 1.0000000
: sample estimates:
: probability of success
:                      1
```

Conclusions

In this case study, we assess whether watching advertisements affects consumers' purchase intention using McNemar's test and the one-sided binomial tests. As there is no unique way to use the provided aggregated data to form a 2×2 table for McNemar's test, we consider the possible range of *p*-values by varying the unknown cell value, which leads to *p*-values less than 0.0016. Therefore, there is sufficient evidence to suggest that watching the ads has a significant effect on consumers' purchase intention. To further assess whether watching the ads increases or decreases the purchase intention, we conduct one-sided binomial tests for the possible extreme cases, which lead to p-values ranging from 9.53×10^{-7} to 0.0011. Therefore, watching the ads has significantly reduced the purchase intention with the p-values ranging from 9.53×10^{-7} to 0.0011 from the one-sided binomial tests. Subjects are randomly selected from the population of interest, and therefore, the result can be generalized to the population of interest. However, this is an observational study, and no causality can be established.

7.1.5 The Cochran-Mantel-Haenszel Test

When a confounding variable is present in a study of the association between a binary predictor and a binary response variable, the Cochran-Mantel-Haenszel (CMH) test can be used to assess the association while controlling for the confounding variable. This section presents a case study that illustrates the application of the CMH test.

Case 26: Drug Effects in Different Hospitals

A pharmaceutical company conducts a study in three hospitals to evaluate a new drug. Data is collected on whether patients received the drug and whether they recovered, with the following details:

- Hospital A: Smaller hospital, fewer patients.
- Hospital B: Medium-sized hospital, moderate volume of patients.
- Hospital C: Large hospital, higher patient volume.

TABLE 5.12: Treatment vs. Recovery Across Hospitals.

	Hospital A		Hospital B		Hospital C	
	Rec.	Unrec.	Rec.	Unrec.	Rec.	Unrec.
Treated	8	2	15	5	25	10
Untreated	3	5	7	10	12	18

Determine if there is a significant association between the treatment and recovery, adjusting for the confounding factor of the hospitals.

For answers to the initial questions and some initial thoughts about this case study, please refer to Section 5.27. We first combine the three contingency tables into a three-dimensional array for further use in the R function `mantelhaen.test()`, which performs the Cochran-Mantel-Haenszel (CMH) test; see Note 7.7 for details of the test.

```
# Create the 2x2x3 array for the stratified data
# Dimensions: Treatment (rows), Outcome (columns), Hospital (layers)
data <- array(c(8, 3, 2, 5, 15, 7, 5, 10, 25, 12, 10, 18), dim = c(2, 2, 3),
                dimnames = list(Treatment = c("Treated", "Untreated"),
                                Outcome = c("Recovered", "Unrecovered"),
                                Hospital = c("A", "B", "C")))
print(data)
: , , Hospital = A
:
:            Outcome
: Treatment   Recovered Unrecovered
:   Treated           8           2
:   Untreated         3           5
:
: , , Hospital = B
:
:            Outcome
: Treatment   Recovered Unrecovered
:   Treated          15           5
:   Untreated         7          10
:
: , , Hospital = C
:
:            Outcome
: Treatment   Recovered Unrecovered
:   Treated          25          10
:   Untreated        12          18
```

The following shows how to make these tables into a long format data, which could be used for logistic regression if the common odds ratio assumption of the CMH test does not hold. Note that, the subject-level data can be recovered by the three contingency tables, and thus a logistic regression that relies on subject-level would be possible.

```
# Convert to long format
df_long <- as.data.frame(as.table(data))

# Expand to subject-level
df_subject <- df_long[rep(1:nrow(df_long), df_long$Freq), 1:3]
rownames(df_subject) <- NULL  # remove row names
head(df_subject, 3)
:   Treatment   Outcome Hospital
: 1   Treated Recovered        A
: 2   Treated Recovered        A
: 3   Treated Recovered        A
```

Data from three hospitals are presented in the form of a table containing counts of treated and untreated individuals for each hospital. The odds ratios calculated for each hospital are the following: Hospital A: 6.67, Hospital B: 4.29, and Hospital C: 3.75.

```
# Calculate odds ratios for each stratum
odds_ratios <- apply(data, 3,
                     function(x) (x[1,1] / x[1,2]) / (x[2,1] / x[2,2]))
names(odds_ratios) <- c("Hospital A", "Hospital B", "Hospital C")

# Odds ratios by hospital:
print(odds_ratios)
: Hospital A Hospital B Hospital C
:   6.666667   4.285714   3.750000
```

We conduct the Breslow-Day test to decide whether the common odds ratio assumption holds; see Note 7.8 for details of the test. Using the R function `DescTools::BreslowDayTest()`, the p-value is 0.891, well above the common significance levels, indicating that there is no significant evidence of heterogeneity in odds ratios, which helps to validate the common odds ratio assumption of the CMH test.

```
DescTools::BreslowDayTest(data)
:
:   Breslow-Day test on Homogeneity of Odds Ratios
:
: data:  data
: X-squared = 0.23189, df = 2, p-value = 0.8905
```

Therefore, we proceed with the Cochran-Mantel-Haenszel test. The p-value is 0.0004, suggesting that there is significant association between treatment and recovery, and the common odds ratio is approximately 4.22, indicating a strong effect of treatment on recovery. (An odds ratio equal to 1 means no association).

```
# Cochran-Mantel-Haenszel test
mantelhaen.test(data)
:
:   Mantel-Haenszel chi-squared test with continuity correction
:
: data:  data
: Mantel-Haenszel X-squared = 12.357, df = 1, p-value = 0.0004394
: alternative hypothesis: true common odds ratio is not equal to 1
: 95 percent confidence interval:
:   1.950638 9.143409
: sample estimates:
: common odds ratio
:          4.223208
```

If the common odds ratio assumption did not hold, then we can use a logistic regression as follows. The analysis of deviance shows that there is no significant interaction between treatment and hospital (p-value = 0.888), indicating that the common odds ratio assumption used in the CMH test may hold. There is significant effect of treatment (p-value = 0.00016), and this is consistent to what the CMH test has concluded. There are no significant differences between hospitals (p-value = 0.942). In Section 7.1.6, we have more discussions on logistic regression and the analysis of deviance.

```
df_subject$Treatment <- as.factor(df_subject$Treatment)
df_subject$Outcome <- as.factor(df_subject$Outcome)
df_subject$Hospital <- as.factor(df_subject$Hospital)
logistic_model <- glm(Outcome ~ Treatment*Hospital, data=df_subject,
                      family = "binomial")
anova(logistic_model, test="Chi")
: Analysis of Deviance Table
:
: Model: binomial, link: logit
:
: Response: Outcome
:
: Terms added sequentially (first to last)
:
:
:                    Df Deviance Resid. Df Resid. Dev  Pr(>Chi)
: NULL                                119     163.01
: Treatment           1  14.2693      118     148.74 0.0001584 ***
: Hospital            2   0.1196      116     148.62 0.9419422
: Treatment:Hospital  2   0.2366      114     148.38 0.8884260
: ---
: Signif. codes:  0 '***' 0.001 '**' 0.01 '*' 0.05 '.' 0.1 ' ' 1
```

Conclusions

In this case study, we apply the Cochran-Mantel-Haenszel (CMH) test to assess the association between treatment and recovery while controlling for the confounding variable of hospital. The Breslow-Day test indicates that the common odds ratio assumption holds, allowing us to proceed with the CMH test. The results show that there is significant association (p-value is 0.0004 from the CMH test) between treatment and recovery across the three hospitals, and the common odds ratio is approximately 4.22. Whether subjects are selected randomly from a population is not specified, and thus generalization to a larger population cannot be determined. This is an observational study, and therefore, there is no causality can be established from this study.

Note 7.7 (The Cochran-Mantel-Haenszel (CMH) Test)**.** The *Cochran-Mantel-Haenszel (CMH) test* ([26, 72]) is a statistical method used to analyze stratified categorical data, and it tests the association between a binary outcome variable and a binary exposure / predictor variable while controlling for a third categorical confounding variable for stratification. It extends the Chi-square test of independence to multiple 2×2 contingency tables. The key assumption of the CMH test is that the odds ratios are consistent across strata (no interaction between the exposure variable and the confounding variable).

Data are stratified into multiple 2×2 tables, according to the categorical confounding variable. Consider K strata, and for stratum k, the data can be summarized in the form of a 2×2 contingency table, as shown below:

	Outcome +	Outcome -
Exposed	a_k	b_k
Unexposed	c_k	d_k

The hypotheses are the following:

H_0 : no association between exposure and outcome across strata, with a common odds ratio equal to 1;
H_a : there is an association with a common odds ratio not equal to 1.

The CMH test statistic is calculated as follows:

$$\chi^2 = \frac{(\sum_k (a_k - \mathrm{E}(a_k)))^2}{\sum_k \mathrm{Var}(a_k)},$$

where $\mathrm{E}(a_k)$ and $\mathrm{Var}(a_k)$ are the expected count and variance of a_k under the null hypothesis. For large sample sizes, the test statistic approximately follows a Chi-square distribution with 1 degree of freedom under the null hypothesis. The common odds ratio can be estimated as follows:

$$\widehat{\mathrm{OR}} = \frac{\sum_k a_k d_k / n_k}{\sum_k b_k c_k / n_k},$$

where $n_k = a_k + b_k + c_k + d_k$, the total sample size in stratum k. The CMH test is a powerful tool for testing for association of stratified 2×2 tables while controlling for confounding variables if the common odds ratio assumption holds. We refer to [26], [72], and [3] for more details.

Note 7.8 (The Breslow-Day Test)**.** The *Breslow-Day test* ([16]) is used to assess the homogeneity of odds ratios across strata in stratified 2×2 contingency tables. This test is commonly used along with the Cochran-Mantel-Haenszel (CMH) test to evaluate whether odds ratios are consistent between different strata. The hypotheses for the Breslow-Day test are as follows:

H_0 : The odds ratios are the same across all strata;
H_a : The odds ratios are different across strata.

The Breslow-Day test statistic is

$$\chi^2 = \sum_{k=1}^{K} \frac{(O_{11k} - E_{11k})^2}{V_{11k}},$$

where K is the number of strata, O_{11k} is the count observed in the (1,1) cell of the kth strata, and E_{11k} and V_{11k} are the expected count and the variance of the count in the (1,1) cell of the kth strata, respectively, under the assumption of the common odds ratio. The degree of freedom for the test is given by: $\mathrm{df} = K - 1$. The expected count E_{11k} is derived from the common odds ratio $\widehat{\mathrm{OR}}$, which is estimated by the CMH test discussed in Note 7.7. The variance V_{11k} is calculated based on a hypergeometric distribution. Refer to [16] for more details.

7.1.6 Logistic Regression

When the response variable is binary and there are multiple predictor variables, logistic regression is widely used to model the relationship between the response and predictors. This section presents a case study that illustrates the application of logistic regression.

Case 27: Job Satisfactory

A company wants to assess job satisfactions among different departments to improve employee satisfaction and retention. A random sample of 70 employees is selected to provide feedback on job satisfaction. In addition, employees' salaries are included. What conclusions can be drawn from the feedback?

For answers to the initial questions and some initial thoughts about this case study, please refer to Section 5.28. Some exploratory analyses can be done to understand the relationship between these variables. For the response variable of Satisfied, "1" is coded for "satisfied" and "0" is for "unsatisfied". There are more satisfied employees in the Sales and Engineering departments and fewer satisfied employees in the Marketing department.

```
library(ggplot2)
data <- read.csv("data/job-satisfactory-department-salary.csv")

data$Department <- as.factor(data$Department)
data$Satisfied <- as.factor(data$Satisfied)

# Center salary for better interpretability
data$Salary <- data$Salary - mean(data$Salary)

table(data$Department, data$Satisfied)
:
:                0  1
:   Engineering  8 13
:   Marketing   14 11
:   Sales        7 17
```

Figure 7.1 suggests that a higher salary may be associated with a higher level of satisfaction, and Figure 7.2 indicates that salaries are different between departments.

```
ggplot(data, aes(x = Salary, y = Satisfied)) + geom_boxplot() +
  stat_summary(fun = "mean", geom = "point", shape = 23, size = 2) +
  labs(x = "Salary (centered)", y = "") + theme_bw()
```

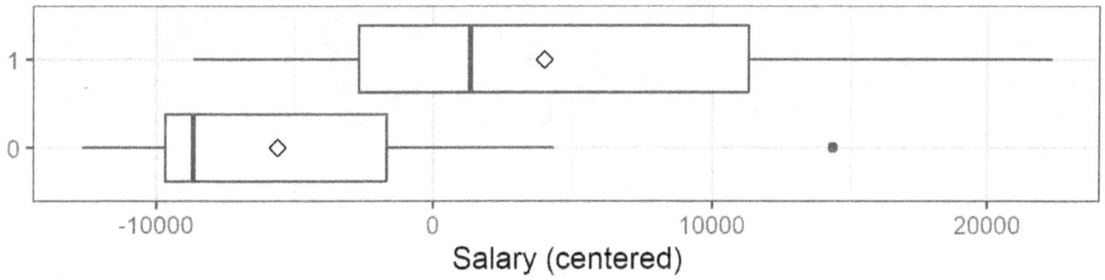

FIGURE 7.1: Box Plots of Salary for Satisfied and Unsatisfied Employees.

```
ggplot(data, aes(x = Salary, y = Department)) + geom_boxplot() +
  stat_summary(fun = "mean", geom = "point", shape = 23, size = 2) +
  labs(x = "Salary (centered)", y = "") + theme_bw()
```

FIGURE 7.2: Box Plots of Salary in Different Departments.

Although the Sales department may have higher levels of job satisfaction, the question is whether the Sales department has the highest salary (in terms of median salary), and after accounting for the effect of salary on job satisfaction, does the Sales department still have higher job satisfaction? To answer these questions, we can fit a logistic regression model with Salary and Department as predictors, and Satisfied as the response variable. We can also include an interaction term between Salary and Department to see whether the effect of Salary on job satisfaction differs among departments.

```
full_model <- glm(Satisfied ~ Department * Salary, data=data,
                  family="binomial")
```

The following likelihood ratio test shows that the interaction term is significant ($p = 0.033$), and thus we keep the interaction and the two main effects in the model.

```
# Analysis of Deviance
drop1(full_model, test="Chi")
: Single term deletions
:
: Model:
: Satisfied ~ Department * Salary
:                   Df Deviance    AIC    LRT Pr(>Chi)
: <none>                 61.590 73.590
: Department:Salary  2   68.413 76.413 6.8228    0.033 *
: ---
: Signif. codes:  0 '***' 0.001 '**' 0.01 '*' 0.05 '.' 0.1 ' ' 1
```

The following summary output shows the estimated coefficients of the fitted model. Note that the baseline department is Engineering, and the coefficients of DepartmentMarketing and DepartmentSales represent the difference in log-odds of satisfaction between these departments and the Engineering department when Salary is at the average level (since we centered Salary to make it more interpretable). The coefficient of Salary represents the effect of Salary on the log-odds of satisfaction for the Engineering department. The interaction terms indicate how the effect of Salary on satisfaction differs for Marketing and Sales compared to Engineering.

```
summary(full_model)
:
: Call:
: glm(formula = Satisfied ~ Department * Salary, family = "binomial",
```

```
:     data = data)
:
: Coefficients:
:                              Estimate Std. Error z value Pr(>|z|)
: (Intercept)                 7.627e-02  5.588e-01   0.136   0.8914
: DepartmentMarketing         2.268e-01  8.012e-01   0.283   0.7771
: DepartmentSales             3.702e+00  1.840e+00   2.011   0.0443 *
: Salary                      1.774e-04  8.015e-05   2.213   0.0269 *
: DepartmentMarketing:Salary -6.030e-05  1.143e-04  -0.528   0.5978
: DepartmentSales:Salary      5.818e-04  3.594e-04   1.619   0.1055
: ---
: Signif. codes:  0 '***' 0.001 '**' 0.01 '*' 0.05 '.' 0.1 ' ' 1
:
: (Dispersion parameter for binomial family taken to be 1)
:
:     Null deviance: 94.973  on 69  degrees of freedom
: Residual deviance: 61.590  on 64  degrees of freedom
: AIC: 73.59
:
: Number of Fisher Scoring iterations: 8
```

After we identify that the interaction term has a significant effect, we may perform a post hoc analysis to find out where the difference is. We can test the slopes of the Salary effect among different departments using the R function `emmeans::emtrends()` as follows. It is possible that even the omnibus test suggests significant interactions, a multiple comparison may still lead to nonsignificant pairwise comparisons, due to reasons such as smaller sample sizes reducing the power of post hoc tests and a much smaller *p*-value is required for post hoc tests, as we have discussed in the notes for multiple comparisons (see Note 6.12).

The following multiple comparison shows that there are no significant differences between the satisfaction-salary slopes among the three departments; note that, the *p*-values for the pairwise comparisons are adjusted for multiple testing. Figure 7.3 will further support the conclusion by visualizing the predicted "slopes" for each department separately.

```
# Estimate the slope of Salary within each department
library(emmeans)
slopes <- emtrends(full_model, ~ Department, var = "Salary")

# Pairwise comparisons of slopes
pairs(slopes)
:  contrast               estimate       SE  df z.ratio p.value
:  Engineering - Marketing 6.03e-05 0.000114 Inf   0.528  0.8578
:  Engineering - Sales    -5.82e-04 0.000359 Inf  -1.619  0.2376
:  Marketing - Sales      -6.42e-04 0.000360 Inf  -1.785  0.1745
:
: P value adjustment: tukey method for comparing a family of 3 estimates
```

Predicted probability of satisfaction for different department and different salaries can be plotted as follows. We notice that although sales department has relatively lower salary, satisfaction quickly catches up after the salary passes certain threshold, and even higher than the other departments at a same relatively higher salary level.

```
newdata <- expand.grid(Salary = seq(min(data$Salary), max(data$Salary),
                       by = 1000), Department = levels(data$Department))

pred <- predict(full_model, newdata, type = "link", se.fit = TRUE)
newdata$fit_link <- pred$fit
newdata$se_link <- pred$se.fit
newdata$lower_link <- newdata$fit_link - 1.96 * newdata$se_link
newdata$upper_link <- newdata$fit_link + 1.96 * newdata$se_link
invlogit <- function(x) exp(x) / (1 + exp(x))
newdata$fit_prob <- invlogit(newdata$fit_link)
newdata$lower_prob <- invlogit(newdata$lower_link)
newdata$upper_prob <- invlogit(newdata$upper_link)
ggplot(newdata, aes(x = Salary, y = fit_prob, color = Department,
                    linetype = Department, fill = Department)) +
  geom_line() + scale_linetype_discrete() +
  geom_ribbon(aes(ymin = lower_prob, ymax = upper_prob), alpha = 0.2) +
  scale_color_grey(start = 0.2, end = 0.8) +
  scale_fill_grey(start = 0.2, end = 0.8) +
  labs(x = "Salary", y = "Predicted Probability of Satisfaction",
    title = "") + theme_bw()
```

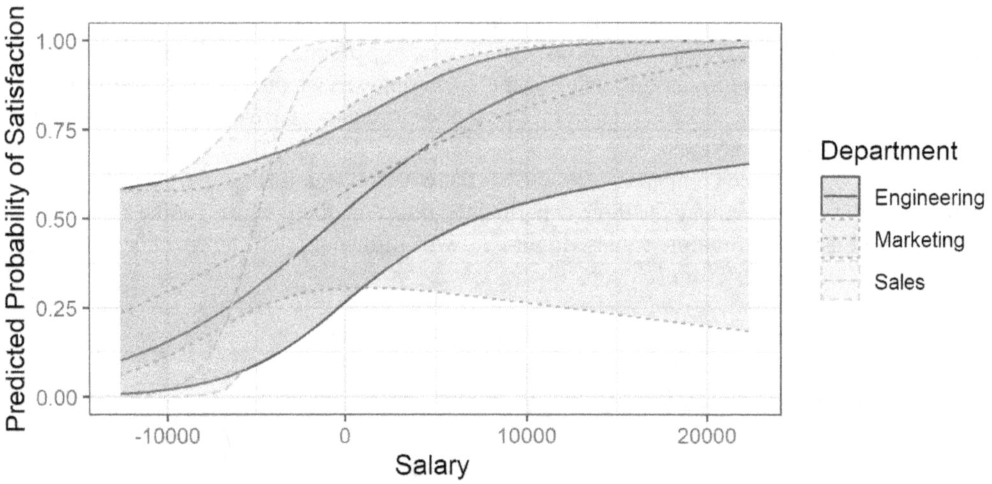

FIGURE 7.3: Predicted Probability of Satisfaction with 95% Confidence Interval.

To find potential outliers and high-leverage points, we can use the R function `influencePlot` in the `car` package; see Figure 7.4. Similarly to leverage in linear regression, leverage (Hat-Values in the plot) in generalized linear regression helps detect observations that disproportionately affect the parameter estimates. Cook's distance can be extended to GLM to help identify those influential data points that combine both residual and leverage information.

The size of each circle represents Cook's distance for that observation. Points that are far from 0 horizontally have high leverage, and points that are far from the center vertically have large residuals. Observations that are both high leverage and have large residuals (i.e., the 55th point in the button right corner) are particularly influential and may warrant further investigation.

In the following plot, the three highlighted points have high leverage and/or large residuals, and they may be influential points that affect the model fit. Further investigation is needed to determine whether these points are data errors, outliers, or high-leverage points. Depending on the context, we may consider removing or re-weighting these points to improve model robustness.

```
car::influencePlot(full_model, fill.col="black")
```

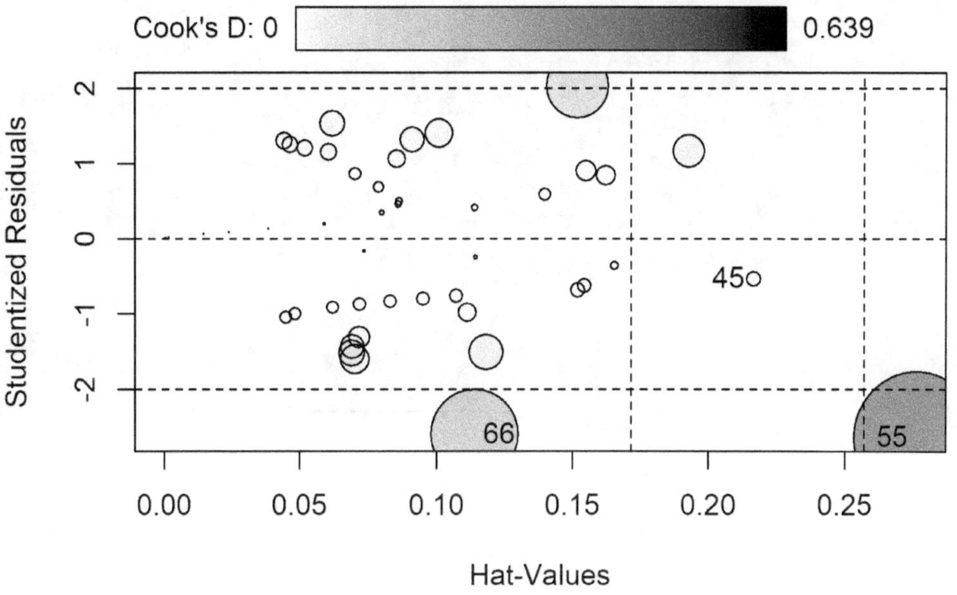

FIGURE 7.4: Regression Influence Plot.

```
:       StudRes       Hat      CookD
: 45 -0.5279008 0.2163084 0.007583919
: 55 -2.6443951 0.2766558 0.638835917
: 66 -2.5992901 0.1140136 0.297575310
```

To evaluate the performance of the fitted model, we can create a confusion matrix based on a threshold of 0.5 for predicted probabilities. The confusion matrix shows the counts of true positives, true negatives, false positives, and false negatives. From the confusion matrix, we can calculate various performance metrics such as accuracy, precision, recall (sensitivity), specificity, and F1 score; see Note 7.10 for details.

```
# Confusion Matrix at threshold 0.5
table(Predicted = as.numeric(predict(full_model, type = "response") > 0.5),
      Actual = full_model$y)
:          Actual
: Predicted  0  1
:         0 21 10
:         1  8 31
```

If we want to evaluate the model performance across different thresholds, we can use ROC curve which balances the sensitivity and specificity of the model. The Area Under the Curve (AUC) should be less than 1 and the closer to 1 the better the performance.

```
# Area Under the Curve (AUC)
library(pROC)
data$pred_prob <- predict(full_model, type = "response")
roc_obj <- roc(data$Satisfied, data$pred_prob)
auc(roc_obj)
: Area under the curve: 0.8701
```

```
ggroc(roc_obj, color = "black", size = 0.5) +
  annotate("segment", x = 1, y = 0, xend = 0, yend = 1, color = "grey",
           linetype = "dashed") + labs(title = "",
                  x = "Specificity (1 - False Positive Rate)",
                  y = "Sensitivity (True Positive Rate)") +
  annotate("text", x = 0.4, y = 0.1, label = paste("AUC =",
           round(auc(roc_obj), 3)), size = 3) + theme_bw()
```

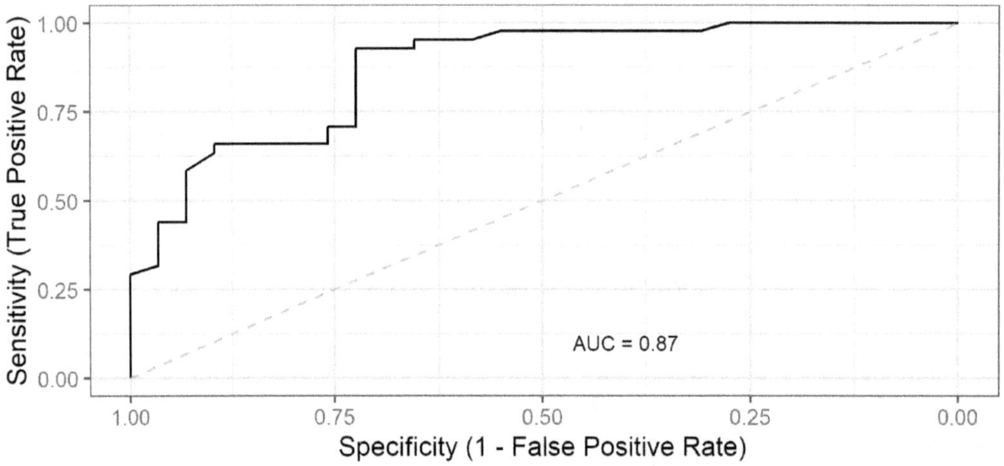

FIGURE 7.5: ROC Curve for Satisfaction Prediction.

Conclusions

In this case study, we fitted a logistic regression model to analyze the relationship between job satisfaction, department, and salary. The results indicate that salary has a significant effect on employees' satisfaction (p-value $< 10^{-6}$ of the analysis of deviance of the logistic regression model), and there is a significant interaction between salary and department (p-value is 0.033 of the analysis of deviance). However, pairwise comparisons suggest that while there is some overall variation, no individual department differs strongly enough from another to be statistically distinct (adjusted p-values are 0.858, 0.238, 0.175, respectively). Subjects are randomly sampled from the three departments of the company, and thus the results can be generalized to these three departments. The fitted model has an AUC of 0.87, indicating good predictive performance. This is an observational study, and therefore, no causality can be established from the findings.

Note 7.9 (Analysis of Deviance for Generalized Linear Models)**.** Analysis of deviance is a key tool for assessing the goodness-of-fit and comparing nested models in generalized linear models (GLMs). *Deviance* of a fitted model measures the discrepancy between the model and the saturated model, a model with a separate parameter for each observation. The saturated model provides a perfect fit: it has as many parameters as data points. The deviance D of the model is defined as

$$D = 2\left[\ell(\text{saturated}) - \ell(\text{model})\right],$$

where $\ell(\text{saturated})$ is the log-likelihood of the saturated model and $\ell(\text{model})$ is the log-likelihood of the fitted model. Similar to ANOVA for linear regression, we can compare nested GLMs using the difference in deviance, which approximately follows a Chi-square distribution under the null hypothesis that the simpler model is sufficient. The degrees of freedom for the test is equal to the difference in the number of parameters between the two models. For more details, see [6, 35].

Note 7.10 (Performance Measures in Binary Classification)**.** Consider a binary classification problem with the following metrics: True Positive (TP): the number of positive instances correctly classified as positive; False Positive (FP): the number of negative instances incorrectly classified as positive; False Negative (FN): the number of positive instances incorrectly classified as negative; True Negative (TN): the number of negative instances correctly classified as negative. These metrics can be organized into a confusion matrix as follows:

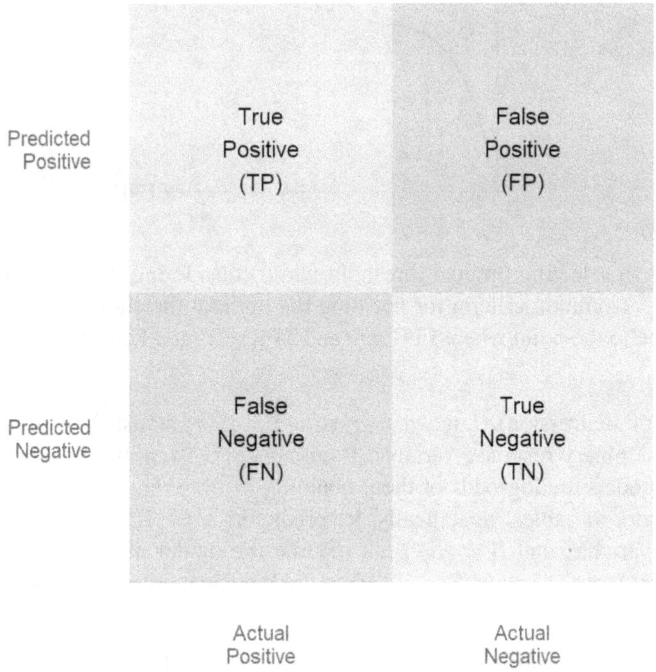

FIGURE 7.6: Confusion Matrix of Binary Classification

Based on the confusion matrix, there are several performance measures can be defined:

Measure	Formula
Accuracy	$(TP+TN)/\text{Total}$
Precision	$TP/(TP+FP)$
Sensitivity / Recall (True Positive Rate)	$TPR = TP/(TP+FN)$
False Negative Rate	$FNR = 1-TPR = FN/(TP+FN)$
Specificity (True Negative Rate)	$TNR = TN/(TN+FP)$
False Positive Rate	$FPR = 1-TNR = FP/(TN+FP)$
F1 score	$F1 = 2\cdot(\text{precision}\cdot\text{recall})/(\text{precision}+\text{recall})$

Another important tool for evaluating binary classifiers is the Receiver Operating Characteristic (ROC) curve, which is a visual tool for evaluating binary classifier performance. It plots True Positive Rate (TPR) against False Positive Rate (FPR). Each point represents a different classification threshold: a lower threshold results in more positive predictions, increasing both TPR and FPR, while a higher threshold results in fewer positive predictions, decreasing both TPR and FPR. The perfect classifier corresponds to the curve that passes the point $(0,1)$, and random classifier corresponds to the diagonal line FPR = TPR. The area under the ROC curve is referred to as AUC, which provides a single scalar value to summarize the ROC curve performance, and a higher AUC indicates better classifier performance. The following plot illustrates a typical ROC.

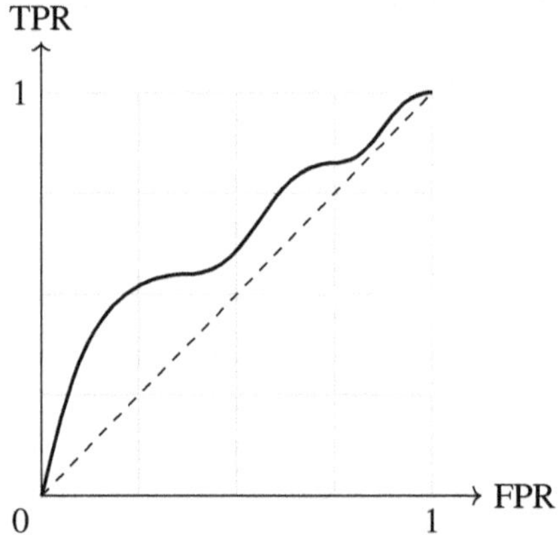

ROC curve helps in selecting optimal threshold while considering the trade-off between sensitivity and specificity. Common criteria for deciding the optimal threshold include $max(\text{TPR}-\text{FPR})$, minimum distance to the point where FPR $= 0$ and TPR $= 1$, and F1 score maximization. For more details, see [41].

Note 7.11 (Logistic Regression)**.** *Logistic regression* ([30]) is a statistical method to perform regression analysis for a binary response variable. It belongs to the family of generalized linear models (GLMs), and it models the log-odds of the probability of the event occurring as a linear combination of the predictor variables. Specifically, let predictors $\mathbf{x} = (1, x_1, x_2, ..., x_p)$, $Y \in \{0,1\}$ be the binary response variable, and $\beta = (\beta_0, \beta_1, ..., \beta_p)$ be the model parameters with β_0 the intercept term, knowing that $\text{logit}(x) = \ln(x/(1-x))$, then the logistic regression model can be expressed as

$$\text{logit}(\text{P}(Y=1|\mathbf{x})) = \ln\left(\frac{\text{P}(Y=1|\mathbf{x})}{\text{P}(Y=0|\mathbf{x})}\right) = \mathbf{x}^\top\beta, \tag{7.1}$$

where the left-hand side is the log-odds of the probability of the event $Y = 1$ given the predictors $\mathbf{x}$, and the right-hand side is a linear combination of the predictors with regression coefficients β.

The standard logistic function, also known as the sigmoid function, is defined as

$$\sigma(\eta) = 1/(1+e^{-\eta}),$$

and it maps any real-valued number to the range $(0, 1)$ as illustrated by the following plot.

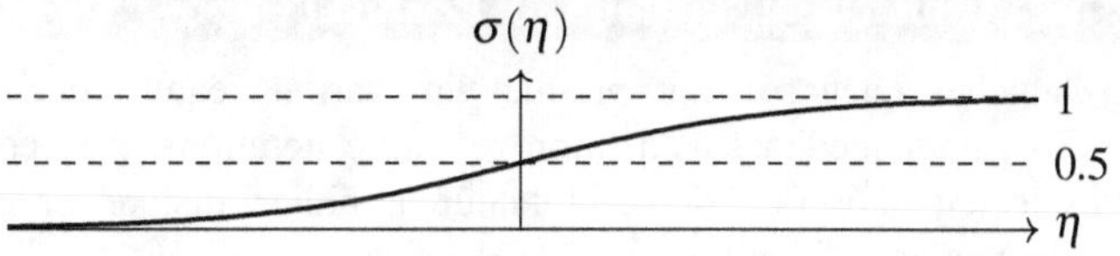

Then the logistic regression model can also be written in terms of the probability of the event occurring as follows:

$$\mathrm{P}(Y = 1|\mathbf{x}) = \sigma(\mathbf{x}^\top \beta).$$

Odds of an event ($Y = 1$) is the ratio of the probability of the event happening ($Y = 1$) to the probability of not happening ($Y = 0$). *Log-odds* is the natural logarithm of the odds. Given $\mathbf{x}$, write the odds of $Y = 1$ as

$$\mathrm{odds}(\mathbf{x}) = \frac{\mathrm{P}(Y = 1|\mathbf{x})}{\mathrm{P}(Y = 0|\mathbf{x})} = \frac{\mathrm{P}(Y = 1|\mathbf{x})}{1 - \mathrm{P}(Y = 1|\mathbf{x})}.$$

Then increase one unit of x_i so that $\mathbf{x} = (1, x_1, \ldots, x_i, \ldots, x_p)$ becomes $\mathbf{x}_i := (1, x_1, \ldots, x_i + 1, \ldots, x_p)$, and then from Equation 7.1, we have

$$\ln(\mathrm{odds}(\mathbf{x}_i)) - \ln(\mathrm{odds}(\mathbf{x})) = \mathbf{x}_i^\top \beta - \mathbf{x}^\top \beta = \beta_i;$$

that is, β_i represents the change in log-odds for a one-unit increase in x_i, keeping other variables constant. Similarly,

$$\frac{\mathrm{odds}(\mathbf{x}_i)}{\mathrm{odds}(\mathbf{x})} = \frac{e^{\mathbf{x}_i^\top \beta}}{e^{\mathbf{x}^\top \beta}} = e^{\beta_i}.$$

So, e^{β_i} represents the "after-to-before" *odds ratio* for a one-unit increase in x_i, keeping other variables constant. Positive coefficients indicate that increasing the feature increases the odds of the event, while negative coefficients indicate the opposite. The magnitude of the coefficient indicates the strength of influence. For more details about logistic regression, see [55, 30], or any standard textbook on generalized linear models such as [76, 4, 35, 40].

Note 7.12 (Complete Separation in Logistic Regression)**.** In practice of logistic regression, *complete separation* refers to a situation where some predictor variables can completely predict the binary response variable; that is, there exists a vector β such that $\beta^\top \mathbf{x}^{(i)} > 0$ with $\mathbf{x}^{(i)}$ the predictors of the observations where $y_i = 1$, and $\beta^\top \mathbf{x}^{(i)} < 0$ with $\mathbf{x}^{(i)}$ the predictors of the observations where $y_i = 0$. Complete separation leads to issues such as nonconvergence in MLE and unreliable parameter estimates as the likelihood function becomes unbounded. This can occur when the dataset is small or some predictors are highly correlated with the response variable. A common practical guideline to decide whether the sample size is sufficient is based on the number of events (i.e., the number of observations with $y_i = 1$) and the number of predictors. A commonly cited rule of thumb is to have at least 10 events per predictor variable to ensure stable estimates in logistic regression; see [87].

Solutions to complete separation include regularization techniques such as L1 penalty used in LASSO regression, L2 penalty used in ridge regression, or a combination of both in elastic net regression, collecting more data to break the separation, and feature selection to remove or combine predictors that cause separation. In R, the `detect_separation()` function from the `detectseparation` package can be used to identify complete separation. For more details about complete separation in logistic regression, see [71, 55].

7.1.7 Cochran's Q Test

When comparing more than two related samples with binary outcomes, Cochran's Q test can be employed to determine if there are significant differences among the samples. This section presents a case study that illustrates the application of Cochran's Q test.

Case 28: Assessing Customer Satisfaction with Product Features

A tech company launches a new software product with various features designed to enhance user experience. To gather feedback and improve future iterations, they conduct a survey asking randomly selected 30 users to rate each feature as either "dislike" or "like", and Table 5.13 summarizes the feedback from the users.

TABLE 5.13: Product Features Feedback.

Feature	Dislike	Like
Feature 1	18	12
Feature 2	26	4
Feature 3	10	20

Determine if there is significant difference between these features according to the responses from the users.

For answers to the initial questions and some initial thoughts about this case study, please refer to Section 5.29. This case is similar to Case 25, except that here we have "dislike" and "like" instead of "no" and "yes", and we have three features instead of "before" and "after". The same group of users rate each of the three features, respectively, so that the responses of "dislike" and "like" among different features can be correlated; that is, we have three related samples with binary outcomes. Similarly to Case 25, here, we also need to use a method that accounts for such dependencies. But we cannot use McNemar's test directly, as it is only applicable to two related samples. Instead, we can use Cochran's Q test, which is an extension of McNemar's test to more than two related samples; see Note 7.13 for more details.

Table 5.13 is not very helpful directly, as we lose information at the user level; that is, we do not know how each user rates the three features, and thus we lose the dependency information between the responses between the features. Unfortunately, this is not an ideal situation and probably can be avoided by keeping more detailed information when one collects the data. The importance of collecting data at its most granular level is never overemphasized if resource allows.

There are many more "Likes" for Feature 3, so it may be worth using Cochran's Q test (see Notes 7.13) for possible subject-level data that are able to lead to the limited information contained in Table 5.13. According to Equation 7.2, only $\sum_{i=1}^{n} R_i^2$ in the test statistic is not provided by the aggregated Table 5.13. The largest p-value will be from the smallest Q value, and thus the smallest value of $\sum_{i=1}^{n} R_i^2$. We know that $\sum_{i=1}^{30} R_i = 36$ and $R_i \in \{0,1,2,3\}$. So, the smallest $\sum_{i=1}^{n} R_i^2$ will be achieved when each R_i takes values around $36/30 = 1.2$. That is, when R_i's take values of 1 or 2 we may achieve the smallest value of $\sum_{i=1}^{n} R_i^2$. Let x_i be the number of users who only give "Like" to the feature $i, i = 1,2,3$, and x_{12}, x_{13}, x_{23} be the number of users who only give "Like" to the feature 1,2, 1,3 and 2,3, respectively. Then the following equations can be easily solved: $x_1 + x_{12} + x13 = 12$, $x_2 + x_{12} + x_{23} = 4$, $x_3 + x_{13} + x_{23} = 20$, $x_1 + x_2 + x_3 + x_{12} + x_{13} + x_{23} = 30$ to get $x_1 = 9, x_2 = 0, x_3 = 15, x_{12} = 1, x_{13} = 2$ and $x_{23} = 3$. Therefore,

$$\min \sum_{i=1}^{n} R_i^2 = 24 \times 1^2 + 6 \times 2^2 = 48,$$

and the smallest test statistic is

$$Q = \frac{(3-1)(3(12^2+4^2+20^2)-36^2)}{3\times 36-48} = 12.8.$$

Then the p-value is 0.0017 that is below the common significance levels, suggesting that there is a significant difference between the features in terms of the proportions of "Like".

```
p_value <- 1-pchisq(12.8, df=2)
print(p_value)
: [1] 0.001661557
```

Alternatively, with the subject-level data constructed above, we can use the R function `DescTools::CochranQTest()` directly to get the same p-value.

```
data <- matrix(c(c(1, 1, 0), rep(c(1, 0, 1), 2), rep(c(0, 1, 1), 3),
                 rep(c(1, 0, 0), 9), rep(c(0, 0, 1), 15)),
               ncol = 3, byrow = TRUE)
colnames(data) <- c("Feature_1", "Feature_2", "Feature_3")

# Run Cochran's Q test
DescTools::CochranQTest(data)
:
:  Cochran's Q test
:
: data:  y
: Q = 12.8, df = 2, p-value = 0.001662
```

Conclusions
In this case study, a Cochran's Q test is conducted to compare the proportions of users who "Like" three different features of a product. The test reveals a significant difference in the proportions of "Like" among the three features (p-value is 0.0017). Although users are randomly sampled, the sample size is small (30 users), so caution should be taken when generalizing the results to the population of all users of the product. This is an observational study, and causality cannot be established.

Note 7.13 (Cochran's Q Test)**.** *Cochran's Q test* ([24]) is a nonparametric statistical test used to determine whether there are significant differences between three or more related samples when the response variable is binary (e.g., "Like" vs. "Dislike"). It is an extension of McNemar's test to the case of more than two conditions. Suppose n subjects each provide binary responses across k conditions. The hypotheses for Cochran's Q test are the following:

$$H_0 : \text{The proportion of successes is the same across all } k \text{ conditions;}$$
$$H_a : \text{At least one condition has a different proportion of successes.}$$

Let y_{ij} denote the response of subject $j, j = 1,\ldots,n$ under condition $i, i = 1,\ldots,k$, where $y_{ij} = 1$ indicates a "success" (e.g., Like) and $y_{ij} = 0$ otherwise. Define

$$T_i = \sum_{j=1}^{n} y_{ij}, \quad R_j = \sum_{i=1}^{k} y_{ij}, \quad G = \sum_{j=1}^{n}\sum_{i=1}^{k} y_{ij}.$$

Here, T_i is the number of successes under condition i, R_j is the number of successes for subject j, and G is the total number of successes in all subjects and conditions. The Cochran's Q statistic is given by

$$Q = \frac{(k-1)\left[k\sum_{i=1}^{k} T_i^2 - G^2\right]}{kG - \sum_{j=1}^{n} R_j^2}. \tag{7.2}$$

A more intuitive way to define the test statistic is the following:

$$Q = (k-1)\frac{\sum_{i=1}^{k}(T_i - G/k)^2}{(1/k)\sum_{j=1}^{n} R_j(k-R_j)},$$

where the numerator is the between-conditions variability of the number of successes, and the denominator is the within-subject variability of the number of successes. Under the null hypothesis that the success probabilities are equal across all k conditions, Q approximately follows a Chi-square distribution with $k-1$ degrees of freedom.

A significant result of Cochran's Q tests indicates that at least one condition differs in the proportion of successes, and post hoc tests such as pairwise McNemar's tests can be used to determine which conditions differ, when appropriate adjustments for multiple comparisons will be needed; see Note 6.12 for notes on multiple comparisons. We refer to [28] for more details on Cochran's Q test and its applications.

7.1.8 Logistic Mixed-Effects Model

When there are repeated measures for each subject and the response variable is binary, logistic mixed-effects models can be used to analyze the data. In this case study, we revisit the customer satisfaction data for three product features discussed in Section 7.1.7. This time, we have subject-level data available, which allows us to explore logistic mixed-effects models.

Case 29: Assessing Customer Satisfaction with Product Features (with Subject-Level Data)

A tech company launches a new software product with various features designed to enhance user experience. To gather feedback and improve future iterations, they conduct a survey asking randomly selected 30 users to rate each feature as either "dislike" or "like", and Table 5.14 summarizes the feedback from the users and subject-level data is provided.

TABLE 5.14: Product Features Feedback.

Feature	Dislike	Like
Feature 1	18	12
Feature 2	26	4
Feature 3	10	20

Determine if there is significant difference between these features according to the responses from the users.

For answers to the initial questions and some initial thoughts about this case study, please refer to Section 5.29 and Section 5.30. This case is the same as Case 28 except that there is subject-level data provided in this case. Of course, we can still use Cochran's Q test for this case. However, with the subject-level data that provides more detailed information, we can explore more advanced methods. In this section, we discuss the logistic mixed-effects model, a popular method for analyzing dependent binary response variables. First, we prepare the data as follows:

```
library(lme4)
library(ggplot2)
data <- read.csv("data/features-like-dislike.csv")

head(data,2); tail(data,2)
:   feature user like
: 1       1    1    0
: 2       1    2    1
:    feature user like
: 89       3   29    1
: 90       3   30    1
## Summary of the feedback of all users:
table(feature=data$feature, like=data$like)
:          like
: feature  0  1
:        1 18 12
:        2 26  4
:        3 10 20
## Likes and dislikes distributed among the three features:
data$feature <- as.factor(data$feature)
data$user <- as.factor(data$user)
data$like <- as.factor(data$like)
```

Figure 7.7 suggests that Feature 3 could have a significantly higher proportion of Likes, and Feature 2 could have a significantly lower proportion of Likes. We can use logistic mixed-effects models to analyze the data more formally.

```
set.seed(123)
ggplot(data, aes(x=feature, y=like)) + theme_bw() +
  geom_point(position = position_jitter(width=0.1, height=0.1)) +
  scale_x_discrete(breaks = c(1,2,3)) + scale_y_discrete(breaks = c(0,1))
```

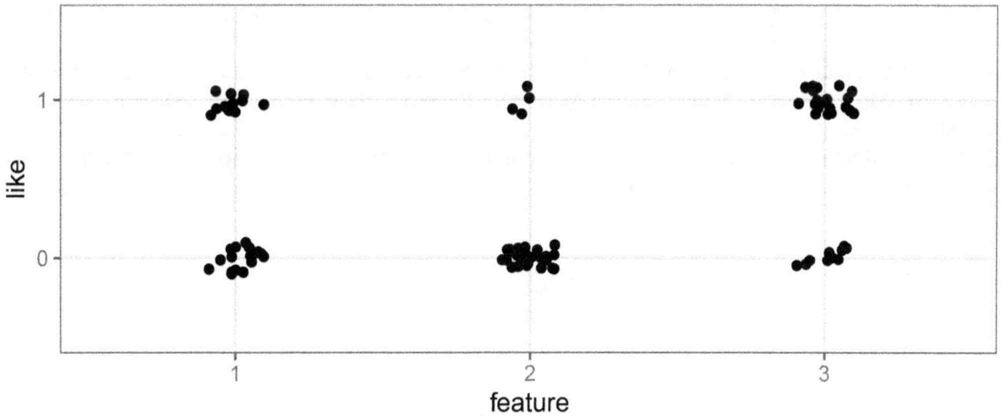

FIGURE 7.7: Distribution of Like Across Features.

As the response variable is binary, we use logistic regression with mixed effects as follows:

$$\ln\left(\frac{P(Y_{ij}=1|V_i=v_i)}{P(Y_{ij}=0|V_i=v_i)}\right)=\mu+\tau_j+v_i,\quad i=1,\ldots,30,\quad j=1,2,3. \tag{7.3}$$

Here, Y_{ij} represents the response variable: $Y_{ij}=1$ for "like" and $Y_{ij}=0$ for "dislike". The ratio $P(Y_{ij}=1)/P(Y_{ij}=0)$ is called the "odds" (of likes), and thus the left-hand side of Equation 7.3 is called the "log-odds" (of likes). μ is the overall intercept, a fixed effect. τ_j's are the fixed effects associated with the features, and we have three constant numbers τ_1, τ_2 and τ_3 for the three features, respectively. V_i is the random effect associated with user i, and $V_1,\ldots,V_{30}$ are assumed to be identically and independently distributed according to a normal distribution with mean 0 and an unknown variance that is to be estimated. We are not aiming to estimate the value of each v_i. Instead, we are interested in estimating the variability of V_i to understand how much the users differ from each other, and the main interest is in the fixed effects τ_j's to understand how the features differ from each other in terms of the odds of being liked. We refer to Note 7.14 for more details on logistic mixed-effects models.

In R, the random effect is represented by `(1|user)` indicating that the data are grouped by users. The 1 here indicates that each user has their own baseline log-odds of liking a feature, and the random intercept v_i captures the deviation of user i's baseline log-odds from the overall intercept μ. The fixed effect of the feature is represented by `feature` in the formula. The following R codes fit the logistic mixed-effects model using the `glmer()` function from the R package `lme4`. Unlike LMMs, which have closed-form expressions for the likelihood, GLMMs require numerical approximations to evaluate the likelihood. The estimates are based on an adaptive Gaussian Hermite approximation of the likelihood, and the parameter "nAGQ" indicates that we use 25 integration points for the numerical integration, which can provide reasonable accuracy for this case.

```
model <- glmer(like ~ feature + (1|user), family = binomial,
               data=data, nAGQ=25)
fixef(model)
: (Intercept)     feature2     feature3
:  -0.5767638  -1.9568817   1.5438043
VarCorr(model)
:  Groups Name        Std.Dev.
:  user   (Intercept) 1.4865
```

To test whether there are significant differences between these three features, we can fit another model excluding the feature and then compare it with the previous model. The following results tell us that there are significant differences with a p-value that is well below the common significance levels. Note that both models are fitted using maximum likelihood estimations, and thus the likelihood ratio test is valid here. The restricted maximum likelihood (REML) method is only applicable to LMMs but not to GLMMs.

```
model_reduced <- glmer(like ~ (1|user), family = binomial,
                       data=data, nAGQ=25)
anova(model, model_reduced)
: Data: data
: Models:
: model_reduced: like ~ (1 | user)
: model: like ~ feature + (1 | user)
:               npar    AIC    BIC  logLik -2*log(L)  Chisq Df Pr(>Chisq)
: model_reduced    2 124.67 129.67 -60.334   120.669
: model            4 105.98 115.98 -48.992    97.985 22.684  2  1.186e-05 ***
```

```
: ---
: Signif. codes:  0 '***' 0.001 '**' 0.01 '*' 0.05 '.' 0.1 ' ' 1
```

For diagnosis of GLMMs, simulation-based residuals using R package DHARMa provide robust and intuitive methods. The Kolmogorov-Smirnov (KS) test is used to check whether the distribution of the residuals deviates from the expected uniform distribution. The outlier test checks whether there are any significant outliers among the residuals. The following R codes simulate residuals using the fitted model, and the QQ plot on uniform scales can be used to check the overall fitting. The plots, the KS test, and the outlier test suggest a good fit; "n.s." in the plot stands for "not significant".

```
library(DHARMa)
set.seed(123)
simres <- simulateResiduals(model, n = 2000)
plotQQunif(simres, testDispersion=FALSE, lcol="black", pch=16)
```

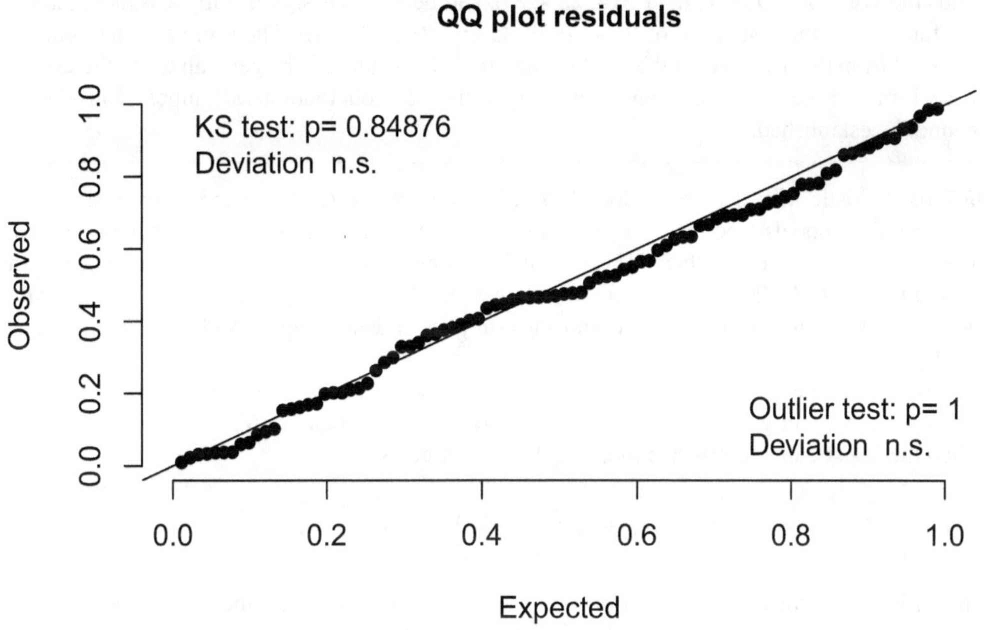

FIGURE 7.8: QQ Plot for Scaled Simulated Residuals.

To interpret a fitted logistic mixed-effects model, we can exponentiate both the left- and right-hand sides of Equation 7.3 to understand how the features affect the odds: $P(Y_{ij} = 1)/P(Y_{ij} = 0)$. The following R outputs are the subject-specific *odds ratios* for Feature 2 vs. Feature 1 and Feature 3 vs. Feature 1. These odds ratios are for subject-specific interpretation: for a typical user, after accounting for their random intercept (i.e., conditional on their random effect v_i), the odds ratio between Feature 2 and Feature 1 is 0.141, and the odds ratio between Feature 3 and Feature 1 is 4.682. Confidence intervals can be constructed using profile confidence intervals or Wald confidence intervals when the sample size is large enough. The results suggest that Feature 3 is significantly better than Feature 1 (as the lower bound of the odds ratios is greater than 1), and Feature 1 is significantly better than Feature 2 (as the upper bound of the odds ratios is less than 1).

```
# Subject-specific odds ratios
OR <- exp(fixef(model))
OR_CI <- exp(confint(model, parm = "beta_", method = "profile"))
cbind(OR[-1],OR_CI[-1,]) #
:                  2.5 %    97.5 %
: feature2 0.1412983 0.02223782  0.610177
: feature3 4.6823697 1.31380094 22.422971
```

Conclusions

In this case study, we analyze the customer satisfaction data for three product features using logistic mixed-effects models. The model uses fixed effects for the features and random effects for individual users to account for the correlation in repeated measures. The analysis reveals significant differences in user preferences among the three features (p-value is 1.2×10^{-5} from the likelihood ratio test). Specifically, for a typical user, conditional on their random effect, Feature 3 is found to be significantly more liked than Feature 1 with the odds ratio of 4.682 (95% C.I.: (1.314, 22.423)), and Feature 2 is significantly less liked than Feature 1 with the odds ratio of 0.141 (95% C.I.: (0.022, 0.610)). The sample is randomly selected from the user base of the product, and thus the results can be generalized to the user base. Furthermore, since this is an observational study without random assignment, causality cannot be established.

Note 7.14 (Logistic Mixed-Effects Model)**.** *Logistic mixed-effects models* are an extension of logistic regression models (see Note 7.11) with random effects added to account for the correlation between binary response variables. These models belong to the broader class of generalized linear mixed models (GLMMs) and are particularly useful when data are collected in a hierarchical or clustered structure, such as repeated measures for subjects or data grouped by clusters (e.g., schools, hospitals).

Let Y_{ij} denote the binary response ($Y_{ij} \in \{0,1\}$) for subject j in group i and $\mathbf{V}_i$ be the random effects associated with group i. Let $p_{ij} = \mathrm{P}(Y_{ij} = 1 | \mathbf{V}_i = \mathbf{v}_i)$ so that $Y_{ij} \big| \mathbf{V}_i = \mathbf{v}_i \sim \text{Bernoulli}(p_{ij})$, and then the logistic mixed-effects model can be expressed as:

$$\text{logit}(p_{ij}) = \ln\left(\frac{p_{ij}}{1-p_{ij}}\right) = \mathbf{x}_{ij}^{\top}\beta + \mathbf{z}_{ij}^{\top}\mathbf{v}_i,$$

where $\mathbf{x}_{ij}$ is the vector of fixed effect covariates with corresponding coefficients β, and $\mathbf{z}_{ij}$ is the design vector for the random effects $\mathbf{V}_i$ that are often assumed to be independent and identically distributed as a multivariate normal distribution:

$$\mathbf{V}_i \overset{iid}{\sim} \mathcal{N}(\mathbf{0}, \Sigma),$$

where Σ is the covariance matrix of the random effects.

In R, the `lme4` package provides the `glmer` function to fit logistic mixed-effects models. The syntax is similar to that of linear mixed-effects models (see Note 6.25), except that the family is set to `binomial` for logistic regression. For more details on logistic mixed-effects models, and GLMMs in general, see [78, 40].

7.2 Multiple Categorical Data

In Section 7.1, methods for analyzing binary data are discussed. Some of those methods such as Chi-square tests can be used directly for categorical data with more than two categories, while others need to be extended. In this section, we present case studies for analyzing categorical data with more than two categories.

7.2.1 Multinomial Logistic Regression

When the response variable is categorical with more than two unordered categories, multinomial logistic regression can be used to model the relationship between the response and a set of predictor variables. This method generalizes binary logistic regression to handle multiple categories. We use the following case study to illustrate the application of multinomial logistic regression.

Case 30: Career Choices Among College Students

A university career center wants to understand how students choose their first job sector after graduation. Graduates typically enter one of three mutually exclusive job sectors: Industry, Academia, Government. The career center randomly collects 300 recently graduated and graduating students and records the following variables: JobSector, GPA, Internship (Yes/No), and Major (STEM/Non-STEM). The goal is to understand how these predictors influence the probability of choosing each job sector.

For answers to the initial questions and some initial thoughts about this case study, refer to Section 5.31. First, we load the data and assign appropriate data types to the variables. In particular, we convert the response variable JobSector to a factor and set Government as the reference category, as in multinomial logistic regression, the choice of reference category affects the interpretation of the model coefficients. The R function `multinom()` from the `nnet` package will be used to fit the multinomial logistic regression model.

```
library(nnet)
data <- read.csv("data/career-choices.csv")
data$Internship <- as.factor(data$Internship)
data$JobSector <- as.factor(data$JobSector)
data$Major <- as.factor(data$Major)

# Set reference category
data$JobSector <- relevel(data$JobSector, ref = "Government")
```

We fit the multinomial logistic regression model using GPA, Internship, and Major as the main effects, and later on we will consider whether adding interaction terms can improve the model fit. An alternative approach, especially when the task is more about prediction than inference, is to start with a full model including all main and interaction effects, and then use `step()` function for a quick variable selection using AIC as the criterion. However, in this case study, we are more interested in understanding the effects of the predictors, so we will likelihood ratio tests to assess the significance of the predictors and interaction terms, which is more straightforward for interpretation than AIC-based model selection. There is no such `drop1()` function for multinomial logistic regression, so we will manually fit the candidate models and perform likelihood ratio tests to compare the models. The initial model with only main effects is fitted as follows:

```
main_mnl <- multinom(JobSector ~ GPA + Internship + Major, data = data,
                     trace = FALSE)
```

Then we add each two-way interaction term separately, and test whether adding these interaction terms significantly improves the model fit using a likelihood ratio test. Three-way interactions can be included if there is a rational basis for doing so, but they are often omitted for simplicity.

```
main_mnl1 <- multinom(JobSector ~ GPA*Internship + Major, data = data,
                      trace = FALSE)
main_mnl2 <- multinom(JobSector ~ GPA + Internship*Major, data = data,
                      trace = FALSE)
main_mnl3 <- multinom(JobSector ~ GPA*Major + Internship, data = data,
                      trace = FALSE)
```

All p-values from the likelihood ratio tests below are greater than 0.05, indicating that the interaction terms do not significantly improve the model fit. In what follows, we will be using the model with only main effects as the candidate model for further analysis and interpretation. However, the p-value for the interaction between Internship and Major is 0.06, which is only slightly larger than the common significance level of 0.05, so we may consider including this interaction term in the model if there is a strong theoretical justification for it.

```
anova(main_mnl1, main_mnl, test = "Chisq")
: Likelihood ratio tests of Multinomial Models
:
: Response: JobSector
:                       Model Resid. df Resid. Dev   Test    Df  LR stat.   Pr(Chi)
: 1 GPA + Internship + Major       592   462.0346
: 2 GPA * Internship + Major       590   461.0432 1 vs 2     2 0.9913752 0.6091519
anova(main_mnl2, main_mnl, test = "Chisq")
: Likelihood ratio tests of Multinomial Models
:
: Response: JobSector
:                       Model Resid. df Resid. Dev   Test    Df LR stat.    Pr(Chi)
: 1 GPA + Internship + Major       592   462.0346
: 2 GPA + Internship * Major       590   456.3923 1 vs 2     2  5.64233 0.05953654
anova(main_mnl3, main_mnl, test = "Chisq")
: Likelihood ratio tests of Multinomial Models
:
: Response: JobSector
:                       Model Resid. df Resid. Dev   Test    Df LR stat.   Pr(Chi)
: 1 GPA + Internship + Major       592   462.0346
: 2 GPA * Major + Internship       590   460.7915 1 vs 2     2 1.243057 0.5371228
```

Then, we assess the overall significance of each predictor variable using the `Anova()` function from the `car` package, which performs a likelihood ratio test for each predictor in the multinomial logistic regression model. The results indicate that `GPA` (p-value: 1.5×10^{-7}) and `Internship` (p-value: 3.4×10^{-4}) are significant predictors of job sector choice, while `Major` is not significant (p-value: 0.77).

```
car::Anova(main_mnl)
: Analysis of Deviance Table (Type II tests)
:
: Response: JobSector
```

```
:              LR Chisq Df Pr(>Chisq)
: GPA           31.3821  2  1.533e-07 ***
: Internship    15.9534  2  0.0003434 ***
: Major          0.5226  2  0.7700573
: ---
: Signif. codes:  0 '***' 0.001 '**' 0.01 '*' 0.05 '.' 0.1 ' ' 1
```

The candidate model is then refit using only the significant predictors GPA and Internship. The summary of the fitted model provides the estimated coefficients and their standard errors for each category relative to the reference category.

```
mnl_model <- multinom(JobSector ~ GPA + Internship,
                      data = data, trace = FALSE)
summary(mnl_model)
: Call:
: multinom(formula = JobSector ~ GPA + Internship, data = data,
:     trace = FALSE)
:
: Coefficients:
:           (Intercept)        GPA InternshipYes
: Academia -2.57057032 1.47654941   -0.04530789
: Industry  0.03142959 0.01325206    1.20056539
:
: Std. Errors:
:          (Intercept)       GPA InternshipYes
: Academia    1.293916 0.4384924     0.4037501
: Industry    1.418377 0.4830149     0.4682041
:
: Residual Deviance: 462.5572
: AIC: 474.5572
```

From the above outputs, we can calculate the z-values and p-values for each coefficient to assess their significance. The results indicate that there are strong evidence that increasing GPA is associated with higher probabilities of choosing Academia over Government (p-value: 0.0008), and that having an internship is associated with higher probabilities of choosing Industry over Government (p-value: 0.01).

```
z_values <- summary(mnl_model)$coefficients /
                             summary(mnl_model)$standard.errors
p_values <- 2 * (1 - pnorm(abs(z_values)))
p_values
:          (Intercept)          GPA InternshipYes
: Academia  0.04696009 0.0007589923    0.91065082
: Industry  0.98232124 0.9781118805    0.01034162
```

We use the exponents of the estimated coefficients to interpret the effects of the predictors. For a one-unit increase in GPA, the relative risk of choosing Academia over Government increases by a factor of $\exp(1.477) \approx 4.38$, holding Internship constant. Similarly, having an internship increases the relative risk of choosing Industry over Government by a factor of $\exp(1.201) \approx 3.32$, holding GPA constant.

Finally, we can use the fitted model to predict the probabilities of choosing each job sector for new students with specified characteristics. For example, the following code predicts the job sector probabilities for a student with a GPA of 2.8 without an internship, and another student with a GPA of 3.8 without an internship. It can be verified that increasing one-unit of GPA is associated with an increase of the relative risk of choosing Academia over Government by a factor of approximately 4.38 ($\approx (0.90932663/0.04348373)/(0.6975728/0.1460339)$), consistent with our earlier interpretation.

```
new_student1 <- data.frame(GPA = 2.8, Internship = "No")
predict(mnl_model, new_student1, type = "probs")
: Government   Academia   Industry
:  0.1460339  0.6975728  0.1563933
new_student2 <- data.frame(GPA = 3.8, Internship = "No")
predict(mnl_model, new_student2, type = "probs")
: Government   Academia   Industry
: 0.04348373 0.90932663 0.04718964
```

Conclusions

In this case study, we use a multinomial logistic regression model to analyze how students choose their first job sector after graduation based on their GPA, internship experience, and majors. The analysis reveals that both GPA and internship experience significantly influence job sector choice, while the major does not have a significant effect. Specifically, higher GPA increases the likelihood of choosing Academia over Government by a factor of 4.38 (p-value: 0.0008) for one-unit increase of GPA, while having an internship increases the likelihood of choosing Industry over Government by a factor of 3.32 (p-value: 0.01). The students are randomly sampled, allowing for generalization of the findings to all recently graduated and graduating students of the university. However, as this is an observational study, causality cannot be established from the results.

Note 7.15 (Multinomial Logit Model)**.** The *multinomial logit model* is used to model a nominal response variable with more than two unordered categories as a function of predictor variables. It is an extension of binary logistic regression model (see Note 7.11) to handle multiple unordered categories in the response variable. Let $Y \in \{1,2,\ldots,J\}$ denote the response variable, and let J be the reference category. For a vector of covariates $\mathbf{x} = (1,x_1,\ldots,x_p)$, the multinomial logit model specifies the log-odds of category j to the reference category as

$$\ln\left(\frac{\mathrm{P}(Y=j\mid\mathbf{x})}{\mathrm{P}(Y=J\mid\mathbf{x})}\right) = \mathbf{x}^\top\beta_j, \quad j=1,\ldots,J-1. \tag{7.4}$$

Equivalently, the probabilities of each non-reference category can be expressed as

$$\mathrm{P}(Y=j\mid\mathbf{x}) = \frac{\exp(\mathbf{x}^\top\beta_j)}{1+\sum_{k=1}^{J-1}\exp(\mathbf{x}^\top\beta_k)}, \quad j=1,\ldots,J-1,$$

and the probability of the reference category is

$$P(Y = J \mid \mathbf{x}) = \frac{1}{1 + \sum_{k=1}^{J-1} \exp(\mathbf{x}^\top \beta_k)}.$$

Given $\mathbf{x}$, write the relative risk of category j to the reference category as

$$\mathrm{RR}_j(\mathbf{x}) = \frac{P(Y = j|\mathbf{x})}{P(Y = J|\mathbf{x})}.$$

Increase one unit of x_i so that $\mathbf{x} = (1, x_1, \ldots, x_i, \ldots, x_p)$ becomes $\mathbf{x}_i := (1, x_1, \ldots, x_i + 1, \ldots, x_p)$, and then from Equation 7.4, we have

$$\ln(\mathrm{RR}_j(\mathbf{x}_i)) - \ln(\mathrm{RR}_j(\mathbf{x})) = \mathbf{x}_i^\top \beta_j - \mathbf{x}^\top \beta_j = \beta_{ji},$$

where β_{ji} is the ith element of the coefficient vector β_j for category j. Therefore, β_{ji} represents the change in log-relative risk for a one-unit increase in x_i, keeping other variables constant. Similarly,

$$\frac{\mathrm{RR}_j(\mathbf{x}_i)}{\mathrm{RR}_j(\mathbf{x})} = \frac{e^{\mathbf{x}_i^\top \beta_j}}{e^{\mathbf{x}^\top \beta_j}} = e^{\beta_{ji}}.$$

So, $e^{\beta_{ji}}$ represents the "after-to-before" relative risk ratio for a one-unit increase in x_i, keeping other variables constant. The multinomial logit model assumes that the odds of choosing one category over another are independent of other alternatives, and maximum likelihood estimation is used to estimate the model parameters. For more details, we refer to [3].

7.2.2 Proportional Odds Model

When the response variable is ordinal, we need to consider the order information in the modeling. A common approach for modeling ordinal response variables is to use the proportional odds model (POM) as described in Note 7.16. In what follows, we illustrate how to fit a POM using a case study on customer satisfaction data.

Case 31: Customer Satisfaction Levels

A company collects survey data from 200 customers about their satisfaction with a service. The goal is to understand how service speed (in minutes) and staff friendliness (rated 1-10) influence satisfaction levels.

For answers to the initial questions and some initial thoughts about this case study, refer to Section 5.32. The response variable has three ordered levels: Dissatisfied, Neutral, and Satisfied. We treat the response as ordinal data and in what follows, we fit a proportional odds model to analyze the effects of waiting time and staff friendliness on customer satisfaction levels.

First, load the data and define the ordinal response variable as an ordered factor with levels from Dissatisfied to Satisfied. Then we can check the distribution of satisfaction levels and the summary statistics of the two predictors: waiting time and staff friendliness.

```
# Predicting Customer Satisfaction Levels
library(MASS)
library(ggplot2)
library(tidyr)
library(dplyr)
data <- read.csv("data/customer-satisfactory.csv")

# Exploratory data analysis
data$satisfaction <- factor(data$satisfaction,
                            levels=c("Dissatisfied","Neutral",
                                     "Satisfied"), ordered = TRUE)

# Check distribution of satisfaction levels
table(data$satisfaction)
:
: Dissatisfied      Neutral    Satisfied
:           93           37           70
summary(data$waiting_time)
:    Min. 1st Qu.  Median    Mean 3rd Qu.    Max.
:   5.400   8.700   9.900   9.985  11.125  16.500
summary(data$staff_friendliness)
:    Min. 1st Qu.  Median    Mean 3rd Qu.    Max.
:    1.00    3.00    6.00    5.51    8.00   10.00
```

For an exploratory data analysis, we plot the satisfaction levels against waiting time and staff friendliness, respectively. Figure 7.9 suggests that dissatisfied customers may be associated with longer waiting time, and Figure 7.10 suggests that satisfied customers may be associated with higher friendliness.

```
# Satisfaction by Waiting Time
ggplot(data, aes(x = waiting_time, y = satisfaction)) +
  geom_boxplot() + theme_bw() +
  stat_summary(fun = "mean", geom = "point", shape = 23, size = 2) +
  labs(title = "", x = "Waiting Time", y = "Satisfaction Level")
```

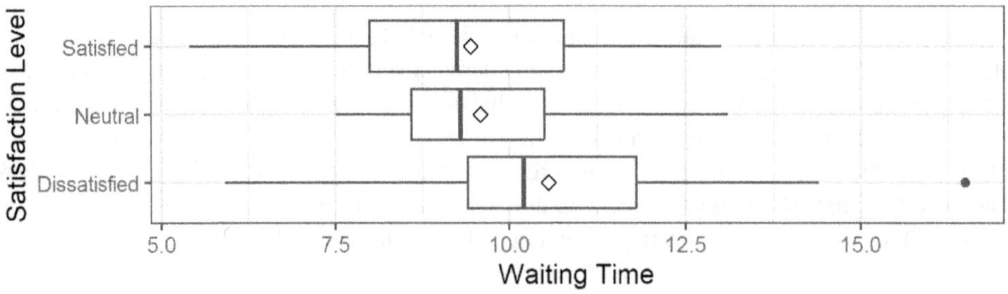

FIGURE 7.9: Satisfaction by Waiting Time.

```
# Satisfaction by Staff Friendliness
ggplot(data, aes(x = staff_friendliness, y = satisfaction)) +
  geom_boxplot() + theme_bw() +
  stat_summary(fun = "mean", geom = "point", shape = 23, size = 2) +
  labs(title = "", x = "Staff Friendliness", y = "Satisfaction Level")
```

FIGURE 7.10: Satisfaction by Staff Friendliness.

Then fit a proportional odds model using the R function `MASS::polr()` including all the main effects and their interaction terms. An analysis of deviance suggests that the interaction term is not significant.

```
# Fit ordinal regression model
model <- polr(satisfaction ~ waiting_time*staff_friendliness, data = data)
car::Anova(model)
: Analysis of Deviance Table (Type II tests)
:
: Response: satisfaction
:                                  LR Chisq Df Pr(>Chisq)
: waiting_time                       35.851  1   2.13e-09 ***
: staff_friendliness                141.375  1  < 2.2e-16 ***
: waiting_time:staff_friendliness     0.000  1     0.9954
: ---
: Signif. codes:  0 '***' 0.001 '**' 0.01 '*' 0.05 '.' 0.1 ' ' 1
```

Therefore, we fit the model again without the interaction term as follows:

```
model <- polr(satisfaction ~ waiting_time+staff_friendliness, data = data)
car::Anova(model)
: Analysis of Deviance Table (Type II tests)
:
: Response: satisfaction
:                     LR Chisq Df Pr(>Chisq)
: waiting_time          35.851  1   2.13e-09 ***
: staff_friendliness   141.375  1  < 2.2e-16 ***
: ---
: Signif. codes:  0 '***' 0.001 '**' 0.01 '*' 0.05 '.' 0.1 ' ' 1
summary(model)
```

```
: Call:
: polr(formula = satisfaction ~ waiting_time + staff_friendliness,
:     data = data)
:
: Coefficients:
:                      Value Std. Error t value
: waiting_time       -0.6314    0.11845   -5.33
: staff_friendliness  0.8848    0.09986    8.86
:
: Intercepts:
:                      Value   Std. Error t value
: Dissatisfied|Neutral -1.7092  1.0404    -1.6428
: Neutral|Satisfied    -0.1617  1.0383    -0.1557
:
: Residual Deviance: 256.401
: AIC: 264.401
```

Therefore, based on Equation 7.5, the fitted model can be expressed as follows:

$$\text{logit}(\text{P}(Y \leq \text{Dissatisfied})) = -1.7 + 0.63 \times \text{waiting} - 0.88 \times \text{friendliness};$$

$$\text{logit}(\text{P}(Y \leq \text{Neutral})) = -0.16 + 0.63 \times \text{waiting} - 0.88 \times \text{friendliness}.$$

The above two equations share the same regression coefficients -0.63 and 0.88, following the parallel assumption, which is why we call such a model a proportional odds model. The proportional odds assumption can be tested using a graphical check that involves plotting logits for each threshold to assess their parallelism, or statistical tests such as the Brant test, which compare POM to a nonproportional model. If the assumption holds, the influence of predictors is consistent across all thresholds.

First, we use graphical methods for checking the parallel assumption. Figures 7.11 and 7.12 illustrate how empirical log-odds change over waiting time and staff friendliness, respectively.

```
# graphical methods for checking the parallel assumption
data$waiting_bin <- cut(data$waiting_time, breaks = 3)  # 5 bins
data$friendliness_bin <- cut(data$staff_friendliness, breaks = 3)

emp_logits_waiting <- data %>%
  group_by(waiting_bin) %>%
  summarise(
    P_Dissat = mean(satisfaction == "Dissatisfied"),
    P_Neutral_or_less = mean(satisfaction %in% c("Dissatisfied",
                                                  "Neutral")),
    Logit_Dissat = log(P_Dissat / (1-P_Dissat)),
    Logit_Neutral_or_less = log(P_Neutral_or_less /
                                   (1-P_Neutral_or_less)))

emp_logits_waiting_long <- pivot_longer(emp_logits_waiting,
                                        cols = starts_with("Logit"),
                                        names_to = "Threshold",
```

```
                                        values_to = "Logit")
ggplot(emp_logits_waiting_long,
       aes(x = waiting_bin, y = Logit, color = Threshold,
           group = Threshold)) + geom_line() + geom_point() +
       scale_color_grey(start = 0.3, end = 0.8) +
       labs(title = "", x = "Waiting Time Bin",
            y = "Empirical Log-Odds") + theme_bw()
```

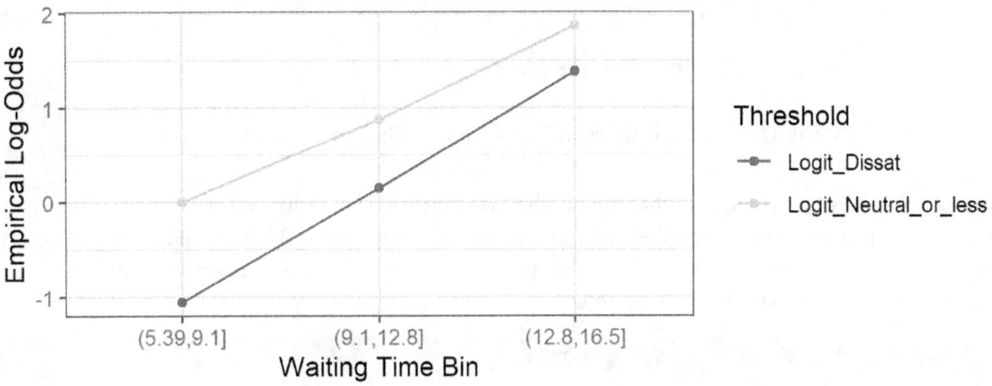

FIGURE 7.11: Empirical Log-Odds by Waiting Time Bin.

```
emp_logits_friendliness <- data %>%
  group_by(friendliness_bin) %>%
  summarise(P_Dissat = mean(satisfaction == "Dissatisfied"),
   P_Neutral_or_less = mean(satisfaction %in% c("Dissatisfied","Neutral")),
   Logit_Dissat = log(P_Dissat / (1-P_Dissat)),
   Logit_Neutral_or_less = log(P_Neutral_or_less / (1-P_Neutral_or_less)))

emp_logits_friendliness_long <- pivot_longer(
  emp_logits_friendliness, cols = starts_with("Logit"),
  names_to = "Threshold", values_to = "Logit")

ggplot(emp_logits_friendliness_long,
       aes(x = friendliness_bin,
           y = Logit, color = Threshold, group = Threshold)) +
  geom_line() + geom_point() + scale_color_grey(start = 0.3, end = 0.8) +
  labs(title = "", x = "Staff Friendliness Bin",
       y = "Empirical Log-Odds") + theme_bw()
```

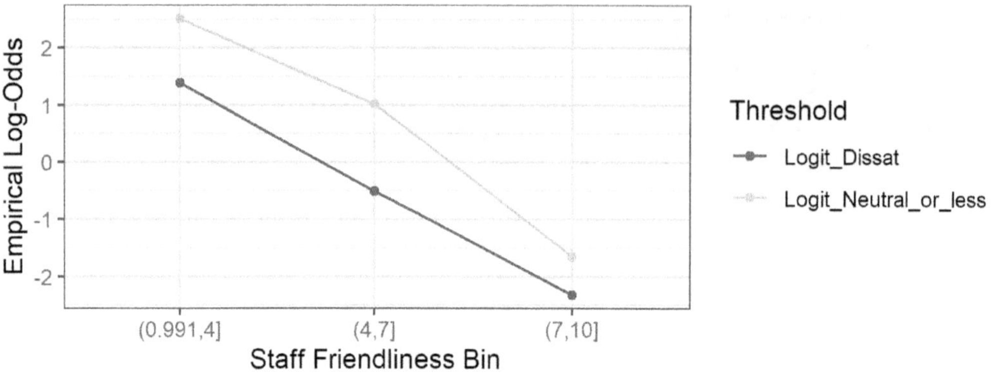

FIGURE 7.12: Empirical Log-Odds by Staff Friendliness Bin.

Both empirical plots suggest that the parallel assumption may hold, and in what follows we use the Brant test to formally check the assumption. The p-values are $0.27, 0.29$ and 0.65 for the omnibus, waiting time, and staff friendliness, respectively, suggesting that there is no strong statistical evidence to reject the parallel assumption.

```
# Brant tests for parallel assumptions
brant::brant(model)
: ----------------------------------------------------
: Test for  X2 df probability
: ----------------------------------------------------
: Omnibus   2.61 2 0.27
: waiting_time  1.11 1 0.29
: staff_friendliness 0.21 1 0.65
: ----------------------------------------------------
:
: H0: Parallel Regression Assumption holds
```

According to Equation 7.5, the regression coefficients β reflects changes of log-odds by one unit increase of **x**. Therefore, we can interpret the results as follows:

- For every one unit increase in waiting time the log-odds of being less satisfied increases 0.63 holding constant all other variables.
- For every one unit increase in staff friendliness the log-odds of being less satisfied decreases 0.88 holding constant all other variables.

Moreover, the 95% confidence intervals for such effects can be derived from the following

```
## use log-odds to interpret
ci <-  confint(model)
cbind(Change_of_Logit = coef(model), ci)
:                       Change_of_Logit       2.5 %      97.5 %
: waiting_time               -0.6313753 -0.8759635 -0.4095981
: staff_friendliness          0.8847518  0.7008079  1.0938727
```

For a given staff friendliness, say the average value of the friendliness, conditional effects of waiting time on log-odds of satisfaction levels can be plotted as follows:

```
# note the negative sign before coef(model)
mean_friendliness <- mean(data$staff_friendliness)
waiting_range <- seq(min(data$waiting_time), max(data$waiting_time),
                     length.out = 100)
logits_waiting <- data.frame(waiting_time = waiting_range,
  "Dissatisfied|Neutral" = model$zeta[1] - coef(model)[1] * waiting_range
  - coef(model)[2] * mean_friendliness,
  "Neutral|Satisfied" = model$zeta[2] - coef(model)[1] * waiting_range
  - coef(model)[2] * mean_friendliness)

logits_waiting_long <- pivot_longer(logits_waiting,
    cols = -waiting_time, names_to = "Threshold", values_to = "Logit")

ggplot(logits_waiting_long, aes(x = waiting_time, y = Logit,
                                color = Threshold)) +
  geom_line() + scale_color_grey(start = 0.3, end = 0.8) +
  labs(title = "", x = "Waiting Time (minutes)", y = "Log-Odds",
       color = "Threshold") + theme_bw()
```

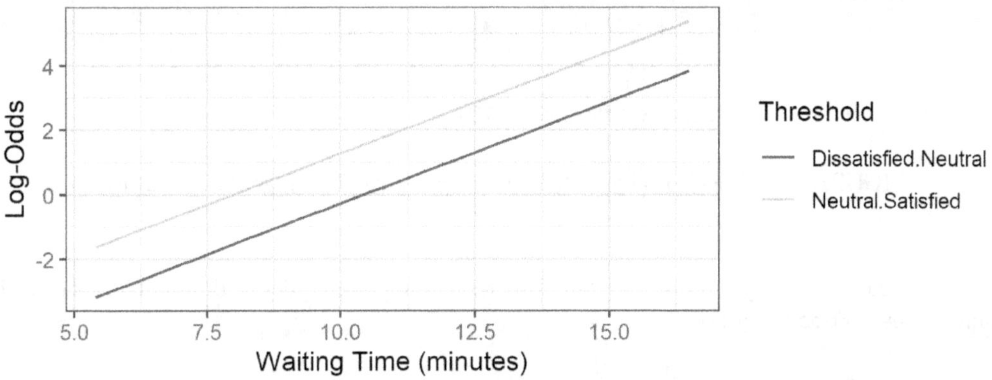

FIGURE 7.13: Log-Odds by Waiting Time (Given Average Staff Friendliness).

For a given waiting time, say the average value of the waiting time, conditional effects of staff friendliness on log-odds of satisfaction levels can be plotted as follows:

```
# note the negative sign before coef(model)
mean_waiting <- mean(data$waiting_time)
friendliness_range <- seq(min(data$staff_friendliness),
                          max(data$staff_friendliness), length.out = 100)
logits_friendliness <- data.frame(Staff_Friendliness = friendliness_range,
  "Dissatisfied|Neutral" = model$zeta[1] - coef(model)[1] * mean_waiting
  - coef(model)[2] * friendliness_range,
  "Neutral|Satisfied" = model$zeta[2] - coef(model)[1] * mean_waiting
  - coef(model)[2] * friendliness_range)
```

```
logits_friendliness_long <- pivot_longer(logits_friendliness,
                                         cols = -Staff_Friendliness,
                                         names_to = "Threshold",
                                         values_to = "Logit")

ggplot(logits_friendliness_long, aes(x = Staff_Friendliness,
                                     y = Logit, color = Threshold)) +
  geom_line() + scale_color_grey(start = 0.3, end = 0.8) +
  labs(title = "", x = "Staff Friendliness (1-10)", y = "Log-Odds",
       color = "Threshold") + theme_bw()
```

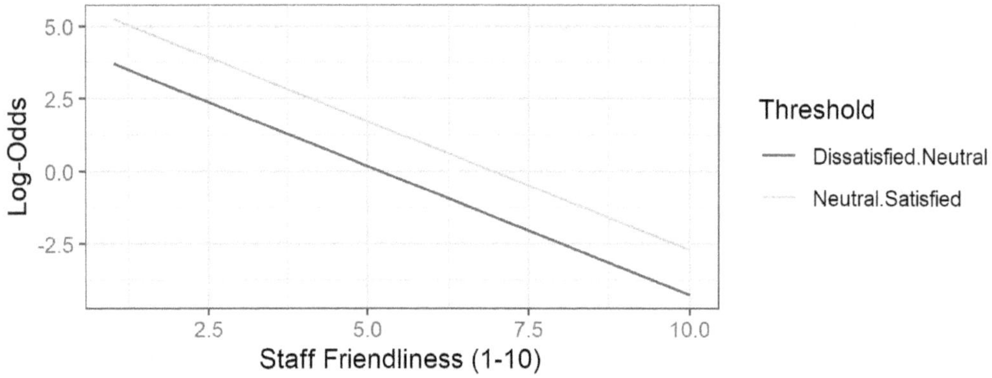

FIGURE 7.14: Log-Odds by Staff Friendliness (Given Average Waiting Time).

If log-odds is not very intuitive, we can also predict the probabilities of different levels directly using `type="probs"` as follows:

```
# Predict probabilities for a new observation
new_data <- data.frame(waiting_time = 8, staff_friendliness = 7)
predict(model, newdata = new_data, type = "probs")
: Dissatisfied      Neutral    Satisfied
:   0.05460664   0.15889390   0.78649946
```

Conclusions

In this case study, we fit a proportional odds model to analyze the effects of waiting time and staff friendliness on customer satisfaction levels. Both waiting time and staff friendliness significantly affect customers' satisfaction: shorter waiting time and higher friendliness are associated with a higher level of satisfaction. Specifically, for every one unit increase in waiting time the log-odds of being less satisfied increases $0.63(95\%\text{ C.I.} : 0.41, 0.88)$ holding constant all other variables; for every one unit increase in staff friendliness the log-odds of being less satisfied decreases $0.88(95\%\text{ C.I.} : 0.70, 1.09)$ holding constant all other variables. This is an observational study, and although there are significant associations no causality can be concluded. How the sample is collected needs to be further verified, and if it is not randomly sampled then the results may not be generalized to the population of interest.

Note 7.16 (Proportional Odds Model)**.** The *proportional odds model (POM)* ([77]) is a type of ordinal regression model to analyze how predictors affect the cumulative probabilities of an ordinal response variable. It is called "proportional odds" because it models the log-odds of being at or below a certain category as a linear function of the predictors, and the odds ratios are assumed to be constant across all thresholds of the response variable, a condition known as the parallel assumption.

Let Y be an ordinal response variable with J ordered categories, and the POM can be expressed as follows:

$$\mathrm{P}(Y \leq j|\mathbf{x}) = \frac{1}{1+\exp(-(\alpha_j - \mathbf{x}^T\beta))}, \quad j = 1, \ldots, J-1,$$

where α_j is the threshold or intercept specific to category j, $\mathbf{x} = (x_1, \ldots, x_p)$ is the vector of predictors, and β is the vector of coefficients, which is assumed constant across thresholds for j's and is the key feature of the proportional odds model. Because the cumulative distribution function of the standard logistic distribution is $F(x) = 1/(1+\exp(-x))$, the model can also be expressed as:

$$\mathrm{P}(Y \leq j|\mathbf{x}) = F(\alpha_j - \mathbf{x}^\top\beta).$$

Therefore, POM is a special case of the *cumulative link model* (CLM) with the logistic link function. In Note 7.22, we discuss CLMMs that extend CLMs by incorporating random effects for dealing with clustered or hierarchical ordinal data.

Given covariates $\mathbf{x}$, the odds of $Y \leq j$ compared to $Y > j$ are defined by:

$$\mathrm{odds}_j(\mathbf{x}) = \frac{\mathrm{P}(Y \leq j|\mathbf{x})}{\mathrm{P}(Y > j|\mathbf{x})} = \exp(\alpha_j - \mathbf{x}^T\beta).$$

Then the log-odds or logit form is expressed as:

$$\ln(\mathrm{odds}_j(\mathbf{x})) = \alpha_j - \mathbf{x}^\top\beta, \tag{7.5}$$

which can also be used to define the POM model, and such an expression is more convenient for interpreting the regression coefficients β.

Increase one unit of x_i so that $\mathbf{x} = (x_1, \cdots, x_i, \cdots, x_p)$ becomes $\mathbf{x}_i := (x_1, \cdots, x_i + 1, \cdots, x_p)$, and then from Equation 7.5, we have

$$\ln(\mathrm{odds}_j(\mathbf{x}_i)) - \ln(\mathrm{odds}_j(\mathbf{x})) = \mathbf{x}_i^\top\beta - \mathbf{x}^\top\beta = -\beta_i; \tag{7.6}$$

that is, $-\beta_i$ represents the change in log-odds of being at or below category j compared to being above category j for a one-unit increase in x_i, keeping other variables constant. Equivalently, we can express the change in odds as:

$$\frac{\text{odds}_j(\mathbf{x}_i)}{\text{odds}_j(\mathbf{x})} = \frac{e^{-\mathbf{x}_i^\top \beta}}{e^{-\mathbf{x}^\top \beta}} = e^{-\beta_i}.$$

So, $e^{-\beta_i}$ represents the "after-to-before" *odds ratio* for a one-unit increase in x_i, keeping other variables constant. Here $e^{-\beta_i}$ does not depend on j due to the parallel assumption, which means that the odds ratio for a one-unit increase in x_i is the same across all thresholds of the response variable. Positive coefficients indicate that increasing the feature increases the odds of being in higher categories (more satisfied), while negative coefficients indicate that increasing the feature increases the odds of being in lower categories (less satisfied). For more details about proportional odds models, see [3, 77].

Note 7.17 (The Brant Test)**.** The *Brant test* ([15]) is used to assess whether the parallel assumption of the proportional odds model holds by fitting binary logistic regressions for each threshold $j = 1, \ldots, J-1$:

$$\ln(\text{odds}_j(\mathbf{x})) = \alpha_j - \mathbf{x}^\top \boldsymbol{\beta}_j,$$

and then testing whether $\boldsymbol{\beta}_1 = \boldsymbol{\beta}_2 = \cdots = \boldsymbol{\beta}_{J-1}$ by comparing the estimates of these $\boldsymbol{\beta}_j$'s. The Brant test can be an omnibus test for the parallel assumption for all predictors, and it can also be used to test the parallel assumption for each predictor separately. The hypotheses for the omnibus Brant test are as follows:

$$\begin{aligned} H_0 &: \ \boldsymbol{\beta}_1 = \boldsymbol{\beta}_2 = \cdots = \boldsymbol{\beta}_{J-1}; \\ H_1 &: \ \text{at least one } \boldsymbol{\beta}_j \text{ is different from the others.} \end{aligned}$$

For the Brant test for each predictor, the hypotheses are similar but only focus on the coefficients of that predictor across different thresholds.

Under the null hypothesis, the maximum likelihood estimates of $\boldsymbol{\beta}_j$'s from the binary logistic regressions should be similar to each other. The Brant test is a Wald-type Chi-square test that compares the estimates of $\boldsymbol{\beta}_j$'s, and a significant result suggests that the parallel assumption is violated. For more details about the Brant test and the test statistics, etc., see [15, 69], and the latter contains detailed calculations of the test statistics.

If the parallel assumption is violated, we can try to use a partial proportional odds model (see Section 7.2.4), which allows some coefficients to vary by threshold, or a multinomial logistic regression (see Note 7.15), which does not account for the ordinal nature of the response.

7.2.3 Partial Proportional Odds Model

In Section 7.2.2, we have introduced the proportional odds model that assumes the parallel assumption holds for all predictors. However, in practice, the parallel assumption may not hold for some predictors. In this case study, we will illustrate how to check the parallel assumption and how to fit a partial proportional odds model when the parallel assumption is violated for some predictors.

Case 32: Analyzing Patient Satisfaction with Healthcare Services

A hospital conducted a survey to assess patient satisfaction with its services. The response variable has three levels: Low Satisfaction, Moderate Satisfaction, High Satisfaction. The hospital collected data on the following predictors: Age: Patient age (years), Wait Time: Time spent waiting before being attended (minutes), and insurance: Whether the patient has private insurance (0 = No, 1 = Yes). The goal is to model how these predictors influence satisfaction levels.

For answers to the initial questions and some initial thoughts about this case study, refer to Section 5.33. First, we need to preprocess the data, including assigning the ordinal response variable and assigning factors for categorical variables.

```
library(MASS)
library(dplyr)
library(tidyr)
library(ggplot2)
library(VGAM)
data <- read.csv("data/patient-satisfaction-healthcare.csv")

# Exploratory data analysis
data$Satisfaction <- factor(data$Satisfaction,
  levels=c("Low","Moderate", "High"), ordered = TRUE)
data$Insurance <- as.factor(data$Insurance)
```

To gain some idea about the data, we plot the satisfaction levels against these predictors, respectively. Figure 7.15 suggests a higher age might associate with a higher level of satisfaction, but this is not very clear.

```
ggplot(data, aes(x = Age, y = Satisfaction)) +
  geom_boxplot() + theme_bw() +
  stat_summary(fun = "mean", geom = "point", shape = 23, size = 2) +
  labs(title = "", x = "Age", y = "Satisfaction Level")
```

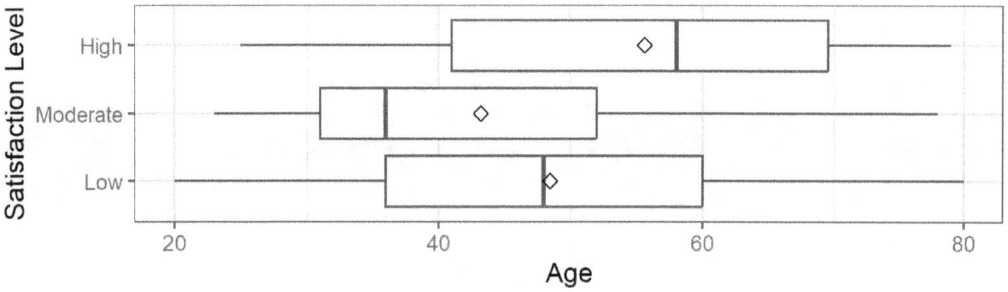

FIGURE 7.15: Satisfaction by Age.

Figure 7.16 indicates that longer waiting time tends to associate with lower level of satisfaction.

```
ggplot(data, aes(x = WaitTime, y = Satisfaction)) +
  geom_boxplot() + theme_bw() +
  stat_summary(fun = "mean", geom = "point", shape = 23, size = 2) +
  labs(title = "", x = "Waiting Time", y = "Satisfaction Level")
```

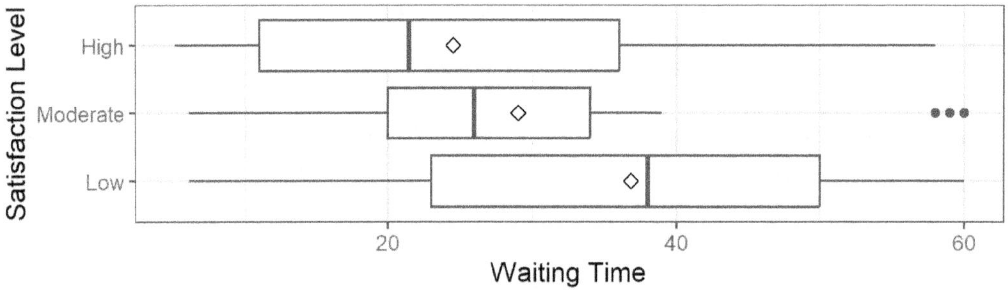

FIGURE 7.16: Satisfaction by Waiting Time.

Figure 7.17 suggests that patients with private insurance tend to have higher level of satisfaction, but the difference is not very large.

```
# Satisfaction by Private Insurance
ggplot(data, aes(x = Insurance, fill = Satisfaction)) +
  geom_bar(position = "fill") + theme_bw() +
  labs(title = "", x = "Insurance", y = "Proportion",
       fill = "Satisfaction") + scale_fill_grey(start = 0.3, end = 0.8)
```

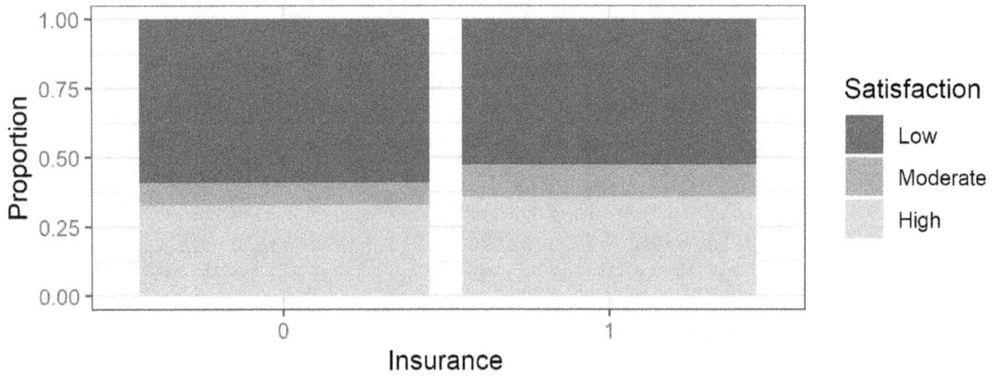

FIGURE 7.17: Satisfaction by Private Insurance.

Then we use graphical methods to check the parallel assumption. Figures 7.18 and 7.19 illustrate how empirical log-odds change over age and waiting time, respectively. The figures suggest that the parallel assumption may not hold for age, but it seems to hold for waiting time.

```
data$age_bin <- cut(data$Age, breaks = 3)
emp_logits_age <- data %>% group_by(age_bin) %>%
    summarise(P_Low = mean(Satisfaction == "Low"),
    P_Moderate_or_less = mean(Satisfaction %in% c("Low", "Moderate")),
    Logit_Low = log(P_Low / (1-P_Low)),
    Logit_Moderate_or_less = log(P_Moderate_or_less /
                                    (1-P_Moderate_or_less)))

emp_logits_age_long <- pivot_longer(emp_logits_age,
  cols = starts_with("Logit"), names_to = "Threshold", values_to = "Logit")
ggplot(emp_logits_age_long, aes(x = age_bin,
  y = Logit, color = Threshold, group = Threshold)) + geom_line() +
  geom_point() + scale_color_grey(start = 0.3, end = 0.8) +
  labs(title = "", x = "Age Bin", y = "Empirical Log-Odds") + theme_bw()
```

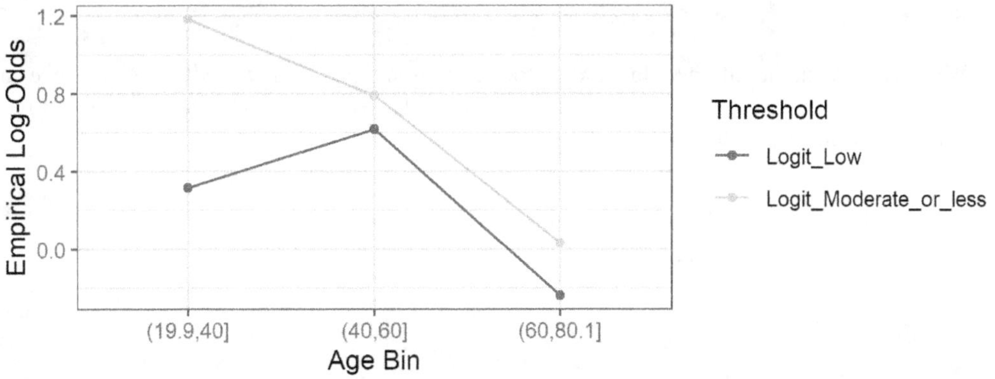

FIGURE 7.18: Empirical Log-Odds by Age Bin.

```
# graphical methods for checking the parallel assumption
data$waiting_bin <- cut(data$WaitTime, breaks = 3)
emp_logits_waiting <- data %>%
  group_by(waiting_bin) %>%
  summarise(P_Low = mean(Satisfaction == "Low"),
    P_Moderate_or_less = mean(Satisfaction %in% c("Low", "Moderate")),
    Logit_Low = log(P_Low / (1-P_Low)),
    Logit_Moderate_or_less = log(P_Moderate_or_less /
                                    (1-P_Moderate_or_less))
  )

emp_logits_waiting_long <- pivot_longer(emp_logits_waiting,
  cols = starts_with("Logit"), names_to = "Threshold", values_to = "Logit")

ggplot(emp_logits_waiting_long, aes(x = waiting_bin, y = Logit,
```

```
  color = Threshold, group = Threshold)) + geom_line() + geom_point() +
  scale_color_grey(start = 0.3, end = 0.8) + theme_bw() +
  labs(title = "", x = "Waiting Time Bin", y = "Empirical Log-Odds")
```

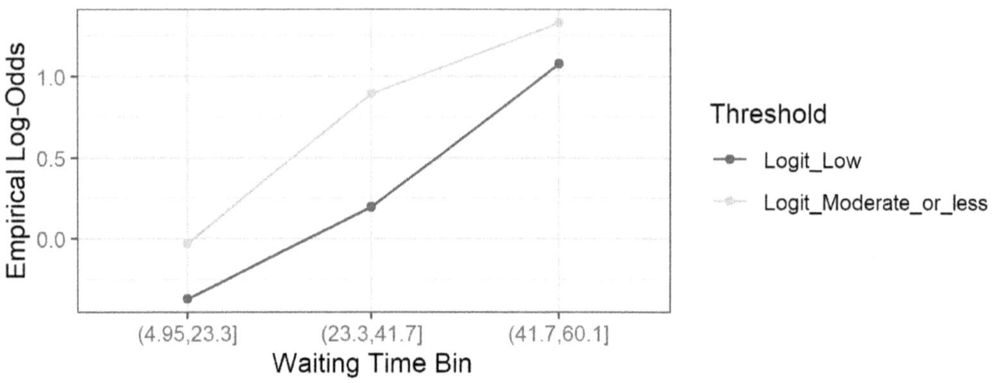

FIGURE 7.19: Empirical Log-Odds by Waiting Time Bin.

We can start with the full model including the interaction terms, and an analysis of deviance can help us quickly remove the Insurance and the interaction terms in the model. So we choose to use only Age and WaitTime in the model.

```
# Fit a standard proportional odds model (for comparison)
full_model <- polr(Satisfaction ~ Age*WaitTime*Insurance, data = data)
car::Anova(full_model)
: Analysis of Deviance Table (Type II tests)
:
: Response: Satisfaction
:                       LR Chisq Df Pr(>Chisq)
: Age                     6.1017  1    0.0135 *
: WaitTime               25.6791  1 4.032e-07 ***
: Insurance               0.4316  1    0.5112
: Age:WaitTime            0.0042  1    0.9485
: Age:Insurance           1.4864  1    0.2228
: WaitTime:Insurance      2.1630  1    0.1414
: Age:WaitTime:Insurance  0.0798  1    0.7776
: ---
: Signif. codes:  0 '***' 0.001 '**' 0.01 '*' 0.05 '.' 0.1 ' ' 1
model <- polr(Satisfaction ~ Age+WaitTime, data = data)
```

Then we use the Brant test for assessing the parallel assumption. The p-value for the omnibus result is 0.06 slightly larger than 5%, however, the p-value for Age is 0.02, suggesting that the parallel assumption does not hold for the Age variable. The p-value is quite large for WaitTime, so the parallel assumption works well here. The results of the Brant tests are consistent to what we have observed from Figures 7.15 and 7.16.

```
brant::brant(model)
: ----------------------------------------------
: Test for X2 df probability
: ----------------------------------------------
: Omnibus  5.48 2 0.06
: Age  5.48 1 0.02
: WaitTime 0.01 1 0.91
: ----------------------------------------------
:
: H0: Parallel Regression Assumption holds
```

When the parallel assumption is violated for some of the variables while holds for the others, we can instead use a partial proportional odds model. The R function `polr()` can not handle the partial proportional odds model directly, and we will be using `VGAM::vglm()` in what follows to fit a partial proportional odds model.

First, use `vglm()` to fit a proportional odds model and a partial proportional odds model, respectively. The latter allows Age to have a non-proportional assumption.

```
# Fit a proportional odds model
pom <- vglm(Satisfaction ~ Age + WaitTime,
            family = cumulative(parallel = TRUE), data = data)

coefficients(summary(pom))
:                   Estimate  Std. Error    z value     Pr(>|z|)
: (Intercept):1 -0.15824994 0.554465144 -0.2854101 7.753300e-01
: (Intercept):2  0.30919310 0.554804071  0.5573014 5.773215e-01
: Age           -0.02146835 0.009023625 -2.3791274 1.735368e-02
: WaitTime       0.04823815 0.009910922  4.8671708 1.132072e-06
```

Then fit a partial proportional odds model as follows:

```
# Allow Age to have non-proportional effects
ppom <- vglm(Satisfaction ~ Age + WaitTime,
             family = cumulative(parallel = FALSE ~ Age, reverse = FALSE),
             data = data)
coefficients(summary(ppom))
:                   Estimate  Std. Error    z value     Pr(>|z|)
: (Intercept):1 -0.36535767 0.567522581 -0.6437764 5.197204e-01
: (Intercept):2  0.61177787 0.588344031  1.0398302 2.984188e-01
: Age:1         -0.01729257 0.009332723 -1.8528970 6.389713e-02
: Age:2         -0.02710402 0.009740282 -2.7826733 5.391306e-03
: WaitTime       0.04831268 0.009919390  4.8705293 1.112997e-06
```

To assess whether the partial proportional odds model significantly improve the modeling, we use a likelihood ratio test and followed by AIC and BIC. As POM is nested in PPOM, we can carry out a likelihood ratio test, and the p-value is 0.048, indicating that there is some evidence that the PPOM is different from the POM.

According to AIC, the POM performs better than PPOM, and according to BIC, PPOM performs better. As BIC encourages simpler model, if a simpler model is preferred, we should use POM, otherwise, PPOM is more appropriate.

```
VGAM::lrtest(ppom, pom)
: Likelihood ratio test
:
: Model 1: Satisfaction ~ Age + WaitTime
: Model 2: Satisfaction ~ Age + WaitTime
:   #Df  LogLik Df Chisq Pr(>Chisq)
: 1 395 -164.31
: 2 396 -166.27  1 3.923    0.04763 *
: ---
: Signif. codes:  0 '***' 0.001 '**' 0.01 '*' 0.05 '.' 0.1 ' ' 1
AIC(pom); AIC(ppom)
: [1] 340.5439
: [1] 338.6209
BIC(pom); BIC(ppom)
: [1] 353.7372
: [1] 355.1125
```

We continue the analysis using PPOM to favor a simpler model. According to Equation 7.6, the regression coefficients reflects changes of log-odds by one unit increase of x. Therefore, we can interpret the results as follows:

- For every one year increase in Age, the log-odds of low satisfaction decreases 0.017 and the log odds of Moderate or Low satisfaction decreases 0.027, holding constant all other variables.
- For every one minute increase in waiting time, the log-odds of being less satisfied (no matter which threshold) increases 0.048, holding constant all other variables.

```
# Extract coefficients
coef_matrix <- coef(ppom, matrix = TRUE)
print(coef_matrix)
:              logitlink(P[Y<=1]) logitlink(P[Y<=2])
: (Intercept)         -0.36535767         0.61177787
: Age                 -0.01729257        -0.02710402
: WaitTime             0.04831268         0.04831268
```

Moreover, the 95% confidence intervals for such effects can be derived from the following.

```
## use log-odds to interpret
ci <-  confint(ppom)
cbind(Change_of_Logit = coef(ppom), ci)
:                 Change_of_Logit       2.5 %        97.5 %
: (Intercept):1      -0.36535767 -1.47768149  0.7469661500
: (Intercept):2       0.61177787 -0.54135524  1.7649109808
```

```
: Age:1              -0.01729257 -0.03558437  0.0009992268
: Age:2              -0.02710402 -0.04619463 -0.0080134215
: WaitTime            0.04831268  0.02887103  0.0677543297
```

As PPOM allows different thresholds to have different slopes against Age, the slopes could intersect at certain ages, and then it creates a problematic situation that at certain ages the odds of low satisfaction could even be higher than the odds of moderate / low satisfaction. In what follows, we create some graphs to check whether such intersecting occurs within the range of ages of interest. Figure 7.20 suggests that there is no intersecting between ages 20 and 80.

```
# Check for intersecting predictors by predicting cumulative probabilities
new_data <- data.frame(Age = seq(20, 80, by = 5), WaitTime = 20)

# Logits for each threshold
pred <- predict(ppom, newdata = new_data, type = "link")

# Convert logits to cumulative probabilities
cum_probs <- plogis(pred)
colnames(cum_probs) <- c("Low", "Low_or_Moderate")

# Plot cumulative probabilities to visualize
plot_data <- data.frame(Age = new_data$Age, `Low` = cum_probs[,1],
  `Low_or_Moderate` = cum_probs[,2]) %>%
  pivot_longer(cols = -Age, names_to = "Threshold",
  values_to = "Cumulative_Probability")

# Plot using ggplot
ggplot(plot_data, aes(x = Age, y = Cumulative_Probability,
                      color = Threshold)) +
  geom_line() + scale_color_manual(
    values = c("Low" = "black", "Low_or_Moderate" = "grey")) +
  labs(title = "", x = "Age", y = "Cumulative Probability", color=NULL) +
  theme_bw() + theme(legend.position.inside = c(0.95, 0.95),
                     legend.justification = c(1,1) )
```

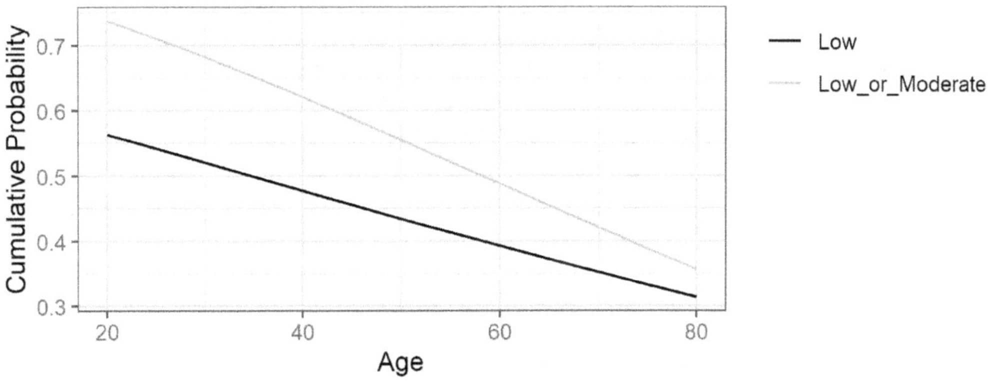

FIGURE 7.20: Cumulative Probabilities by Age for Different Thresholds.

Conclusions

In this case study, a partial proportional odds model is used for modeling patient satisfaction data as the parallel assumption for the proportional odds model is violated for Age but holds for Waiting Time. For every one year increase in Age, the log-odds of low satisfaction decreases $0.017(95\%\ \text{C.I.}: -0.001, 0.036)$ and the log-odds of Moderate or Low satisfaction decreases $0.027(95\%\ \text{C.I.}: 0.008, 0.046)$, holding constant all other variables. For every one minute increase in waiting time, the log-odds of being less satisfied (no matter which threshold) increases $0.048(95\%\ \text{C.I.}: 0.029, 0.068)$, holding constant all other variables. Having private insurance does not significantly affect satisfaction. This is an observational study, and no causality of the effects can be established. How the sample is collected needs to be further clarified to determine whether the result can be generalized to all patients of the hospital.

7.2.4 Extended McNemar's Tests

When comparing the marginal proportions of paired categorical data with more than two categories, extensions of McNemar's test can be used. This section presents a case study that illustrates the application of such extended McNemar's tests.

Case 33: Effect of a Dietary Intervention on Cholesterol Levels

A nutritionist assesses the impact of a dietary intervention on cholesterol levels. The study involves measuring cholesterol levels in patients before and after a dietary intervention. The cholesterol level of each patient is classified into three levels: Low, Medium, and High. The following contingency table (Table 5.15) summarizes the cholesterol levels of 100 patients before and after the dietary intervention:

TABLE 5.15: Cholesterol Levels.

Before \ After	Low	Medium	High
Low	30	10	5
Medium	8	20	10
High	2	5	10

Determine if there is significant difference in cholesterol levels before and after the dietary intervention.

For answers to the initial questions and initial thoughts, see Section 5.34. This case is quite similar to Case 24 in terms of data structures, except that here we have three categories, instead of two categories in Case 24. First, we prepare the contingency table as follows:

```
library(DescTools)
tab <- matrix(c(30, 10, 5, 8, 20, 10, 2, 5, 10), nrow = 3, byrow = TRUE)
dimnames(tab) <- list(Before = c("Low", "Medium", "High"),
                      After = c("Low", "Medium", "High"))
tab
:         After
: Before   Low Medium High
:   Low     30     10    5
:   Medium   8     20   10
:   High     2      5   10
```

A mosaic plot to visualize the data is shown in Figure 7.21 which suggests that the proportions of most levels remain the same by looking at the diagonal cells (top left to bottom right), but we need formal tests to assess whether there are significant differences in the marginal proportions before and after the dietary intervention.

```
# Visualize the data with a mosaic plot
mosaicplot(tab, xlab = "Before", ylab = "After", main = "",
           col = c(grey(0.3), grey(0.55), grey(0.8)))
```

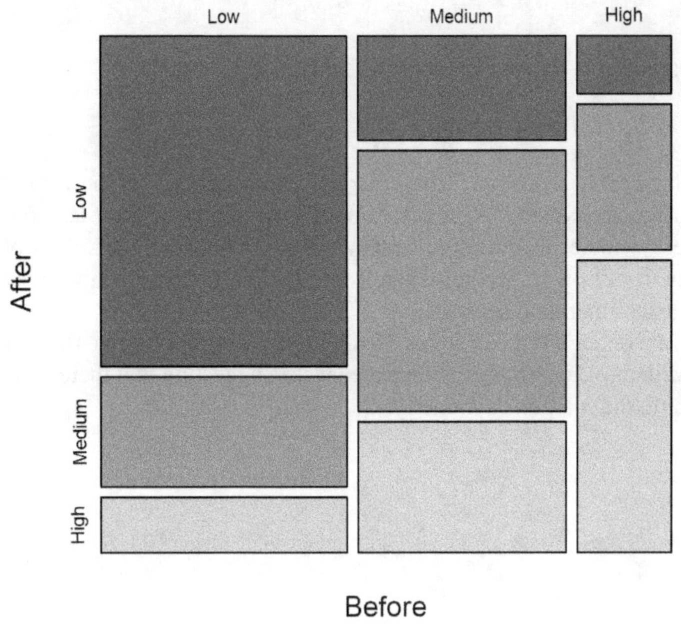

FIGURE 7.21: Mosaic Plot of Before vs. After.

To test whether marginal proportions are significantly different before and after the dietary intervention, we can use extensions of McNemar's test for paired nominal or ordinal data with more than two categories. Specifically, the null hypothesis can be

$$H_0 : \pi_{i+} = \pi_{+i} \text{ for all } i = 1,2,3,$$

where π_{i+} and π_{+i} denote the marginal probabilities of the ith row (Before) and the ith column (After), respectively. For assessing such marginal homogeneity of paired nominal data, we use the Stuart-Maxwell test and / or the Bhapkar test, and these two are very similar except for how they estimate the variance components in their test statistics. Both tests are extensions of McNemar's test to marginal homogeneity for $k \times k$ squared contingency tables for paired nominal data. For a large sample size, they provide very similar results; see Note 7.18 and 7.19 for details. The p-values of these two tests are 0.205 and 0.194, respectively, suggesting that we fail to reject the null hypothesis at the common significance levels.

```
StuartMaxwellTest(tab)
:
:  Stuart-Maxwell test
:
```

```
: data:  tab
: chi-squared = 3.1737, df = 2, p-value = 0.2046
```

```
BhapkarTest(tab)
:
:  Bhapkar test
:
: data:  tab
: chi-squared = 3.2777, df = 2, p-value = 0.1942
```

Another extension of McNemar's test is the McNemar-Bowker test and

$$H_0 : \pi_{ij} = \pi_{ji} \text{ for all } i \neq j.$$

That is, it tests marginal symmetry and is more strict than marginal homogeneity as the former implies the latter but not the opposite; see Note 7.20 for details on the McNemar-Bowker test. For this case study, if we want to assess whether there is significant asymmetry between transitions of levels, say, there are more changes from Low to Medium than from Medium to Low, then we may use the McNemar-Bowker test to assess marginal symmetry.

In R, the same function as for McNemar's test can be used and the following McNemar-Bowker test leads to the p-value (0.366) well above common significance levels, suggesting that there is no significant evidence to reject the null, the marginal symmetry.

```
mcnemar.test(tab)
:
:  McNemar's Chi-squared test
:
: data:  tab
: McNemar's chi-squared = 3.1746, df = 3, p-value = 0.3655
```

Note that all three tests discussed above are large-sample tests and their results may not be reliable when the sample size is small. In addition, they do not take into account the ordinal nature of the response variable (cholesterol levels: Low, Medium, High). For testing marginal homogeneity for paired ordinal data, we may consider modeling the data using a cumulative link mixed model (CLMM); see Note 7.22 for details on CLMMs.

First, we need to restructure the data from Table 5.15 into a long format of subject-level data with pseudo IDs.

```
library(tidyr)
library(lme4)
df <- as.data.frame(as.table(tab))
df_wide <- df[rep(seq_len(nrow(df)), df$Freq), 1:2]
rownames(df_wide) <- NULL
df_wide$ID <- seq_len(nrow(df_wide))
df_long <- df_wide %>% pivot_longer(cols = c(Before, After),
                                    names_to = "Time", values_to = "Level")
head(df_long)
: # A tibble: 6 x 3
:      ID Time   Level
:   <int> <chr>  <fct>
```

```
: 1       1 Before Low
: 2       1 After  Low
: 3       2 Before Low
: 4       2 After  Low
: 5       3 Before Low
: 6       3 After  Low
```

Then a cumulative link mixed model can be used for the ordinal response variable `Level`. A likelihood ratio test comparing models with and without the fixed effect `Time` leads p-value (0.1124) larger than the common significance levels, and therefore `Time`, is not significant, suggesting no significant difference in cholesterol levels before and after the dietary intervention, consistent with what we have obtained from the other tests.

```
library(ordinal)
df_long$Level <- factor(df_long$Level, levels=c("Low","Medium", "High"),
                        ordered = T)
model_ordinal <- clmm(Level ~ Time + (1|ID), data = df_long)
model_ordinal_reduced <- clmm(Level ~ (1|ID), data = df_long)
anova(model_ordinal, model_ordinal_reduced)
: Likelihood ratio tests of cumulative link models:
:
:                         formula:                  link: threshold:
: model_ordinal_reduced Level ~ (1 | ID)            logit flexible
: model_ordinal         Level ~ Time + (1 | ID) logit flexible
:
:                         no.par    AIC  logLik LR.stat df Pr(>Chisq)
: model_ordinal_reduced        3 408.13 -201.06
: model_ordinal                4 407.61 -199.80  2.5204  1     0.1124
```

Conclusions

In this case study, we have compared several extensions of McNemar's test for paired nominal data with more than two categories, including the Stuart-Maxwell test (p-value: 0.205), the Bhapkar test (p-value: 0.194), and the McNemar-Bowker test (p-value: 0.366). All tests suggest no significant difference in the marginal distributions of cholesterol levels before and after the dietary intervention. Additionally, a cumulative link mixed model for the ordinal response variable also indicates no significant effect of time (before vs. after) on cholesterol levels (p-value: 0.112). The population of interest and how the patients are selected need to be further clarified to decide whether the study can generalize to the population. This is an observational study and no causality can be established.

Note 7.18 (The Stuart-Maxwell Test). The *Stuart-Maxwell test* ([105, 74]) is a generalization of McNemar's test for the case of a square $k \times k$ contingency table that arises from paired categorical data. It is used to test the null hypothesis of *marginal homogeneity*, i.e.

$$H_0 : \pi_{i+} = \pi_{+i} \quad \text{for all } i = 1, \ldots, k,$$

where π_{i+} and π_{+i} denote the marginal probabilities of the ith row and the ith column, respectively. Let n_{ij} denote the observed frequency in cell (i, j) and define the difference between the totals of the ith row and the totals of the ith column as

$$d_i = n_{i+} - n_{+i}, \quad i = 1, \ldots, k.$$

Let $\mathbf{d} = (d_1, d_2, \ldots, d_{k-1})^\top$ of length $k-1$ (since $\sum_i d_i = 0$), and then under H_0, the Stuart-Maxwell test statistic can be written as

$$\chi^2_{SM} = \mathbf{d}^\top \mathbf{V}_{SM}^{-1} \mathbf{d},$$

where $\mathbf{V}_{SM}$ is the estimated covariance matrix of $\mathbf{d}$, with entries given by

$$v_{ii} = n_{i+} + n_{+i} - 2n_{ii}, \quad v_{ij} = -(n_{ij} + n_{ji}), \quad i \neq j,$$

and χ^2_{SM} is asymptotically distributed as a Chi-square random variable with $k-1$ degrees of freedom. See [105] and [74] for more details on the Stuart-Maxwell test. A similar test for marginal homogeneity is the Bhapkar test (see Note 7.19), which uses a different estimation of the covariance matrix and is large-sample equivalent to the Stuart-Maxwell test; see [59] for more details.

Note 7.19 (The Bhapkar Test). The *Bhapkar test* ([11]) is another large sample test for *marginal homogeneity* in a $k \times k$ square contingency table with paired categorical responses, and it has the same hypothesis as the Stuart-Maxwell test but uses a different estimation of the covariance matrix of the difference vector $\mathbf{d}$ compared to the Stuart-Maxwell test (see Note 7.18).

Define the difference vector same as in the Stuart-Maxwell test:

$$\mathbf{d} = (d_1, d_2, \ldots, d_{k-1})^\top, \quad \text{where } d_i = n_{i+} - n_{+i},$$

and the statistic of the Bhapkar test is given by

$$\chi^2_B = \mathbf{d}^\top \mathbf{V}_B^{-1} \mathbf{d},$$

where entries of the covariance matrix $\mathbf{V}_B$ are estimated as

$$v_{ii} = n_{i+} + n_{+i} - 2n_{ii} + (n_{i+} - n_{+i})^2/n, \quad v_{ij} = -(n_{ij} + n_{ji}) - (n_{i+} - n_{+i})(n_{j+} - n_{+j})/n, \quad i \neq j.$$

Note that

$$\mathbf{V}_B = \mathbf{V}_{SM} - \frac{1}{n}\mathbf{d}\mathbf{d}^\top,$$

which is an adjustment to $\mathbf{V}_{SM}$. Under the null hypothesis of marginal homogeneity, the statistic χ^2_B is asymptotically distributed as a Chi-square random variable with $k-1$ degrees of freedom. The Bhapkar test and the Stuart-Maxwell test are asymptotically equivalent under large samples, and we refer to [11, 59, 3] for more details.

Note 7.20 (The McNemar-Bowker Test). The *McNemar-Bowker test* ([80, 14]) is a generalization of McNemar's test for *marginal symmetry* in a $k \times k$ contingency table, where $k \geq 2$. It is used to test the following null hypothesis:

$$H_0 : \pi_{ij} = \pi_{ji} \quad \text{for all } i \neq j,$$

where π_{ij} denotes the probability that a paired observation falls in cell (i, j). Let n_{ij} be the observed frequency in cell (i, j). The test statistic is defined as

$$\chi^2_{\text{MB}} = \sum_{i<j} \frac{(n_{ij} - n_{ji})^2}{n_{ij} + n_{ji}},$$

which compares the off-diagonal elements of the table. Under the null hypothesis of marginal symmetry, χ^2_{MB} is asymptotically distributed as a Chi-square random variable with $k(k-1)/2$ degrees of freedom. For the special case of $k = 2$, the McNemar-Bowker test reduces to the classical McNemar test for 2×2 tables.

Note 7.21 (Relationships Among McNemar and Its Extensions)**.** McNemar's test (Note 7.6), and three of its extensions that we have discussed, the Stuart-Maxwell test (Note 7.18), the Bhapkar test (Note 7.19), and the McNemar-Bowker test (Note 7.20), are closely related but differ in their hypotheses and scopes. McNemar's test applies to 2×2 tables and evaluates marginal homogeneity by testing whether the probabilities of discordant cells are equal. The McNemar-Bowker test generalizes this idea to $k \times k$ tables by comparing the symmetry of off-diagonal cell frequencies. In contrast, the Stuart-Maxwell and Bhapkar tests both target marginal homogeneity in $k \times k$ tables by comparing row and column marginal distributions rather than off-diagonal symmetry, and they are asymptotically equivalent but differ in their derivations. For more discussions on other methods for analyzing paired categorical data, see [6].

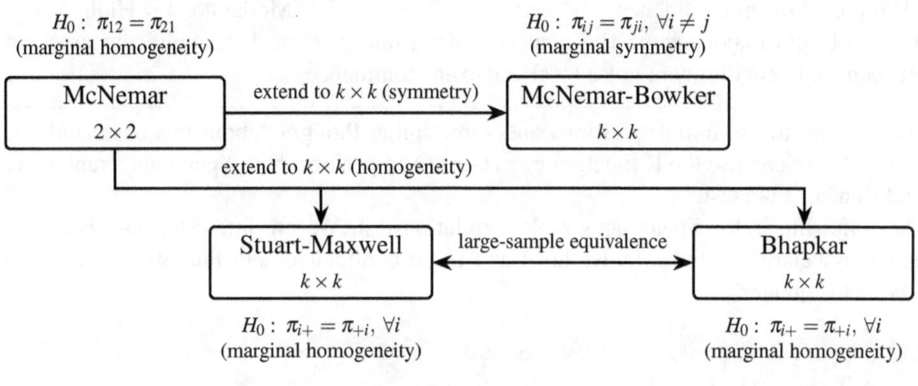

FIGURE 7.22: Relationships among McNemar, McNemar-Bowker, Stuart-Maxwell, and Bhapkar Tests.

Note 7.22 (Cumulative Link Mixed Model)**.** The *cumulative link mixed model (CLMM)* extends the cumulative link model (CLM) by incorporating random effects. Without the random effects, CLMM reduces to the standard cumulative link model (CLM) for ordinal data, and the proportional odds model (POM) discussed in Note 7.16 is a special case of CLM with the logistic link function.

Let Y_{ij} denote the ordinal response for subject j in cluster i, then CLMM assumes that the cumulative probabilities follow

$$P(Y_{ij} \leq k \mid \eta_{ij}) = F(\alpha_k - \eta_{ij}), \qquad k = 1, \ldots, K-1,$$

where $F(\cdot)$ is a specified link function such as the logistic CDF, $\alpha_1 < \cdots < \alpha_{K-1}$ are unknown threshold parameters, and η_{ij} is the linear predictor. A typical mixed-effects structure is

$$\eta_{ij} = \mathbf{x}_{ij}^\top \beta + \mathbf{z}_{ij}^\top \mathbf{v}_i,$$

where $\mathbf{x}_{ij}$ is the vector of fixed effect covariates with corresponding coefficients β, and $\mathbf{z}_{ij}$ is the design vector for the random effects $\mathbf{V}_i$ that are often assumed to be independent and identically distributed as a multivariate normal distribution:

$$\mathbf{V}_i \overset{iid}{\sim} \mathcal{N}(\mathbf{0}, \Sigma),$$

where Σ is the covariance matrix of the random effects, capturing cluster-specific deviations from the population-level effects. For more details on cumulative link mixed models, see [22] and [2].

7.2.5 Nonlinear Correlation Tests

When data do not follow a normal distribution or when the relationship between the variables is not linear, some alternative correlation tests can be used to assess the association between two variables as long as the data can be ranked. These tests do not assume any specific distribution for the data and are robust to outliers. In this section, we discuss two commonly used nonlinear correlation tests: Spearman's rank correlation and Kendall's tau tests through a case study involving study environment noise levels and exam confidence.

Case 34: Environment Noise Level and Exam Confidence

An education researcher investigates whether the level of noise in a student's study environment is associated with their self-reported confidence before an exam. While excessive noise may be distracting, students may adapt differently, so only a weak association is expected. The researcher collects data from 30 student volunteers, recording their study environment noise level (1 = Very Quiet, 2 = Quiet, 3 = Moderate, 4 = Noisy, 5 = Very Noisy) and their self-reported exam confidence (1 = Very Low, 2 = Low, 3 = Moderate, 4 = High, 5 = Very High) before a major exam. The goal is to determine if there is a significant association between study environment noise level and exam confidence.

For answers to the initial questions and some initial thoughts about this case study, refer to Section 5.35. We can use the R function `cor.test()` to perform both Spearman's rank correlation test and Kendall's tau test.

The following is for Spearman's rank correlation test. We are interested in whether there is a positive association between noise level and exam confidence, and thus we set the alternative hypothesis to "greater".

```
data <- read.csv("data/noise-exam.csv")
cor.test(data$noise_level, data$exam_confidence, method = "spearman",
         alternative ="greater")
:
:  Spearman's rank correlation rho
:
: data:  data$noise_level and data$exam_confidence
: S = 4221.4, p-value = 0.3747
: alternative hypothesis: true rho is greater than 0
: sample estimates:
:        rho
: 0.06085806
```

There is a warning message above indicating that there are ties in the data (The warning message about ties is suppressed in the code chunk above to keep the output clean). When there are ties of ranks in either variable, the p-value reported by `cor.test()` is inaccurate. In such cases, the R function `SuppDists::pSpearman()` can be used to compute the p-value more accurately. The following leads to p-value of 0.376, indicating insufficient evidence to conclude a positive association in terms of Spearman's rank correlation between noise level and exam confidence.

```
rho <- cor(data$noise_level, data$exam_confidence, method = "spearman")
1 - SuppDists::pSpearman(rho, r=30, lower.tail = TRUE)
: [1] 0.3758481
```

The following is for Kendall's tau test. Again, we set the alternative hypothesis to "greater" to test for a positive association.

```
cor.test(data$noise_level, data$exam_confidence, method = "kendall",
         alternative ="greater")
:
```

```
:  Kendall's rank correlation tau
:
: data:  data$noise_level and data$exam_confidence
: z = 0.33127, p-value = 0.3702
: alternative hypothesis: true tau is greater than 0
: sample estimates:
:        tau
: 0.05198996
```

There is also a warning message about ties above (The warning message about ties is suppressed in the code chunk above to keep the output clean), and we can use `SuppDists::pKendall()` to compute the p-value more accurately. The following leads to a p-value of 0.336, again indicating insufficient evidence to conclude a positive association in terms of Kendall's tau between noise level and exam confidence.

```
tau <- cor(data$noise_level, data$exam_confidence, method = "kendall")
1 - SuppDists::pKendall(tau, N=30, lower.tail = TRUE)
: [1] 0.3355791
```

Conclusions

In this case study, we explore the use of rank-based correlation tests to assess the association between study environment noise levels and exam confidence among students. Both Spearman's rank correlation test and Kendall's tau test are applied to the ordinal data collected from a survey of 30 students. The results from both tests indicate insufficient evidence to conclude a positive association between noise level and exam confidence, as reflected by the p-values of 0.376 and 0.336, respectively. The subjects are selected based on volunteering and not randomly, which may limit the generalizability of the findings. This is an observational study and thus causality cannot be established.

Note 7.23 (Spearman's Rank Correlation)**.** *Spearman's rank correlation* ([103]) is a nonparametric measure of association between two random variables. It is based on ranks rather than the raw data, making it a common choice for measuring associations in ordinal data or data that do not follow a normal distribution, or when the relationship between the variables is not linear. For a continuous bivariate random vector (X,Y), let F_X and F_Y denote their respective cumulative distribution functions. Then the Spearman's rank correlation coefficient is the Pearson's correlation between $F_X(X)$ and $F_Y(Y)$, that is, $\rho = \text{Corr}(F_X(X), F_Y(Y))$.

To test the null hypothesis $H_0 : \rho = 0$ with a sample of paired observations (X_i, Y_i), $i = 1, \ldots, n$, where there are no ties in the ranks, the test statistic is given by

$$r_s = \frac{12\sum_{i=1}^{n}[(R_i - (n+1)/2)(S_i - (n+1)/2)]}{n(n^2-1)},$$

where R_i and S_i denote the respective ranks of the paired observations X_i and Y_i, $i = 1, \ldots, n$. For small sample sizes, the distribution of r_s under H_0 can be determined exactly, and for sufficiently large sample sizes, the distribution of r_s can be approximated by a normal distribution. When there are ties in the ranks, permutation tests can be used to compute p-values more accurately; see Comments 46 in Chapter 8 of [53] for more details. For more details on Spearman's rank correlation and other measures of association and dependency between random variables, see [61, 53].

Note 7.24 (Kendall's Tau). *Kendall's tau* ([62]) is a nonparametric measure of association between two random variables. It is based on the concept of concordant and discordant pairs, making it a common choice for measuring associations in ordinal data or data that do not follow a normal distribution, or when the relationship between the variables is not linear. For a continuous bivariate random vector (X,Y), Kendall's tau is defined as the difference between the probability of concordance and the probability of discordance:

$$\tau = \mathrm{P}[(X_1 - X_2)(Y_1 - Y_2) > 0] - \mathrm{P}[(X_1 - X_2)(Y_1 - Y_2) < 0],$$

where (X_1, Y_1) and (X_2, Y_2) are independent and identically distributed as (X,Y).

Given n paired observations (X_i, Y_i), $i = 1, \ldots, n$, if there are no ties in the data, we can test the null hypothesis $H_0 : \tau = 0$ using the following test statistic:

$$K = \frac{C - D}{n(n-1)/2},$$

where C is the number of concordant pairs, D is the number of discordant pairs, and $n(n-1)/2$ is the total number of possible pairs. Under H_0 and for large n, asymptotic normality holds for the test statistic, and p-values can be approximated accordingly. For small n, exact distributions of K under H_0 can be used to compute p-values. When there are ties in the data, some adjustments are needed; see Section 8.1 of [53] for more details. For more details on Kendall's tau and other measures of association and dependency between random variables, see [61, 53].

7.3 Count Data

When the response variable is a count, we can use methods such as binomial, Poisson, and negative binomial regression to analyze the data. In this section, we present case studies that have counts as a response variable.

7.3.1 Poisson and Negative Binomial Regression

We first present a case study that illustrates the application of Poisson regression and negative binomial regression.

Case 35: Modeling Property Crimes with Multiple Predictors

Local government officials and community planners want to evaluate how socioeconomic conditions are associated with property crimes across counties. They have collected data from 30 counties over a single calendar year. For each county, the following variables are recorded: the total number of reported property crimes, total population, the poverty percentage, the unemployment rate, population density (persons per square mile), and median household income (in thousands of dollars). The goal is to understand what factors are associated with the number of property crimes across counties.

For answers to the initial questions and some initial thoughts about this case study, refer to Section 5.36.

First, we draw a histogram of the response variable, and Figure 7.23 shows that the distribution of crime counts is right-skewed, which is common for count data. If we consider crimes as the number of "successes" in a large number of "trials", where the trials can be considered as the potential targets of crimes (e.g., properties in the county), then in this case, the number of crimes can be considered relatively small compared to potentially larger number of trials, and therefore, a Poisson regression might be a reasonable starting point for modeling; see Note 7.25 for when Poisson approximation is appropriate for count data.

```
library(ggplot2)
data <- read.csv("data/county-property-crimes.csv")
ggplot(data, aes(x = crimes)) +
  geom_histogram(binwidth = 1000, fill = "grey50") + theme_bw() +
  labs(title = "Distribution of Property Crime Counts",
       x = "Number of Crimes in Counties", y = "Frequency")
```

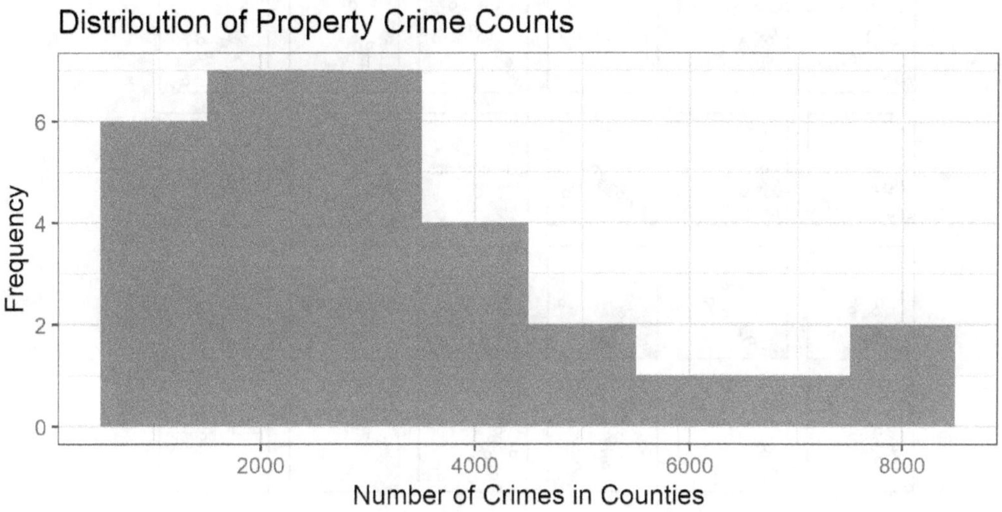

FIGURE 7.23: Histogram of Property Crime Counts.

Next, we examine the pairwise relationships between the response variable and each predictor, as well as the relationships among the predictors using a scatter plot as shown in Figure 7.24. It is clearly that higher crime counts are observed in counties with higher population, and it justifies the need to use population as an offset term in the model; for details on the offset term in Poisson regression, see Note 7.26. We also observe that the predictors are correlated with each other, and we will need to check for multicollinearity when fitting the model; see Note 6.15 for details on variance inflation factors (VIF) for checking multicollinearity.

```
pairs(data[,2:7])
```

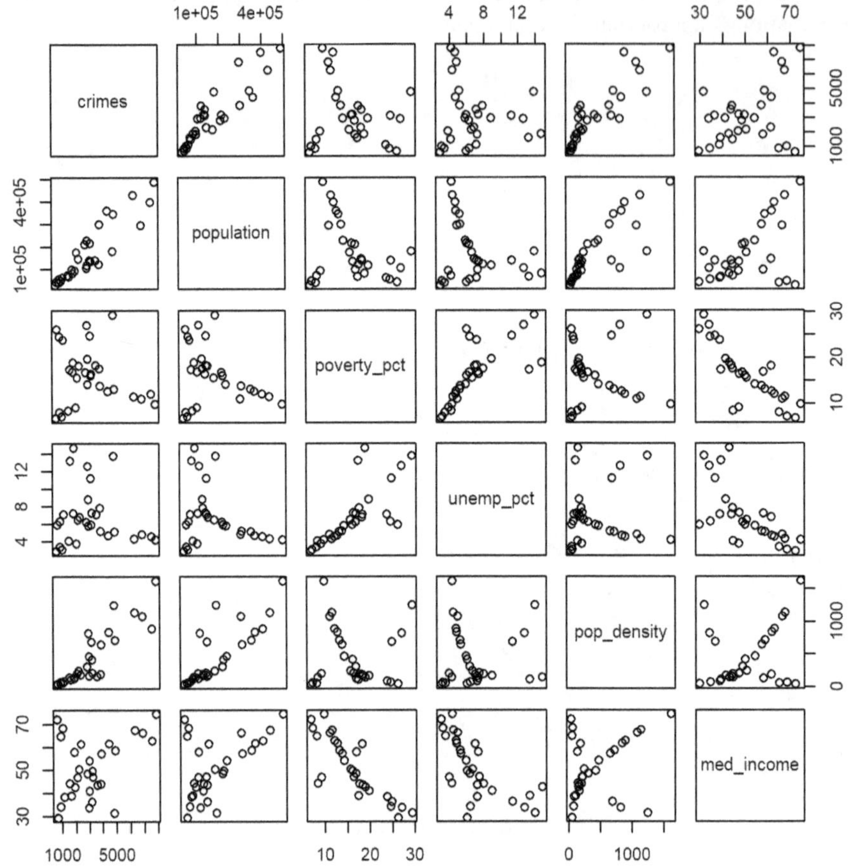

FIGURE 7.24: Scatter Plots of the Variables.

A Poisson regression model is fitted with all predictors included, where we take logarithm of the population density as a predictor as the raw data is heavily right-skewed. The logarithm of population is included as an offset term; the logarithm is needed, and please refer to Note 7.26 for details on the offset term in Poisson regression.

```
fit_pois_full <- glm(crimes ~ poverty_pct + unemp_pct + med_income +
                         log(pop_density) + offset(log(population)),
                         data = data, family = poisson(link = 'log'))
```

The VIF values are calculated to check for multicollinearity. The results show that the VIF values for poverty percentage and median household income are above 5, indicating potential multicollinearity issues. Therefore, we remove the median income from the model, and refit the model. Then the VIF values are all below 5, suggesting that multicollinearity is not a concern for the remaining predictors.

```
car::vif(fit_pois_full)
:       poverty_pct         unemp_pct        med_income log(pop_density)
:          6.091311          3.589453          6.227804          1.623319
fit_pois <- glm(crimes ~ poverty_pct + unemp_pct + log(pop_density) +
```

```
                  offset(log(population)), data = data,
                  family = poisson(link = 'log'))
car::vif(fit_pois)
:      poverty_pct          unemp_pct log(pop_density)
:         3.382073           3.382525         1.042193
```

The next line of code helps to check for overdispersion, and the value is much greater than 1, indicating that the data are overdispersed for the Poisson regression model; see Note 7.26 for the concept of overdispersion and methods to handle it.

```
sum(residuals(fit_pois, type="pearson")^2) / fit_pois$df.residual
: [1] 186.8659
```

To address the issue of overdispersion, we first use a quasi-Poisson regression model which includes a dispersion parameter to adjust the standard errors of the estimates. Then we use the R function `drop1()` to select predictors based on the *F* test. Note that the *F* test is required for the quasi-Poisson model as the likelihood ratio test is not valid for the quasi-likelihood models; see Note 7.27 for more details.

```
# Quasi-Poisson (robust SEs accounting for overdispersion)
fit_qpois <- glm(crimes ~ poverty_pct + unemp_pct + log(pop_density) +
                  offset(log(population)), data = data,
                  family = quasipoisson(link = 'log'))
drop1(fit_qpois, test="F")
: Single term deletions
:
: Model:
: crimes ~ poverty_pct + unemp_pct + log(pop_density) + offset(log(population))
:                  Df Deviance F value  Pr(>F)
: <none>                4733.3
: poverty_pct       1   4791.6  0.3204 0.57623
: unemp_pct         1   5696.0  5.2881 0.02975 *
: log(pop_density)  1   4908.1  0.9605 0.33610
: ---
: Signif. codes:  0 '***' 0.001 '**' 0.01 '*' 0.05 '.' 0.1 ' ' 1
```

The results show that only the unemployment rate is statistically significant, and therefore, we refit the model with only the unemployment rate as the predictor. The dispersion parameter is estimated to be 185. The exponentiated coefficient for the unemployment rate is 1.05, which means that a one percentage increase in the unemployment rate is associated with a 5% increase in the crime rate, holding other variables constant.

```
fit_qpois <- glm(crimes ~ unemp_pct + offset(log(population)),
    data = data, family = quasipoisson(link = 'log'))

summary(fit_qpois)$dispersion
: [1] 184.9115
exp(coef(fit_qpois))[2]
: unemp_pct
:  1.054215
exp(confint(fit_qpois))[2,]
:    2.5 %   97.5 %
: 1.023862 1.084154
```

Diagnostics are performed to check the model fit, and the residual plots do not show any obvious patterns and concerns.

```
# Model Diagnostics: Residual plots
par(mfrow = c(2, 2))
plot(fit_qpois)
```

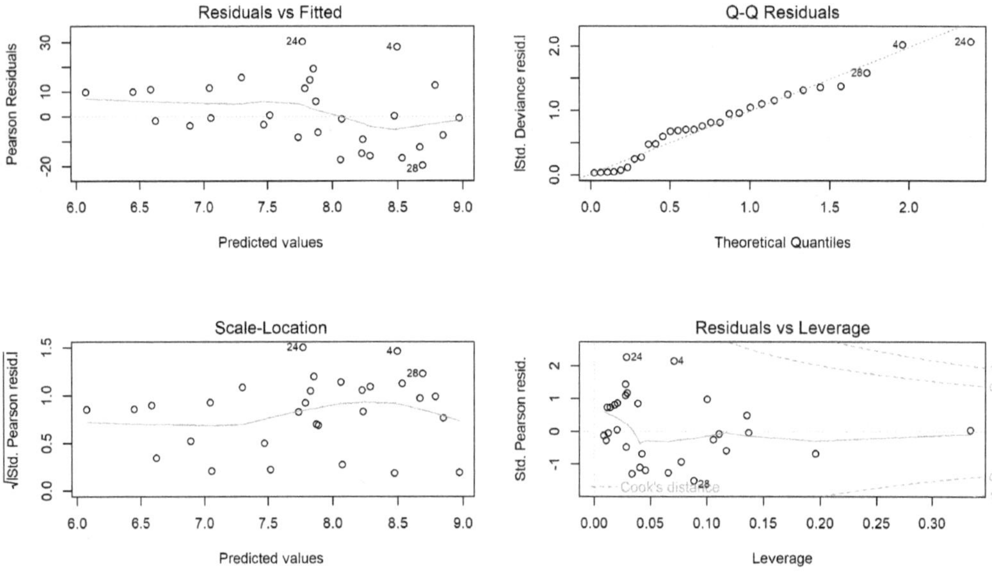


```
par(mfrow = c(1, 1))
```

The observed–predicted plot can be used to check the overall fit of the model, and Figure 7.25 shows that the predicted crime rates are generally close to the observed crime rates. Although there are some outliers, overall, the model seems to fit the data reasonably well.

```
data$pred_qpois <- predict(fit_qpois, type = 'response')
data$rate_obs <- data$crimes / data$population * 1e5
data$rate_qpois <- data$pred_qpois / data$population * 1e5

# Plot observed vs. predicted
ggplot(data, aes(x = rate_obs, y = rate_qpois, label = county_id)) +
  geom_abline(slope = 1, intercept = 0, linetype = 'dashed') +
  geom_point(color = "grey50", size = 2) +
  ggrepel::geom_text_repel(size = 2.8) +
  labs(title = "Observed vs. Predicted Crime Rates (Quasi-Poisson)",
    x = "Observed Crimes per 100k", y = "Predicted Crimes per 100k") +
    theme_bw()
```

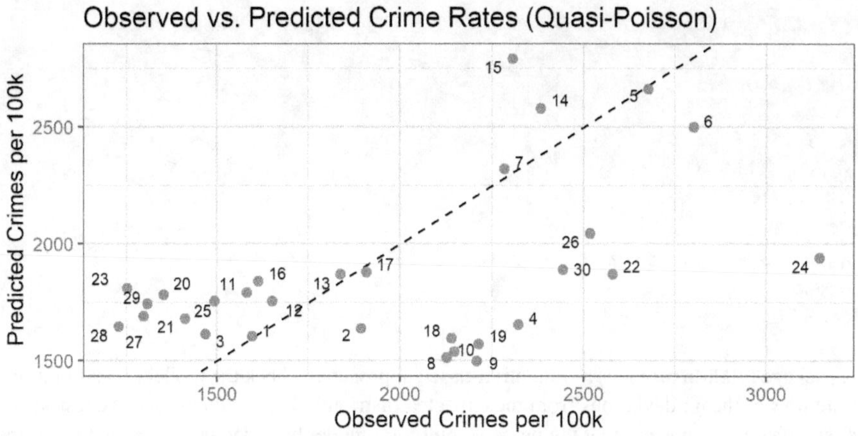

FIGURE 7.25: Observed vs. Predicted Crime Counts (Quasi-Poisson).

Another approach to handle overdispersion is to use negative binomial regression, which includes a dispersion parameter to model the variance separately from the mean; see Note 7.28 for details on negative binomial regression. We use the R function `MASS::glm.nb()` to fit the negative binomial regression model as the function can automatically estimate the dispersion parameter. Again, due to the multicollinearity issue, we remove the median income from the model, and then use the likelihood ratio test to select predictors using the R function `drop1()`. The following results show that only the unemployment rate is statistically significant, which is consistent with the results from the quasi-Poisson model.

```
# Negative binomial regression
library(MASS)
fit_nb <- glm.nb(crimes ~ poverty_pct + unemp_pct + log(pop_density) +
                  offset(log(population)),  data = data)

drop1(fit_nb, test="LRT")
: Single term deletions
:
: Model:
: crimes ~ poverty_pct + unemp_pct + log(pop_density) + offset(log(population))
:                  Df Deviance    AIC    LRT Pr(>Chi)
: <none>              30.146 472.56
: poverty_pct       1   31.886 472.31 1.7405 0.187081
: unemp_pct         1   39.306 479.72 9.1599 0.002474 **
: log(pop_density)  1   33.096 473.52 2.9506 0.085848 .
: ---
: Signif. codes:  0 '***' 0.001 '**' 0.01 '*' 0.05 '.' 0.1 ' ' 1
```

We refit the model using only the unemployment rate as the predictor. The exponentiated coefficient for the unemployment rate is 1.04, which means that a one percentage increase in the unemployment rate is associated with a 4% increase in the crime rate, holding other variables constant. The estimated dispersion parameter θ is around 18.4, which corresponds to $k = 1/18.4$ in Note 7.28, indicating that the variance is greater than the mean and thus overdispersion is present in the data. From Note 7.28, the extra variance introduced by k is $1 + k\mu$ multiples of the mean μ, and therefore, the variance can be much larger than the mean when μ is large even if k is small, which is the case here.

```
fit_nb <- MASS::glm.nb(crimes ~ unemp_pct + offset(log(population)),
                       data = data)
exp(coef(fit_nb))[2]
: unemp_pct
:  1.036147
exp(confint(fit_nb))[2,]
:     2.5 %   97.5 %
: 1.010013 1.063753
# Estimated dispersion parameter theta
fit_nb$theta
: [1] 18.38751
```

The R package **DHARMa** provides a simulation-based approach to check the model fit, and Figure 7.26 show that there are no significant deviations from the expected distribution of residuals, and the test for overdispersion is not significant, suggesting that the negative binomial model has adequately accounted for the overdispersion in the data.

```
library(DHARMa)
set.seed(123)
simres <- simulateResiduals(fit_nb, n = 2000)
plotQQunif(simres, testDispersion=TRUE, lcol="black", pch=16)
```

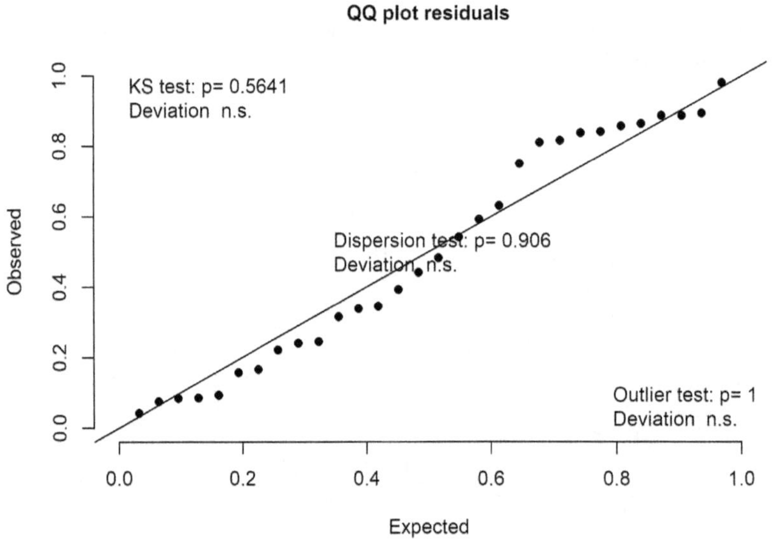

FIGURE 7.26: Diagnostic Plot for Negative Binomial Model

The observed–predicted plot can be used to check the overall fit of the model, and Figure 7.27 shows that the predicted crime rates are generally close to the observed crime rates, and the performance of the negative binomial model is similar to that of the quasi-Poisson model.

```
data$pred_nb <- predict(fit_nb, type = 'response')
data$rate_nb <- data$pred_nb / data$population * 1e5

# Plot observed vs. predicted
```

```
ggplot(data, aes(x = rate_obs, y = rate_nb, label = county_id)) +
  geom_abline(slope = 1, intercept = 0, linetype = 'dashed') +
  geom_point(color = "grey50", size = 2) +
  ggrepel::geom_text_repel(size = 2.8) +
  labs(title = "Observed vs. Predicted Crime Rates (Negative Binomial)",
       x = "Observed Crimes per 100k",
       y = "Predicted Crimes per 100k") + theme_bw()
```

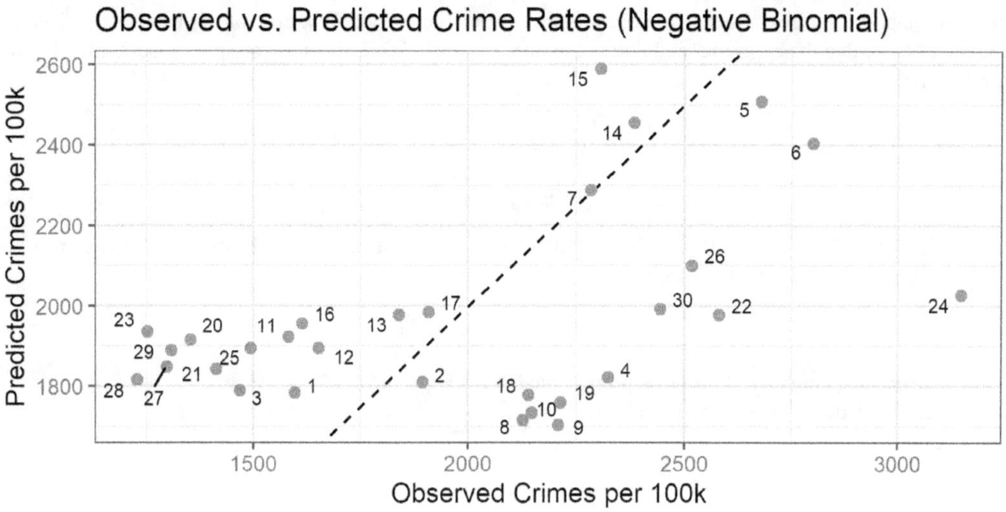

FIGURE 7.27: Observed vs. Predicted Crime Rates (Negative Binomial).

Conclusions

In this case study, we use both quasi-Poisson regression and negative binomial regression to model the crime counts among counties, while using the population of each county as an offset term to account for different risk exposures. The results from both models suggest that the unemployment rate is a significant predictor of crime rates, while other factors may not be significant or their effects may be confounded by multicollinearity and thus need to be removed from the model. In particular, the two models suggest that a one percentage increase in the unemployment rate is associated with a 4% to 5% increase in the property crime rate, holding other variables constant. How the counties are selected for the analysis need to be clarified to consider the generalizability of the results, and the analysis is based on a cross-sectional data, and therefore, no causal conclusions can be established from the results.

Note 7.25 (Poisson Approximation to the Binomial Distribution)**.** When the number of trials is large and the probability of success is small, the Poisson distribution provides an approximation to the Binomial distribution. Let $X \sim \text{Binomial}(n, p)$, where n is the number of independent trials and p is the probability of success on each trial. The probability mass function of the Binomial distribution is

$$\mathrm{P}(X = k) = \binom{n}{k} p^k (1-p)^{n-k}, \quad k = 0, 1, \ldots, n.$$

If n is large, p is small, and $\lambda = np$ is a finite constant, then it can be shown that as $n \to \infty$ and $p \to 0$ with $np = \lambda$ fixed, the Binomial probability mass function converges to the Poisson probability mass function:

$$P(X = k) \to e^{-\lambda} \frac{\lambda^k}{k!}.$$

Therefore, in practice, it is more convenient to use the Poisson regression model to analyze count data when the counts are relatively small compared to the number of trials, especially in many cases the number of trials is not explicitly known but expected to be large. For details on the derivation of the Poisson approximation, see standard textbooks on probability, such as [91].

Note 7.26 (Poisson Regression)**.** *Poisson regression* is commonly used to perform regression analysis when the response variable is a count. Let Y denote the number of events observed. Under a Poisson regression model,

$$Y|\mathbf{x} \sim \text{Poisson}(\mu),$$

where $\mu = \text{E}(Y|\mathbf{x})$ is the expected count. When observations correspond to different exposures (e.g., time duration, population size, or area), it is often more appropriate to model the *rate* of events instead of the raw counts. Let t denote the exposure, and define the rate $\lambda = \mu/t$. The Poisson regression model with a log link can then be written as

$$\ln(\lambda) = \ln(\text{E}(Y|\mathbf{x})/t) = \mathbf{x}^\top \beta,$$

where $\mathbf{x} = (1, x_1, \ldots, x_i, \ldots, x_p)$, or equivalently,

$$\ln(\mu) = \ln(\text{E}(Y|\mathbf{x})) = \mathbf{x}^\top \beta + \ln(t),$$

where, $\ln(t)$ is included as an *offset*, a term with a fixed coefficient of 1 that adjusts for different exposures across subjects. Thus, the expected count is

$$\mu = \text{E}(Y|\mathbf{x}) = t \exp\left(\mathbf{x}^\top \beta\right),$$

that is, the expected number of events is proportional to the exposure t.

Increase one unit of x_i so that $\mathbf{x} = (1, x_1, \ldots, x_i, \ldots, x_p)$ becomes $\mathbf{x}_i := (1, x_1, \ldots, x_i + 1, \ldots, x_p)$, and then we have

$$\ln(\text{E}(Y|\mathbf{x}_i)/t) - \ln(\text{E}(Y|\mathbf{x})/t) = \mathbf{x}_i^\top \beta - \mathbf{x}^\top \beta = \beta_i; \tag{7.7}$$

that is, β_i represents the change in log-rates (or log-counts) for a one-unit increase in x_i, keeping other variables constant. Similarly,

$$\frac{\text{E}(Y|\mathbf{x}_i)/t}{\text{E}(Y|\mathbf{x})/t} = \frac{e^{\mathbf{x}_i^\top \beta}}{e^{\mathbf{x}^\top \beta}} = e^{\beta_i}.$$

That is, $\exp(\beta_i)$ represents the ratio of rates (or counts for a given exposure) associated with a one-unit increase in the predictor x_i, holding other variables constant.

When different subjects have the same exposure t, for example, consider the number of crimes in different cities where the population size is the same, then the offset term $\ln(t)$ becomes a constant and can be absorbed into the intercept. In this case, the Poisson regression model reduces to

$$\ln(\text{E}(Y|\mathbf{x})) = \mathbf{x}^\top \beta,$$

and the interpretations of β_i and $\exp(\beta_i)$ are similar, but only for the expected counts instead of rates.

A key assumption of the Poisson regression model is that the mean and variance of the response variables are equal, that is, $\text{Var}(Y|\mathbf{x}) = \text{E}(Y|\mathbf{x})$. However, in practice, the data may exhibit overdispersion (variance greater than the mean) or underdispersion (variance less than the mean). In such cases, alternative models such as quasi-Poisson model (Note 7.27) or negative binomial regression (Note 7.28) can be used to account for such dispersion. For more details on Poisson regression, see some standard references on generalized linear models [76, 4, 35, 40].

Note 7.27 (Quasi-Poisson Regression)**.** The *quasi-Poisson regression model* is an extension of the Poisson regression model, and it is used when the equidispersion assumption in the Poisson model does not hold. Let Y denote the count response variable. In a quasi-Poisson model, the mean structure is the same as in a Poisson regression:

$$E(Y|\mathbf{x}) = \mu, \qquad \ln(\mu) = \mathbf{x}^\top \beta,$$

Unlike the Poisson model, the quasi-Poisson approach does not assume a full probability distribution for Y. Instead, it specifies only the relationship between the mean and variance:

$$\text{Var}(Y|\mathbf{x}) = \phi \mu,$$

where $\phi > 0$ is a dispersion parameter. When $\phi = 1$, the model reduces to the standard Poisson regression; when $\phi > 1$, the data exhibit overdispersion; and when $\phi < 1$, the data exhibit underdispersion. Parameter estimates for β are identical to those from the Poisson regression, but the estimated variance–covariance matrix is inflated by the estimated dispersion parameter $\hat{\phi}$ accordingly, so that the estimated standard errors will reflect the non-equidispersion in the data. See [40] for more details on quasi-Poisson regression.

Note 7.28 (Negative Binomial Regression)**.** *Negative binomial regression* is commonly used for performing regression analysis on count data when overdispersion is present in the Poisson regression model (Note 7.26). Let Y denote the count response variable. In the negative binomial regression model,

$$Y|\mathbf{x} \sim \text{NB}(\mu, k),$$

where $\mu = \text{E}(Y|\mathbf{x})$ is the mean and $k > 0$ is a dispersion parameter controlling the degree of overdispersion. The mean and variance of Y given covariates $\mathbf{x}$ are

$$\text{E}(Y|\mathbf{x}) = \mu, \qquad \text{Var}(Y|\mathbf{x}) = \mu + k\mu^2,$$

and therefore, the variance can be greater than the mean and thus overdispersion can be captured by the parameter k. When $k \to 0$, the negative binomial regression model reduces to the Poisson regression model. Note that, negative binomial regression cannot handle the underdispersion case. The way to relate the mean to explanatory variables is the same as in Poisson regression, that is,

$$\ln(\text{E}(Y|\mathbf{x})) = \mathbf{x}^\top \beta.$$

The interpretation of the regression coefficients is also similar to that in Poisson regression; see Note 7.26 for details on the interpretation. For more details on negative binomial regression, see [20, 40].

7.3.2 Binomial Regression

When the response variable is a count of successes out of a certain number of trials, we can use binomial regression to model the relationship between the predictors and the probability of success. We use the following case study to illustrate the application of binomial regression, and beyond that, we will discuss a common situation when the number of trials may affect the probability of success as well.

Case 36: Assessing Factors Affecting Customer Complaints

A telecommunications company serves many customers through phone calls, and customers have the option to leave a complaint about the service they received from the service representative right after their service interaction. The company wants to understand factors associated with the number of complaints received by service representatives during a workday, which can help the company to improve customer satisfaction and service quality. The company collects data from 40 service representatives randomly selected from multiple call centers for a single day. For each representative, the following variables are recorded: number of complaints, number of customers served, and years of experience of the service representative. The goal is to understand what factors influence the number of complaints about the service provided by service representatives.

For answers to the initial questions and some initial thoughts about this case study, refer to Section 5.37.

First, we visualize the data using scatter plots, and Figure 7.28 shows that the number of complaints increases with the number of customers served, which is expected. The number of customers served can be considered as the number of trials, and the number of complaints can be treated as the number of successes (each customer can either make a complaint or not). Therefore, a binomial regression model may be a reasonable starting point for modeling. Note that, simply using the number of complaints as the response variable and the number of customers served as a predictor would not be appropriate, as their relationship is not what we actually want to model. Although the description of the case study does not make it explicit, what we really care about is the complaint rate, or in other words, the probability of a complaint being made by a customer.

```
library(ggplot2)
library(MASS)
library(GGally)

data <- read.csv("data/customer-complaints.csv")
ggpairs(data, progress=F, lower=list(continuous="smooth")) + theme_bw()
```

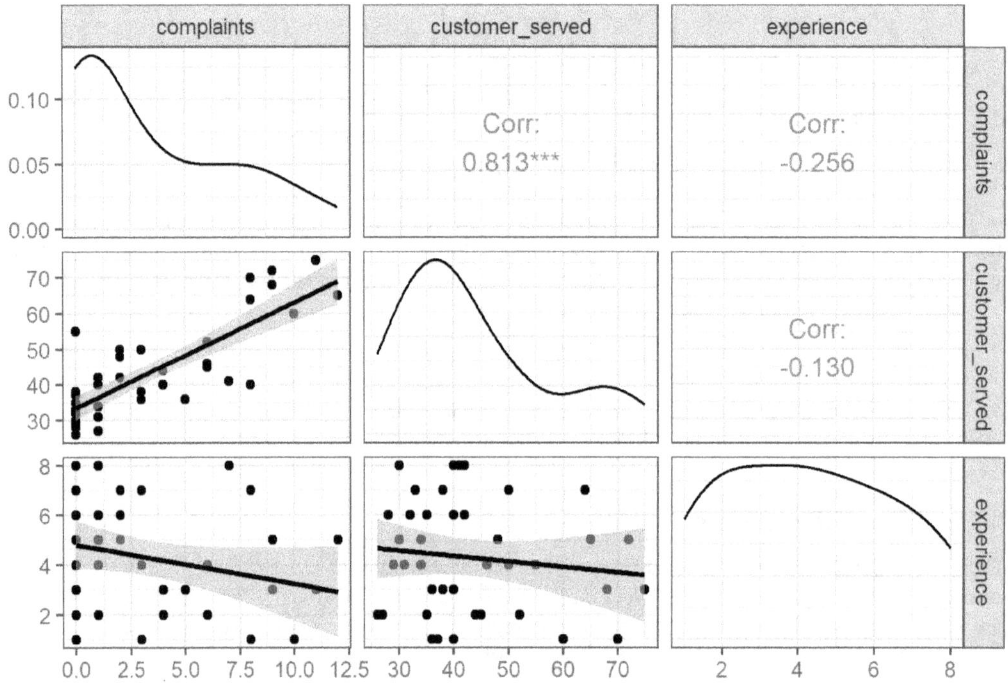

FIGURE 7.28: Scatter Plots of the Data.

We fit a binomial regression model as follows, and note how the response variable is specified using the **cbind()** function to indicate the number of successes and failures. There is an alternative way to specify the model using the complaint rate as the response variable and the number of customers served as weights, but the first approach is more straightforward and thus is generally recommended.

```
model_bn <- glm(cbind(complaints, customer_served-complaints) ~
                      experience, data=data, family = binomial)
model_bn_alt <- glm(complaints/customer_served ~ experience,
                weights = customer_served, data=data, family = binomial)
```

Binomial distribution has a fixed relationship between the mean and variance (Notes 7.29, 7.30), which may not hold for real data. A more flexible model is the quasi-binomial regression model (see Note 7.30), where a dispersion parameter is introduced to account for overdispersion or underdispersion. The following R code suggests that the dispersion parameter is greater than 1, indicating overdispersion in the data, and thus the quasi-binomial regression model may be more appropriate for this case study.

```
sum(residuals(model_bn, type = "pearson")^2) / df.residual(model_bn)
: [1] 2.433746
```

We fit a quasi-binomial model as follows, and use F tests to examine the significance of the predictors; note that the F test is needed as there is a free dispersion parameter. The results suggest that there is no strong evidence that experience is associated with the log-odds of complaints (p-value: 0.1).

```
# quasi-binomial
model_qbn <- glm(cbind(complaints, customer_served - complaints) ~
                 experience, family = quasibinomial, data = data)
drop1(model_qbn, test = "F")
: Single term deletions
:
: Model:
: cbind(complaints, customer_served - complaints) ~ experience
:            Df Deviance F value  Pr(>F)
: <none>          111.64
: experience  1   120.04  2.8598 0.09901 .
: ---
: Signif. codes:  0 '***' 0.001 '**' 0.01 '*' 0.05 '.' 0.1 ' ' 1
```

However, we should not stop here, as the complaint rate may be associated with workload, which could be reflected by the number of customers served. Figure 7.29 shows that the complaint rate seems to increase with either the number of customers served or the logarithms of the number of customers served, and thus it may be necessary to include the number of customers as a predictor in the model.

```
data$complaint_rate <- data$complaints / data$customer_served
p1 <- ggplot(data, aes(x = customer_served, y = complaint_rate)) +
  geom_point(size = 1) + theme_bw() +
  geom_smooth(method = "lm", se = FALSE, formula = 'y ~ x',
              color = "grey50", linewidth = 0.5, linetype = "dashed") +
  scale_color_grey(start = 0.3, end = 0.8) +
  labs(title = "", x = "Customer Served",
       y = "Complaint Rates")

p2 <- ggplot(data, aes(x = log(customer_served), y = complaint_rate)) +
  geom_point(size = 1) + theme_bw() +
  geom_smooth(method = "lm", se = FALSE, formula = 'y ~ x',
              color = "grey50", linewidth = 0.5, linetype = "dashed") +
  scale_color_grey(start = 0.3, end = 0.8) +
  labs(title = "", x = "log(Customer Served)",
       y = "Complaint Rates")
gridExtra::grid.arrange(p1, p2, ncol = 2)
```

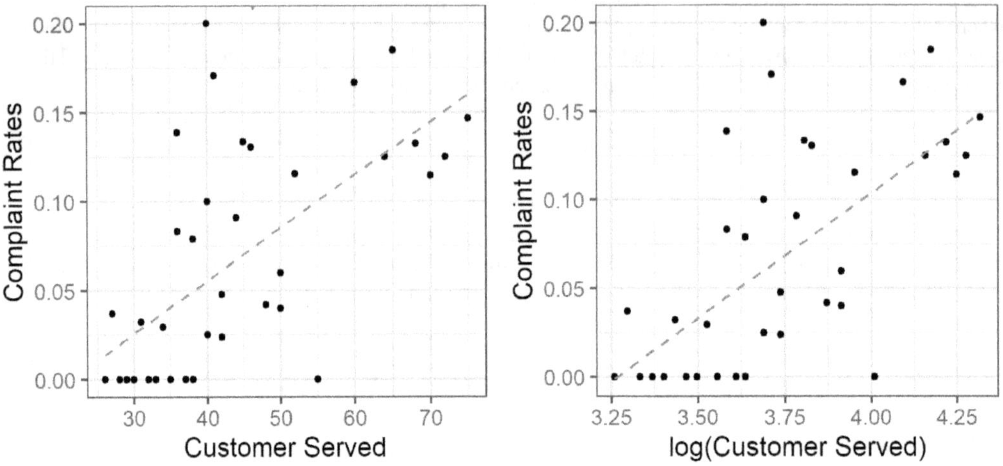

FIGURE 7.29: Complaint Rates vs. Customer Served.

We refit the quasi-binomial model, adding the number of customers served and the logarithm of it as a predictor, respectively. Then we use 10-fold cross-validation to compare the two models, and the results suggest that the model with the logarithm of the number of customers served has a smaller cross-validated error.

```
model_qbn <- glm(cbind(complaints, customer_served - complaints) ~
                  experience + customer_served,
                  family = quasibinomial, data = data)
model_qbn_log <- glm(cbind(complaints, customer_served - complaints) ~
                     experience + log(customer_served),
                     family = quasibinomial, data = data)

library(boot)
set.seed(123)
cv.glm(data, model_qbn, K = 10)$delta
: [1] 0.002920057 0.002888958
cv.glm(data, model_qbn_log, K = 10)$delta
: [1] 0.002734875 0.002708178
```

We use `model_qbn_log` as the final model, and use F tests to examine the significance of the predictors. The results suggest that the experience is not significant (p-value: 0.20), while the logarithm of the number of customers served is significant (p-value: 0.00014).

```
drop1(model_qbn_log, test = "F")
: Single term deletions
:
: Model:
: cbind(complaints, customer_served - complaints) ~ experience +
:     log(customer_served)
:                      Df Deviance F value    Pr(>F)
: <none>                  75.110
: experience            1  78.547  1.6929 0.2012574
: log(customer_served)  1 111.639 17.9944 0.0001422 ***
: ---
```

```
: Signif. codes:  0 '***' 0.001 '**' 0.01 '*' 0.05 '.' 0.1 ' ' 1
```

Diagnostics are performed using the residual plots, and there are no obvious violations of the model assumptions.

```
# Model Diagnostics: Residual plots
par(mfrow = c(2, 2))
plot(model_qbn_log)
```

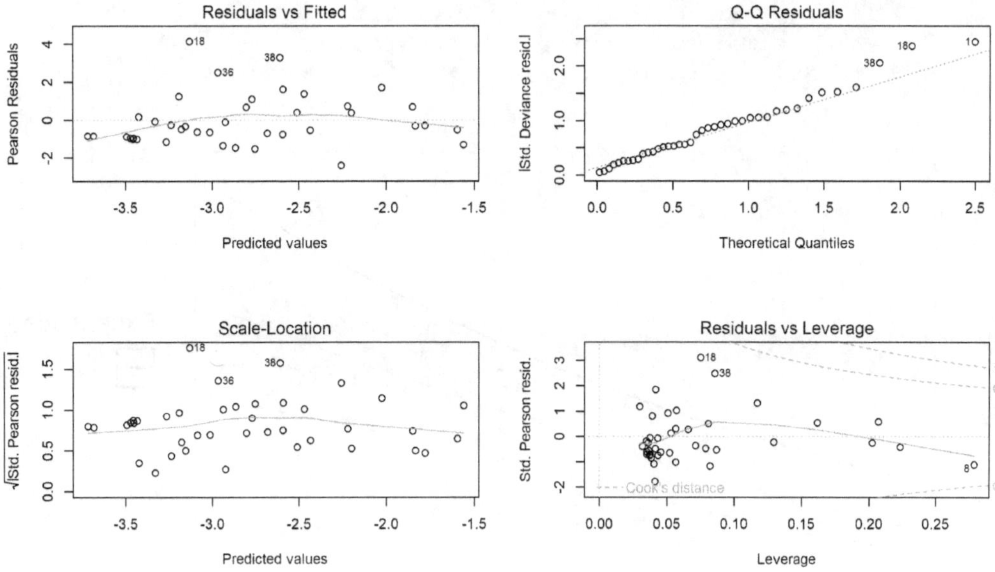

FIGURE 7.30: Model Diagnostics: Residual Plots.

```
par(mfrow = c(1, 1))
```

Although the experience is not significant, we keep it in the model as it is a variable of interest, and the following are the estimated regression coefficients. A larger number of years of experience seems to lead to a smaller log-odds of complaints, although the evidence is not statistically significant at common significance levels. However, the logarithm of the number of customers served is significantly associated with the log-odds of complaints at common significance levels, with one unit increase in the logarithm of the number of customers served being associated with an increase of about 1.87 (95% CI: 1.02, 2.76) in the log-odds of complaints, holding experience constant.

```
coef(model_qbn_log)[2:3]
:           experience log(customer_served)
:          -0.08165773           1.86927547
confint(model_qbn_log)[2:3,]
:                            2.5 %     97.5 %
: experience           -0.2034127 0.03720795
: log(customer_served)  1.0194954 2.75599269
```

A plot showing the predicted probability of complaints across the number of customers served for two different levels of experience is shown in Figure 7.31, and such a plot can be useful for communicating the results. Here, only two levels of experience are shown for better visualization.

```
library(ggeffects)
pred <- ggpredict(model_qbn_log, terms = c("customer_served [all]",
                  "experience [1,8]"))
pred <- data.frame(pred)
names(pred)[names(pred) == "group"] <- "Experience"
names(pred)[names(pred) == "x"] <- "Customers"

ggplot(pred, aes(x = Customers, y = predicted, color = Experience,
                linetype = Experience, group = Experience)) + geom_line() +
       coord_cartesian(ylim = c(0,0.3))+
       geom_ribbon(aes(ymin = conf.low, ymax = conf.high,
                       fill = Experience), alpha = 0.2) +
       scale_color_manual(values = c("black", "grey70")) +
       scale_fill_manual(values = c("black", "grey70")) + theme_bw()
```

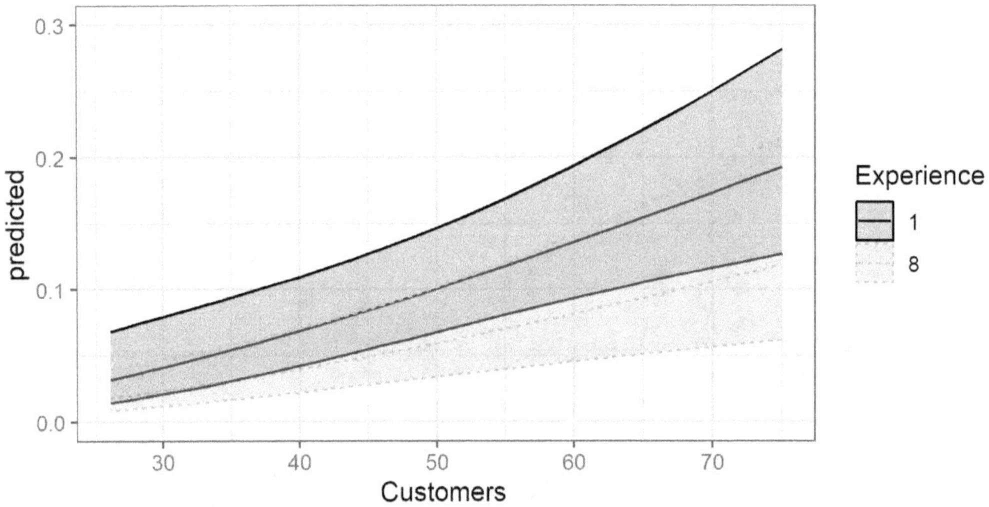

FIGURE 7.31: Predicted Probability of Complaints.

Using the total number of customers served for that day as a variable to reflect the workload may not be ideal, as it may not directly capture the workload experienced by the customer representative when the customer is served. For example, if a customer is served at the beginning of the day, then the cumulative workload (fatigue) experienced by the representative at that time may be very low. Therefore, if possible, a variable that better reflects the fatigue level when each customer is served should be collected, and models such as logistic mixed-effects model (Section 7.1.8) can then be used to account for the correlation among customers served by the same representative.

Conclusions

In this case study, we use a quasi-binomial regression model to analyze the relationship between the complaint rate and the predictors, including the years of experience of the customer representative and the number of customers served. The results suggest that the logarithm of the number of customers served is significantly associated with the log-odds of complaints (p-value: 0.00014), with one unit increase in the logarithm of the number of customers served being associated with an increase of about 1.87 (95% CI: $1.02, 2.76$) in the log-odds of complaints, holding experience constant. However, the experience is not significant at common significance levels (p-value: 0.20). We also discuss the situation that the number of customers served variable, used as a predictor, should be replaced by another variable that can capture the fatigue level experienced by the customer representative when each customer is served, and then a model with more granular data and more appropriate modeling approach can be used to better understand the relationship between the complaint rate and workload or fatigue. The subjects are randomly sampled, and thus the results can be generalized to the customer representatives of the company. However, this is an observational study, and thus causal conclusions cannot be drawn from the results.

Note 7.29 (Binomial Regression)**.** *Binomial regression* is used to perform regression analysis on the number of successes in a fixed number of trials. Let $Y|\mathbf{x} \sim \text{Binomial}(n, p)$ denote the number of successes out of n trials, where $p = \text{P(success)}$. The model relates the mean $\text{E}(Y|\mathbf{x}) = np$ to a linear predictor via a logit link function of p:

$$\ln\left(\frac{\text{E}(Y|\mathbf{x})}{n - \text{E}(Y|\mathbf{x})}\right) = \ln\left(\frac{p}{1-p}\right) = \mathbf{x}^\top \beta,$$

so that

$$p = \frac{\exp(\mathbf{x}^\top \beta)}{1 + \exp(\mathbf{x}^\top \beta)}.$$

Increase one unit of x_i so that $\mathbf{x} = (1, x_1, \ldots, x_i, \ldots, x_p)$ becomes $\mathbf{x}_i := (1, x_1, \ldots, x_i + 1, \ldots, x_p)$, and then we have

$$\ln\left(\frac{\text{E}(Y|\mathbf{x}_i)}{n - \text{E}(Y|\mathbf{x}_i)}\right) - \ln\left(\frac{\text{E}(Y|\mathbf{x})}{n - \text{E}(Y|\mathbf{x})}\right) = \mathbf{x}_i^\top \beta - \mathbf{x}^\top \beta = \beta_i; \tag{7.8}$$

that is, β_i represents the change in log-odds of success for a one-unit increase in x_i, keeping other variables constant. Therefore, $\exp(\beta_i)$ represents the odds ratio of success associated with a one-unit increase in the predictor x_i, holding other variables constant.

Note from Equation 7.8 that the probability of success p is a function of the predictors $\mathbf{x}$, but not the number of trials n. However, sometimes, the number of trials n may also affect the probability of success p. For example, in the case study, the total number of customers served reflects the workload, and a higher workload may lead to a higher probability of complaints. In practice, we can consider including n as a predictor in the model. However, we should be cautious when interpreting β_i, as the probability of success would depend on the value of n as well as the other predictors. This situation is not ideal, and if possible, a proxy variable that better reflects the workload should be collected and included in the model instead of n. For more details on binomial regression, see some standard references on generalized linear models [76, 4, 35, 40].

Note 7.30 (Quasi-Binomial Regression)**.** The *quasi-binomial regression model* is an extension of the binomial regression model, and it is used when the variance of the response exceeds or falls below the variance suggested by the binomial distribution. Let Y denote the number of successes out of n trials with success probability p. The expected value is $\text{E}(Y) = np$, and the variance is $\text{Var}(Y) = \phi np(1-p)$. Unlike the standard binomial model, the quasi-binomial model introduces a dispersion parameter ϕ to allow extra variability,

$$\text{Var}(Y) = \phi np(1-p).$$

When $\phi > 1$ the data exhibit overdispersion, and when $\phi < 1$ the data are underdispersed relative to the binomial assumption. We refer to [40] for more details about quasi-binomial regression.

8

Appendix

8.1 R Packages Used in the Case Studies

The R software version 4.5.3 is used to generate the results, and the following R packages are used in the case studies:

Package	Version	Package	Version
boot	1.3-32	lme4	2.0-1
car	3.1-5	lsr	0.5.2
corrplot	0.95	MASS	7.3-65
DescTools	0.99.60	metafor	4.8-0
DHARMa	0.4.7	multcomp	1.4-30
dplyr	1.2.0	nnet	7.3-20
emmeans	2.0.2	ordinal	2025.12-29
ez	4.5-0	pROC	1.19.0.1
GGally	2.4.0	pwr	1.3-0
ggeffects	2.3.2	rstatix	0.7.3
ggplot2	4.0.2	semPlot	1.1.8
gridExtra	2.3	tidyr	1.3.2
lavaan	0.6-21	VGAM	1.1-14

DOI: 10.1201/9781003683827-8

8.2 Worksheet for Initial Questions

Case: ____________ Title: ______________________________________

Initial Questions
1. What are the research objectives?
2. What are the statistical questions?
3. What is the response variable, and what is the data type of the response variable?
4. What are the explanatory variables of interest?
5. Are there covariates?
6. What is the population of interest?
7. What is the subject, and what is the number of distinct subjects?
8. Are there subject-level data?
9. Are response variables dependent (repeated measures / clustered subjects)?
10. Are the subjects selected randomly?
11. Are the subjects randomly assigned to different groups?

Initial Thoughts

8.3 Other Resources

A companion website for this book is available at `https://larryleihua.github.io/PSCBook/`, where datasets and printable worksheets including case descriptions and initial questions and thoughts for all the case studies are available.

Bibliography

[1] Statistical inference in the 21st century: A world beyond $p < 0.05$. *The American Statistician*, 73(sup1):1–19, 2019.

[2] Alan Agresti. *Analysis of Ordinal Categorical Data*. John Wiley & Sons, 2010.

[3] Alan Agresti. *Categorical Data Analysis*. Wiley, 3rd edition, 2013.

[4] Alan Agresti. *Foundations of Linear and Generalized Linear Models*. John Wiley & Sons, 2015.

[5] Alan Agresti. *Statistical Methods for the Social Sciences*. Pearson, Boston, 5th edition, 2018.

[6] Alan Agresti. *An Introduction to Categorical Data Analysis*. Wiley, 3rd edition, 2019.

[7] Leona S. Aiken and Stephen G. West. *Multiple Regression: Testing and Interpreting Interactions*. Sage Publications, 1991.

[8] Douglas Bates, Martin Mächler, Ben Bolker, and Steve Walker. Fitting linear mixed-effects models using lme4. *Journal of Statistical Software*, 67(1):1–48, 2015.

[9] David A. Belsley, Edwin Kuh, and Roy E. Welsch. *Regression Diagnostics*. Wiley, 1980.

[10] Yoav Benjamini and Yosef Hochberg. Controlling the false discovery rate: A practical and powerful approach to multiple testing. *Journal of the Royal Statistical Society Series B*, 57(1):289–300, 1995.

[11] Vasant P. Bhapkar. A note on the equivalence of two test criteria for hypotheses in categorical data. *Journal of the American Statistical Association*, 61(313):228–235, 1966.

[12] Carlo Emilio Bonferroni. Teoria statistica delle classi e calcolo delle probabilità. *Pubblicazioni del R Istituto Superiore di Scienze Economiche e Commerciali di Firenze*, 8:3–62, 1936.

[13] Michael Borenstein, Larry V. Hedges, Julian P. T. Higgins, and Hannah R. Rothstein. *Introduction to Meta-Analysis*. John Wiley & Sons, 2nd edition, 2021.

[14] A. H. Bowker. A test for symmetry in contingency tables. *Journal of the American Statistical Association*, 43:572–574, 1948.

[15] Rollin Brant. Assessing proportionality in the proportional odds model for ordinal logistic regression. *Biometrics*, pages 1171–1178, 1990.

[16] Norman E. Breslow and Nicholas E. Day. *Statistical Methods in Cancer Research, Volume I: The Analysis of Case-Control Studies*. Number 32 in IARC Scientific Publications. International Agency for Research on Cancer, Lyon, France, 1980.

[17] Morton B. Brown and Alan B. Forsythe. Robust tests for the equality of variances. *Journal of the American Statistical Association*, 69(346):364–367, 1974.

[18] Timothy A. Brown. *Confirmatory Factor Analysis for Applied Research*. Guilford Publications, 2015.

[19] Warren S. Browner, Thomas B. Newman, Steven R. Cummings, and Deborah G. Grady. *Designing Clinical Research*. Lippincott Williams & Wilkins, 5th edition, 2022.

[20] A. Colin Cameron and Pravin K. Trivedi. *Regression Analysis of Count Data*. Cambridge University Press, 2nd edition, 2013.

[21] George Casella and Roger L. Berger. *Statistical Inference*. Duxbury, 2nd edition, 2002.

[22] Rune H. B. Christensen. *ordinal—Regression Models for Ordinal Data*, 2025. R package version 2025.12-29.

[23] Norman Cliff. Dominance statistics: Ordinal analyses to answer ordinal questions. *Psychological Bulletin*, 114(3):494, 1993.

[24] William G. Cochran. The comparison of percentages in matched samples. *Biometrika*, 37(3/4):256–266, 1950.

[25] William G. Cochran. The combination of estimates from different experiments. *Biometrics*, 10(1):101–129, 1954.

[26] William G. Cochran. Some methods for strengthening the common χ^2 tests. *Biometrics*, 10(4):417–451, 1954.

[27] Jacob Cohen. *Statistical Power Analysis for the Behavioral Sciences*. Routledge, New York, 2nd edition, 1988.

[28] William Jay Conover. *Practical Nonparametric Statistics*. John Wiley & Sons, 1999.

[29] R. Dennis Cook. Detection of influential observation in linear regression. *Technometrics*, 19(1):15–18, 1977.

[30] David R. Cox. The regression analysis of binary sequences. *Journal of the Royal Statistical Society Series B: Statistical Methodology*, 20(2):215–232, 1958.

[31] Harald Cramér. *Mathematical Methods of Statistics*. Princeton University Press, 1946.

[32] Lee J. Cronbach. Coefficient alpha and the internal structure of tests. *Psychometrika*, 16:297–334, 1951.

[33] Rebecca DerSimonian and Nan Laird. Meta-analysis in clinical trials. *Controlled Clinical Trials*, 7(3):177–188, 1986.

[34] Jayant V. Deshpande, Uttara Naik-Nimbalkar, and Isha Dewan. *Nonparametric Statistics: Theory and Methods*. World Scientific, 2017.

[35] Annette J. Dobson and Adrian G. Barnett. *An Introduction to Generalized Linear Models*. Chapman and Hall/CRC, 2018.

[36] Olive Jean Dunn. Multiple comparisons using rank sums. *Technometrics*, 6:241–252, 1964.

[37] Charles W. Dunnett. A multiple comparison procedure for comparing several treatments with a control. *Journal of the American Statistical Association*, 50(272):1096–1121, 1955.

[38] Bradley Efron. Better bootstrap confidence intervals. *Journal of the American Statistical Association*, 82(397):171–185, 1987.

[39] Paul D. Ellis. *The Essential Guide to Effect Sizes: Statistical Power, Meta-Analysis, and the Interpretation of Research Results*. Cambridge University Press, 2010.

[40] Julian J. Faraway. *Extending the Linear Model with R: Generalized Linear, Mixed Effects and Nonparametric Regression Models*, volume 124. CRC Press, 2016.

[41] Tom Fawcett. An introduction to roc analysis. *Pattern Recognition Letters*, 27(8):861–874, 2006.

[42] Ronald A. Fisher. On the interpretation of χ^2 from contingency tables, and the calculation of p. *Journal of the Royal Statistical Society*, 85(1):87–94, 1922.

[43] Garrett M. Fitzmaurice, Nan M. Laird, and James H. Ware. *Applied Longitudinal Analysis*. John Wiley & Sons, 2012.

[44] John Fox and Georges Monette. Generalized collinearity diagnostics. *Journal of the American Statistical Association*, 87(417):178–183, 1992.

[45] Milton Friedman. The use of ranks to avoid the assumption of normality implicit in the analysis of variance. *Journal of the American Statistical Association*, 32(200):675–701, 1937.

[46] Ellen R. Girden. *ANOVA: Repeated Measures*, volume 07-084 of *SAGE University Papers Series on Quantitative Applications in the Social Sciences*. SAGE Publications, Newbury Park, CA, 1992.

[47] Phillip I. Good. *Permutation Tests: A Practical Guide to Resampling Methods for Testing Hypotheses*. Springer, New York, 3rd edition, 2013.

[48] Ulrich Halekoh and Søren Højsgaard. A kenward-roger approximation and parametric bootstrap methods for tests in linear mixed models–the r package pbkrtest. *Journal of Statistical Software*, 59:1–32, 2014.

[49] David A. Harville. Maximum likelihood approaches to variance component estimation and to related problems. *Journal of the American Statistical Association*, 72(358):320–338, 1977.

[50] Larry V. Hedges. Distribution theory for glass's estimator of effect size and related estimators. *Journal of Educational Statistics*, 6(2):107–128, 1981.

[51] Larry V. Hedges, Elizabeth Tipton, Rrita Zejnullahi, and Karina G. Diaz. Effect sizes in ancova and difference-in-differences designs. *British Journal of Mathematical and Statistical Psychology*, 76(2):259–282, 2023.

[52] Robert V. Hogg, Joseph W. McKean, and Allen T. Craig. *Introduction to Mathematical Statistics*. Pearson, Boston, MA, 8th edition, 2019.

[53] Myles Hollander, Douglas A. Wolfe, and Eric Chicken. *Nonparametric Statistical Methods*. John Wiley & Sons, 2013.

[54] Sture Holm. A simple sequentially rejective multiple test procedure. *Scandinavian Journal of Statistics*, 6(2):65–70, 1979.

[55] David W. Hosmer, Stanley Lemeshow, and Rodney X. Sturdivant. *Applied Logistic Regression*. John Wiley and Sons, 3rd edition, 2013.

[56] Jason C. Hsu. *Multiple Comparisons: Theory and Methods*. Chapman and Hall, 1996.

[57] Bradley E. Huitema. *The Analysis of Covariance and Alternatives*. Wiley, 2nd edition, 2011.

[58] Huynh Huynh and Leonard S. Feldt. Conditions under which mean square ratios in repeated measurements designs have exact f-distributions. *Journal of the American Statistical Association*, 65(332):1582–1589, 1970.

[59] CT Ireland, HH Ku, and S Kullback. Symmetry and marginal homogeneity of an r$\times$ r contingency table. *Journal of the American Statistical Association*, 64(328):1323–1341, 1969.

[60] Susan Jamieson. Likert scales: how to (ab)use them. *Medical Education*, 38(12):1217–1218, 2004.

[61] Harry Joe. *Multivariate Models and Dependence Concepts*, volume 73 of *Monographs on Statistics and Applied Probability*. Chapman & Hall, London, 1997.

[62] Maurice G. Kendall. A new measure of rank correlation. *Biometrika*, 30(1-2):81–93, 1938.

[63] Michael G. Kenward and James H. Roger. Small sample inference for fixed effects from restricted maximum likelihood. *Biometrics*, pages 983–997, 1997.

[64] Clyde Y. Kramer. Extension of multiple range tests to group means with unequal numbers of replications. *Biometrics*, 12(3):307–310, 1956.

[65] William H. Kruskal and W. Allen Wallis. Use of ranks in one-criterion variance analysis. *Journal of the American Statistical Association*, 47:583–621, 1952.

[66] Daniël Lakens. Calculating and reporting effect sizes to facilitate cumulative science: a practical primer for t-tests and anovas. *Frontiers in Psychology*, 4:863, 2013.

[67] Howard Levene. Robust tests for equality of variances. *Contributions to Probability and Statistics: Essays in Honor of Harold Hotelling*, pages 278–292, 1960.

[68] Rensis Likert. A technique for the measurement of attitudes. *Archives of Psychology*, 140:1–55, 1932.

[69] J Scott Long. Regression models for categorical and limited dependent variables (vol. 7). *Advanced Quantitative Techniques in the Social Sciences*, 219, 1997.

[70] Steven G. Luke. Evaluating significance in linear mixed-effects models in r. *Behavior Research Methods*, 49(4):1494–1502, 2017.

[71] Mohammad Ali Mansournia, Angelika Geroldinger, Sander Greenland, and Georg Heinze. Separation in logistic regression: causes, consequences, and control. *American Journal of Epidemiology*, 187(4):864–870, 2018.

[72] Nathan Mantel and William Haenszel. Statistical aspects of the analysis of data from retrospective studies of disease. *JNCI: Journal of the National Cancer Institute*, 22(4):719–748, 04 1959.

[73] Donald W. Marquardt. Generalized inverses, ridge regression, biased linear estimation, and nonorthogonal designs. *Technometrics*, 12(3):591–612, 1970.

[74] A. E. Maxwell. Comparing the classification of subjects by two independent judges. *British Journal of Psychiatry*, 116(535):651–655, 1970.

[75] Scott E. Maxwell and Harold D. Delaney. *Designing Experiments and Analyzing Data.* Psychology Press, 2nd edition, 2004.

[76] P. McCullagh and J.A. Nelder. *Generalized Linear Models.* Chapman and Hall/CRC Monographs on Statistics and Applied Probability Series. Chapman & Hall, 2nd edition, 1989.

[77] Peter McCullagh. Regression models for ordinal data. *Journal of the Royal Statistical Society: Series B (Methodological)*, 42(2):109–127, 1980.

[78] Charles E. McCulloch, Shayle R. Searle, and John M. Neuhaus. *Generalized, Linear, and Mixed Models.* Wiley, 2nd edition, 2008.

[79] Roderick P. McDonald. *Test Theory: A Unified Treatment.* Lawrence Erlbaum Associates, Mahwah, NJ, 1999.

[80] Quinn McNemar. Note on the sampling error of the difference between correlated proportions or percentages. *Psychometrika*, 12(2):153–157, 1947.

[81] Bengt O. Muthén. A general structural equation model with dichotomous, ordered categorical, and continuous latent variable indicators. *Psychometrika*, 49(1):115–132, 1984.

[82] Dorothy Newman. The distribution of range in samples from a normal population, expressed in terms of an independent estimate of standard deviation. *Biometrika*, 31(1/2):20–30, 1939.

[83] Geoff Norman. Likert scales, levels of measurement and the "laws" of statistics. *Advances in Health Sciences Education*, 15(5):625–632, 2010.

[84] Jum C. Nunnally and Ira H. Bernstein. *Psychometric Theory.* McGraw-Hill, New York, 3rd edition, 1994.

[85] Robert M. O'Brien. A caution regarding rules of thumb for variance inflation factors. *Quality & Quantity*, 41(5):673–690, 2007.

[86] Stephen Olejnik and James Algina. Generalized eta and omega squared statistics: Measures of effect size for some common research designs. *Psychological Methods*, 8(4):434–447, 2003.

[87] Peter Peduzzi, John Concato, Elizabeth Kemper, Theodore R. Holford, and Alvan R. Feinstein. A simulation study of the number of events per variable in logistic regression analysis. *Journal of Clinical Epidemiology*, 49(12):1373–1379, 1996.

[88] Louis M. Rea and Richard A. Parker. *Designing and Conducting Survey Research: A Comprehensive Guide*. John Wiley & Sons, 2014.

[89] Alvin C. Rencher and G Bruce Schaalje. *Linear Models in Statistics*. John Wiley & Sons, 2008.

[90] Jeanine Romano, Jeffrey D. Kromrey, Jesse Coraggio, and Jeff Skowronek. Appropriate statistics for ordinal level data: Should we really be using t-test and cohen'sd for evaluating group differences on the nsse and other surveys. In *Annual Meeting of the Florida Association of Institutional Research*, volume 177, 2006.

[91] Sheldon M. Ross. *A First Course in Probability*. Pearson, 9th edition, 2014.

[92] Yves Rosseel. lavaan: An R package for structural equation modeling. *Journal of Statistical Software*, 48(2):1–36, 2012.

[93] Kenneth J. Rothman, Sander Greenland, and Timothy L. Lash. *Modern Epidemiology*, volume 4. Wolters Kluwer, 2020.

[94] Henry Rouanet and Dominique Lepine. Comparison between treatments in a repeated-measurement design: Anova and multivariate methods. *British Journal of Mathematical and Statistical Psychology*, 23(2):147–163, 1970.

[95] F. E. Satterthwaite. An approximate distribution of estimates of variance components. *Biometrics Bulletin*, 2(6):110–114, 1946.

[96] Henry Scheffé. A method for judging all contrasts in the analysis of variance. *Biometrika*, 40(1-2):87–104, 1953.

[97] Fabian Scheipl, Sonja Greven, and Helmut Kuechenhoff. Size and power of tests for a zero random effect variance or polynomial regression in additive and linear mixed models. *Computational Statistics & Data Analysis*, 52(7):3283–3299, 2008.

[98] Steven G. Self and Kung-Yee Liang. Asymptotic properties of maximum likelihood estimators and likelihood ratio tests under nonstandard conditions. *Journal of the American Statistical Association*, 82(398):605–610, 1987.

[99] S Shaphiro and MBJB Wilk. An analysis of variance test for normality. *Biometrika*, 52(3):591–611, 1965.

[100] David J. Sheskin. *Handbook of Parametric and Nonparametric Statistical Procedures*. Chapman and Hall/CRC, 5th edition, 2011.

[101] Zbyněk Šidák. Rectangular confidence regions for the means of multivariate normal distributions. *Journal of the American Statistical Association*, 62(318):626–633, 1967.

[102] George W. Snedecor and William G. Cochran. *Statistical Methods*. Iowa State University Press, 8th edition, 1989.

[103] Charles Spearman. The proof and measurement of association between two things. *The American Journal of Psychology*, 15:72–101, 1904.

[104] Daniel O. Stram and Jae Won Lee. Variance components testing in the longitudinal mixed effects model. *Biometrics*, 50(4):1171–1177, 1994.

[105] Alan Stuart. A test for homogeneity of the marginal distributions in a two-way classification. *Biometrika*, 42(3/4):412–416, 1955.

[106] Student. The probable error of a mean. *Biometrika*, 6(1):1–25, 1908.

[107] The R Core Team. *R: A Language and Environment for Statistical Computing*. R Foundation for Statistical Computing, 2026.

[108] Henry C. Thode. *Testing for Normality*. Marcel Dekker, New York, 2002.

[109] Bruce Thompson. *Exploratory and Confirmatory Factor Analysis: Understanding Concepts and Applications*. American Psychological Association, 2004.

[110] John W. Tukey. Comparing individual means in the analysis of variance. *Biometrics*, 5(2):99–114, 1949.

[111] John W. Tukey. The problem of multiple comparisons. In *The Collected Works of John W. Tukey, Volume VIII: Multiple Comparisons*, pages 1–300. Chapman & Hall, 1994. Originally circulated as a 1953 technical report.

[112] Geert Verbeke and Emmanuel Lesaffre. A linear mixed-effects model with heterogeneity in the random-effects population. *Journal of the American Statistical Association*, 91(433):217–221, 1996.

[113] Ronald L. Wasserstein and Nicole A. Lazar. The ASA statement on p-values: Context, process, and purpose. *The American Statistician*, 70(2):129–133, 2016.

[114] B. L. Welch. The generalization of 'student's' problem when several different population variances are involved. *Biometrika*, 34(1/2):28–35, 1947.

[115] Frank Wilcoxon. Individual comparisons by ranking methods. *Biometrics Bulletin*, 1:80–83, 1945.

Index

Note: Page numbers followed by f and t indicate figures and tables, respectively.

For Product Safety Concerns and Information please contact our EU
representative GPSR@taylorandfrancis.com
Taylor & Francis Verlag GmbH, Kaufingerstraße 24, 80331 München, Germany

www.ingramcontent.com/pod-product-compliance
Lightning Source LLC
LaVergne TN
LVHW060401240926
848761LV00034B/305
* 9 7 8 1 0 4 1 1 6 3 0 5 3 *